ADVANCES IN CELL AGING AND GERONTOLOGY
VOLUME 7

Interorganellar Signaling in Age-Related Disease

ADVANCES IN CELL AGING AND GERONTOLOGY
VOLUME 7

Interorganellar Signaling in Age-Related Disease

Volume Editor:
Mark P. Mattson,
PhD, National Institute on Aging,
NIH
Baltimore, MD
USA

2001

ELSEVIER
Amsterdam - London - New York - Oxford - Paris - Shannon - Tokyo

ELSEVIER SCIENCE B.V.
Sara Burgerhartstraat 25
P.O. Box 211, 1000 AE Amsterdam, The Netherlands

First edition 2001

Library of Congress Cataloging in Publication Data
A catalog record from the Library of Congress has been applied for.

ISSN: 1566 - 3124
ISBN: 0-444-50495-8

∞ The paper used in this publication meets the requirements of ANSI/NISO Z39.48-1992 (Permanence of Paper).
Printed in The Netherlands.

TABLE OF CONTENTS

PREFACE

The purpose of this volume is to integrate previously distinct fields to elucidate the cellular and molecular mechanisms underlying the pathophysiology of different age-related disorders. Specifically, the chapters consider how alterations in the function of one organelle lead to abnormal regulation of functions in other organelles, and how these interorganellar "chain reactions" ultimately lead to overall cell dysfunction and degeneration.

Molecular genetic studies have identified a number of gene products in which mutations cause age-related disease. Several of these mutations have been shown, in experimental studies of cultured cells and transgenic animals, to disrupt the function of a specific aspect of organellar signaling. Two important homeostatic systems that involve multiple organelles are those regulating intracellular free calcium levels and oxyradical metabolism. These two regulatory systems, in particular, appear to be prone to dysfunction in age-related diseases ranging from Alzheimer's disease to cancer to diabetes. Reading through the chapters of this volume reveals the latter underlying theme. The chapters, therefore, provide the most recent insight into how disruptions of interorganellar signaling lead to disease.

The first chapter considers how alterations in function of peroxisomes can lead to a variety of disorders. Peroxisome dysfunction apparently results in enhanced free radical production and perturbed energy metabolism. This then leads to mitochondrial dysfunction and, ultimately, cell damage. Jihan Youseef and Mostafa Badr provide a detailed description of cutting edge research in the field of peroxisomal disorders. The next chapter provides the synthesis of molecular/genetic and experimental data documenting alterations in neuronal calcium regulation associated with the pathogenesis of Alzheimer's disease. Some cases of Alzheimer's disease are caused by mutations in presenilins or the amyloid precursor protein. Mutations in presenilin appear to have, as their primary effect, alteration in endoplasmic reticulum calcium regulation, which secondarily leads to altered proteolytic processing of amyloid precursor protein and production of neurotoxic forms of amyloid β-peptide. Perturbed ER function secondarily leads to mitochondrial dysfunction and activation of apoptotic signaling cascades involving caspases and related proteins.

Diabetes mellitus is an age-related disorder involving damage to pancreatic β-cells and altered energy metabolism. Matthias Bauer and colleagues describe the evidence supporting an important role for mitochondrial dysfunction in the pathogenesis of diabetes. In addition to playing a role in damage to β cells, mitochondrial dysfunction throughout insulin responsive cells may occur in diabetes and contribute to the widespread vascular damage and other secondary alterations. Thomas Preston and Gurmit Singh describe the role of mitochondrial signaling in cancer. Many cancers are caused by lack of appropriate cell death. It has been shown that a common mechanistic feature of many of these cancers is stabilization of mitochondrial function. For example, the anti-apoptotic protein, Bcl-2 appears to prevent cell death and promote cancer by stabilizing mitochondrial function. Accordingly, recent clinical approaches to treating cancer patients are focusing on drugs that promote mitochondrial permeabililty transition and apoptosis. Proper protein folding and export, and degradation of damaged proteins, in the endoplasmic reticulum and proteosome is critical for normal cell function. A set of proteins called "chaperones" play a central role in controlling proper protein production and turnover. An increasing literature suggests many different age-related diseases are associated with abnormal protein turnover. Alberto Macario and Everly Conway de Macario describe the chaperone system and how alterations in this system lead to age-related disease in a number of different organ systems. Recent years have witnessed an explosion of research on an enzyme activity called telomerase. Michael Fossel covers the field of telomerase and its role in cancer and age-related degenerative disease. Emerging data are identifying specific mechanisms whereby telomerase controls chromosome shortening and novel mechanisms whereby specific protein components of telomerase, such as TERT, promote cell survival and prevent cellular senescence.

Jeff Keller and Gordon Glazner cover the now quite large field of the role of disturbances in mitochondrial function in age-related neurodegenerative disorders. From Alzheimer's disease to Huntington's disease to Parkinson's disease and amyotrophic lateral sclerosis, disturbances in mitochondrial calcium homeostasis and free radical metabolism are increasingly implicated as pivotal events in the disease process. Preventative and therapeutic approaches to neurodegenerative disorders are being developed based upon stabilizing mitochondrial function and reducing mitochondrial oxyradical production.

The final chapter of *InterOrganellar Signaling in Age-Related Disease* focuses on abnormalities in proteosome-mediated proteolysis and the ubquitin/proteasome pathway in cancer. Vito Spataro and Chris Norbury describe clinical data and data form experimental studies of cancer in cell culture and animal models that point to abnormalities in the ubquitin/proteasome pathway in various cancers. Collectively, the chapters in this volume will provide the reader with novel insight into the inter-relationships of the function of different organelles in the sequences of events that lead to cellular dysfunction and degeneration in the aging human population. The chapters are rich in information for cell and molecular biologists pursuing studies of the different diseases covered. In addition, the clinician will find value in understanding mechanisms underlying age-related disease as such an understanding will lead to novel therapeutic approaches for an array of age-related diseases.

PEROXISOMAL ALTERATIONS IN AGING AND AGE-RELATED DISEASE

JIHAN YOUSSEF & MOSTAFA BADR

University of Missouri-Kansas City, Kansas City, MO 64108-2792

Table of Contents

Introduction

Numerous evidence indicates that cells accumulate damage over a lifetime. This slow incremental damage results in gradual loss of differentiated functions and growth rate accompanied by an increased probability for the development of cancer and ultimately death (Dice, 1993; Rubin, 1997). Although such changes are encountered in most organs, they are normally held to a minimum by the organized state of the tissue and homeostatic regulation of the organism (Cutler, 1982; Rubin, 1997).
Despite a rising interest in the health problems of the elderly, information on senescence-related alterations in essential metabolic pathways, as well as their responses to various chemicals, are scarce. This is particularly true for peroxisomal pathways that are involved in a multitude of cellular functions, yet little attention has been given to their alterations in aging and age-related disease.

Interorganellar Signaling in Age-Related Disease, Edited by Mark P. Mattson, 1—28

Cellular Aging

Theories of aging are generally classified into organ theories, physiological theories and genome-based theories. However, this classification can obscure obvious overlaps (Perez-Campo et al., 1998). A hypothesis has been advanced proposing that cellular aging represents active, genetically programmed events in the same sense as do early development and morphogenesis (Dice and Goff, 1988). Another hypothesis is based on the passive accumulation of errors in cellular constituents, such as DNA, RNA, protein and lipids, due to a variety of insults coupled with imperfect repair mechanisms (Dice and Goff, 1988). These types of explanations for cellular aging are not necessarily mutually exclusive (Dice and Goff, 1988). For example, production and elimination rates of detrimental free radicals depend on a multiplicity of peptides and proteins which are encoded for in the genome (Perez-Campo et al., 1998).

General Effects

The theory most frequently mentioned in the current literature to explain aging is the free radical theory which suggests the involvement of free radical production as a cause of aging (Rubin, 1997; Perez-Campo et al., 1998). This proposal stipulates that aging results from the deleterious effects of free radicals produced in the course of cellular metabolism. These free radicals oxidize enzyme proteins at specific sites which leads to loss of function (Masters and Crane, 1995). Remarkable reversal of aging effects have been observed in experimental animals given compounds which combat oxygen free radicals (Masters and Crane, 1995). Furthermore, increased expression of antioxidant enzymes significantly increased the life span of fruit flies (Orr and Sohal, 1994). In animals, a strong negative exponential correlation exists between liver oxygen radical production and maximum life span (Perez-Campo et al., 1998). Mitochondrial, microsomal (Mehlhorn and Cole, 1985), as well as peroxisomal metabolic pathways, are sources of these free radicals.

Non-enzymatic glycosylation has also been proposed as one of the fundamental mechanisms of aging (Martin, 1992). Although enzymatic glycosylation is a normal metabolic process in which a sugar is bound to a specific site of a protein through an enzyme-mediated reaction, non-enzymatic glycosylation is a random unregulated process (Roffe, 1998). Non-enzymatic glycosylation occurs, at any site sterically available, in the presence of a reducing sugar and is independent on enzymatic catalysis (Roffe, 1998), the result is both structural and physicochemical changes in properties of involved proteins. These changes include reduced solubility and increased resistance to proteolytic digestion, and have been proposed to have a particular impact on the heart (Roffe, 1998). The likelihood of spontaneous non-enzymatic glycosylation increases with increasing levels of reducing sugars and with time (Roffe, 1998). It is, therefore, most pronounced in long-lived proteins such as collagen and elastin (Roffe, 1998). Advanced glycosylation end-products of arterial collagen, as a consequence of contact with circulating glucose, is implicated in the generation of atherosclerotic plaque (Sakata et al., 1995).

Evidence also suggests that changes in apoptosis may be an important factor in aging. Senescent cells are remarkably resistant to stimuli which normally induce apoptosis in young cells (Warner, 1997). Although such a characteristic may extend the life span of terminally differentiated cells, it may also hinder the elimination of damaged cells leading to tissue dysfunction and possibly cancer (Warner, 1997). Further investigation of the relationship between apoptosis and cellular aging may reveal factors and mechanisms important to the issue of controlling aging and cancer.

The most consistent and general changes in cellular behavior in aging animals, with only a few exceptions, is a reduction in the rate of cell proliferation (Rubin, 1997; Roffe, 1998). There is an age-dependent decline in the rate of cell proliferation in most if not all organs, thus age-related changes are not confined to non-renewable cell populations such as nerve and muscle (Rubin, 1997). The rate of cell proliferation in all organs is diminished greatly by early middle age and then remains fairly constant or declines slowly (Rubin, 1997). Although one study showed an enhanced proliferation in the gastrointestinal tract of aged versus young mature Fischer 344 rats, the weight of evidence indicates a reduction with age in cellular proliferation rate in many organs in mammals (Rubin, 1997).

Tissue-Specific Effects

Common mechanisms of aging may operate at the cellular level in most organs and cause wide-spread decline in physiological processes eventually culminating in death (Table 1). For example, the aged liver has a reduced ability to regenerate (Tauchi and Sato, 1978; Popper, 1986). Increases in mitosis in response to hepatectomy or to chemicals is diminished in aged animals, compared with their young counterparts (Bucher, 1963; Tauchi and Sato, 1978; Schapiro et al., 1982; Popper, 1986). Furthermore, the pool of proliferating hepatocytes is smaller in old animals, and is more sharply localized to periportal hepatocytes (Popper, 1986). The importance of these changes in aging are unclear. Studies have produced conflicting data on the effect of aging on the size of hepatocytes. While most studies described conspicuous enlarged as well as small hepatocytes in old rats (Popper, 1986), others found far fewer variations of cell size, especially in female rats (Popper, 1986).

Table 1 Altered characteristics of senescent cells

Parameter	Changes with Aging
Transcription	↓
Translation	↓
Proteolysis	↓
Responsiveness to Hormones	↓
Lipofuscin	↑
Abnormal Nuclei	↑
Chromosomal Abnormalities	↑

↓, decline; ↑, increase. Data are derived from Makarides, 1983 and Dice, 1985.

Interestingly, hepatocellular organelles age faster than the cell itself. Aged Fischer 344 rats had clear nuclear variations in that the normally prominent diploid cells, which frequently contain smaller nuclei, were replaced by polyploid, multinucleated cells (Popper, 1986). Mitochondria appear to generally decrease in number and size as a function of aging, with the smallest mitochondria with shortest cristae observed in the oldest rats (Dice, 1985; Popper, 1986). Similar observations have been documented in both mice and humans (Wilson and Franke, 1975; Herbener, 1976; Popper, 1986). Overall, mitochondrial phosphorylation as well as turnover of mitochondrial proteins are reduced in the aged liver (Weindruch et al. 1980; Popper, 1986). Similarly, the endoplasmic reticulum seems to be diminished in aged animals (Schmucker and Wang, 1980; Popper, 1986). In rats, studies have documented the existence of an age-related decline in several mixed-function oxidase activities without any evidence of significant changes in cytochrome P-450 or NADPH cytochrome reductase content (Birnbaum, 1980). In contrast to results obtained in rats and despite a decline in several mixed-function oxidase activities with age, benzo(a)pyrene hydroxylase and 7-ethoxycoumarin de-ethylase showed significant increases in senescent mice (Birnbaum, 1980). Conversely, activity of liver cytosolic alcohol dehydrogenase increased with age in male Fischer 344 rats (Rikans and Moore, 1987). However, activity of this enzyme was not altered with age in female rats, but was higher than in males in both young and old rats (Rikans and Moore, 1987). In contrast, hepatic aldehyde dehydrogenase activity was similar in both male and female rats and was unchanged with age (Rikans and Moore, 1987). Lysosomes are conspicuously increased in aged animals with their volume density reaching about five times that at birth (Popper, 1986).

The aged heart also exhibits characteristics which are different from those observed in young hearts. Although a net decrease in the number of cardiac myocytes occurs with age, the total myocyte mass increases through enlargement (hypertrophy) of the remaining myocytes (Roffe, 1998). Furthermore, there is an age-related decrease in maximal cardiac oxygen consumption (Hossack and Bruce, 1982; Higgintham et al., 1985), but cardiac output at rest and with exercise is similar in young and older individuals free of cardiovascular disease (Rodeheffer et al., 1984). In isolated rat cardiac myocytes, the contractile response to noradrenaline declines with age (Roffe, 1998). This attenuation of adrenergic responsiveness in the older organism is due to less avid binding of catecholamines by the receptor and a reduction in intracellular response to receptor activation (Xiao and Lakatta, 1991).

The main sources of energy in the myocardium are high-energy phosphates such as ATP and creatine phosphate. ATP levels are reduced in senescent myocardium due to reduced mitochondrial function (Roffe, 1998). Also, creatine phosphate, which is generated from creatine by phosphorylation, is reduced in senescent rat myocardium (Roffe, 1998). In hearts of aged (24-month-old) rats, oxidative phosphorylation was found to be markedly reduced compared with younger animals, due to a reduction in mitochondrial complex I activity (Genova, M.et al., 1995). However, other studies, both in humans and rats, have shown an age-related decrease in cytochrome oxidase, an enzyme that is dependent on the integrity of mitochondrial membranes, particularly their cardiolipin content (McMillin et al., 1993). A reduction of coenzyme Q (ubiquinone) in senescent rat heart has also been considered as a cause of lower respiratory function in aged hearts (Roffe, 1998).

In addition to the above changes, age-related reductions in the activities of cardiac enzymes of the citric acid cycle and those associated with β-oxidation of fatty acids have also been reported (Hansford and Castro, 1985). Levels of carnitine, which binds to fatty acids and facilitates their transfer across mitochondrial membranes, and of acetyl carnitine transferase are also reduced in aged rat hearts (Singh, 1973).

Peroxisomes: Function and Proliferation

The peroxisome is a ubiquitous single membrane-limited cytoplasmic organelle present in animal, plant and fungal cells. In animals, peroxisomes are most abundant in the liver and kidney (Bentley et al., 1993). Morphometric studies show that there are between 370 and 620 peroxisomes per rat hepatocyte, occupying 1.5 to 2% of the cell volume (Staubli et al., 1969). The average peroxisome is approximately one fifth the volume of a mitochondrion (Weibel et al., 1969). Peroxisomal proteins are synthesized on free polyribosomes with most of them made at their final size. These proteins are posttranslationally imported from the cytosol into peroxisomes and have half-lives ranging from 1-15 minutes or longer (Lazarow and Fujiki, 1985).

Function of Peroxisomes

Peroxisomes are respiratory organelles which play a role in cellular active oxygen metabolism (Del Rio et al., 1990). They are involved in a number of essential cellular functions, including, but not limited to, cellular respiration, gluconeogenesis, lipid metabolism, thermogenesis, and synthesis of plasmalogens (Tolbert, 1981; Masters, 1998; Youssef and Badr, 1998). Among the oxidases identified in peroxisomes are: urate oxidase, acyl-CoA oxidase, amino acid oxidase, polyamine oxidase, trihydroxycholestanoyl-CoA oxidase and pristanoyl-CoA oxidase (Van den Bosch et al., 1992). Peroxisomal β-oxidation of fatty acids is among the best-characterized peroxisomal metabolic pathways. This system is not a functional duplicate of the mitochondrial system, as it is now clear that peroxisomes are involved in the degradation of a distinct set of compounds such as the very long-chain fatty acids and branched-chain fatty acids (Van den Bosch et al., 1992). Oxidation of erucic acid, arachidonic acid and tetracosatetraenoic acid is initiated in peroxisomes (Van den Bosch et al., 1992). Mammalian peroxisomes are also a site for cholesterol and ether lipid synthesis (Van den Bosch et al., 1992).

Peroxisome Proliferation

Various chemicals, conditions and factors cause peroxisome proliferation (Tables 2 & 3). This phenomenon was first described in the 1960s. It was observed that numerous electron dense single membrane-limited structures appeared in rat liver following clofibrate feeding (Paget, 1963). Since that time, a myriad of therapeutic agents, industrial chemicals and environmental pollutants of diverse chemical structure have been shown to cause peroxisome proliferation, particularly in rodent livers (Bentley et al., 1993; Youssef and Badr, 1998). This group of chemicals has thus been referred to collectively as peroxisome proliferators.

In addition to increasing the number of peroxisomes, they induce the activities of peroxisomal enzymes and cause hepatomegaly (Bentley et al., 1993). These effects are not necessarily coupled, but are dose- and time-dependent (Bentley et al., 1993). Furthermore, these effects are tissue- and species-specific (De la Iglesia et al., 1982; Eacho et al., 1986; Oesch et al., 1988; Lake et al., 1989; Watanabe et al., 1989; Sakuma et al., 1992), with the liver being the most responsive tissue (Bentley et al., 1993; De Craemer et al., 1994). Starvation, experimental diabetes, administration of peroxisome proliferators and high-fat diets have also been shown to provoke a peroxisome proliferative effect in myocardium and kidneys of rodents, albeit to a lesser extent than in the liver (De Craemer et al., 1994).

Table 2 Age-related diseases associated with altered peroxisomal metabolism

Disease	References
Diabetes	Horie et al., 1981; Thomas et al., 1989
Hypothyroidism/Hyperthyroidism	Fringes and Reith, 1982; Fringes et al.,1982
Obesity	Bentely et al., 1993
Inflammation	Levy et al., 1968

Table 3 Representative peroxisome-proliferating chemicals

	Chemical
I.	Fibrates
	Clofibrate
	Bezafibrate
	Ciprofibrate
	Nafenopin
II.	Nonfibrates
a)	Phthalic acid esters
	Di (2-ethylhexyl) phthalate
	Mono (2-ethylhexyl) phthalate
b)	Perfluorinated fatty acids
	Perfluorooctanoic acid Perfluorodecanoic acid
	Perfluorooctane sulfonic acid
c)	Miscellaneous
	Aspirin
	Ethanol
	Tibric acid
	WY- 14,643
	Cetaben
	Dehydroepiandrosterone
	Valproic acid

Species differences in the response to peroxisome proliferators have been documented. While rats and mice are extremely responsive to the pleiotophic effects of these chemicals, guinea pigs are not and hamsters display an intermediate response (Bentley et al., 1993; Eacho et al., 1986; Oesch et al., 1988; Lake et al., 1989). Humans and other primates are either insensitive or respond marginally to peroxisome proliferators (Eacho et al., 1986; Lake et al., 1989; Watanabe et al., 1989; Bentley, 1993). The potencies of various agents as peroxisome proliferators vary considerably (Bentley, 1993).

Peroxisome Proliferator-Activated Receptors (PPARs)

Mechanisms involved in the response to such a diverse group of compounds, factors and conditions are unclear. However, a receptor-based mechanism for the pleiotropic response to peroxisome proliferators in rodents has been suggested, where a nuclear peroxisome proliferator receptor belonging to the steroid hormone receptor superfamily has been identified (Issemann and Green, 1990; Tsai and O'Malley, 1994). This receptor is activated by several peroxisome proliferators and is termed the peroxisome proliferator- activated receptor (PPAR). Activation of PPAR stimulates the expression of genes encoding peroxisomal and nonperoxisomal proteins (Gottlicher et al., 1992; Youssef and Badr, 1998). Recently, several isoforms of the PPAR have been found and cloned (Kliewer et al., 1994; Vidal-Puig et al., 1997).

It has been demonstrated that the prostaglandin 15-deoxy - $\Delta^{12,14}$ - PGJ_2 and the antidiabetic thiazolidinediones are ligands for the $PPAR_{\gamma}$ subtype of this receptor (Figure 1; Kliewer et al., 1997; Tontonoz et al., 1997). Additionally, leukotriene B_4 and several known peroxisome proliferating agents including WY-14,643 and fibrates have been found to bind specifically to $PPAR_{\alpha}$ (Figure 1; Devchand et al., 1996; Forman et al., 1997). Fatty acids and eicosanoids also bind to $PPAR_{\alpha}$, with varying degrees of specificity (Figure 1; Kliewer et al., 1997; Forman et al., 1997; Wolf, 1998; Gustafsson, 1998).

Studies have documented the existence of a human form of $PPAR_{\alpha}$ ($hPPAR_{\alpha}$; Sher et al., 1993; Mukherjee et al., 1994) and PPARγ ($hPPAR_{\gamma}$; Greene et al., 1995). The tissue distribution pattern of $hPPAR_{\alpha}$ mRNA is similar to that of the rat $PPAR_{\alpha}$. Both are highly expressed in the skeletal muscle, liver and kidneys, while expressed at low levels in the brain and lung (Mukherijee et al., 1994; Su et al., 1998). The principal site of expression of PPARγ is the adipose tissue, but this receptor is also expressed, albeit at lower levels, in many other tissues and cell types such as mammary and colonic epithelium (Mueller et al., 1998).

Effect of Aging and Age-Related Diseases on PPAR-Involved Pathways

Information on the relationship between aging and peroxisomal metabolism is generally scarce. Since the liver is the most responsive organ to the effects of peroxisome proliferators, the vast majority of available data deals with the liver (Youssef and Badr, 1999). Factors responsible for the reported decline in peroxisomal enzyme activities and diminution of their response to peroxisome proliferating chemicals in

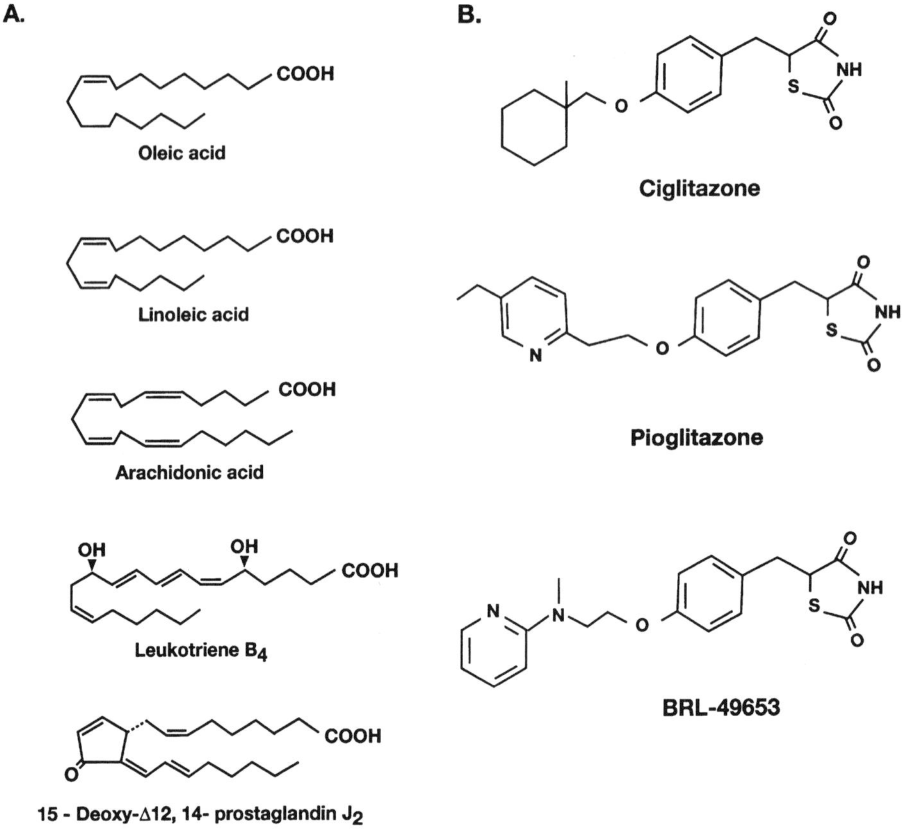

Figure 1. Chemical Structures of Known Representative Physiological (A) and Xenobiotic (B) PPAR ligands.

aged animals are unclear. Aged organs suffer from defects in gene transcription and translation of certain messengers (Roth, 1997). These defects may be the result of deficiencies in important receptors and/or impaired signal transduction pathways (Figure 2).

Convergence of several members of the steroid receptor family has been proposed as a means to explain the effect of peroxisome proliferators. An overlap between the gene networks regulated by peroxisome proliferators and those for retinoids, thyroid hormones and farnesol has been documented (Masters and Crane, 1995, O'Brien et al., 1996). Peroxisome proliferator activated receptors (PPAR), thyroid receptors (TR) and retinoid receptors (RXR) recognize the same DNA response sequence and cross talk among these receptors takes place (Masters and Crane, 1995; O'Brien et al., 1996). PPAR binds to Peroxisome Proliferator-Response Elements (PPRES) as heterodimers with RXR and as such activate gene transcription in response to activators (Mukherjee et al., 1998).

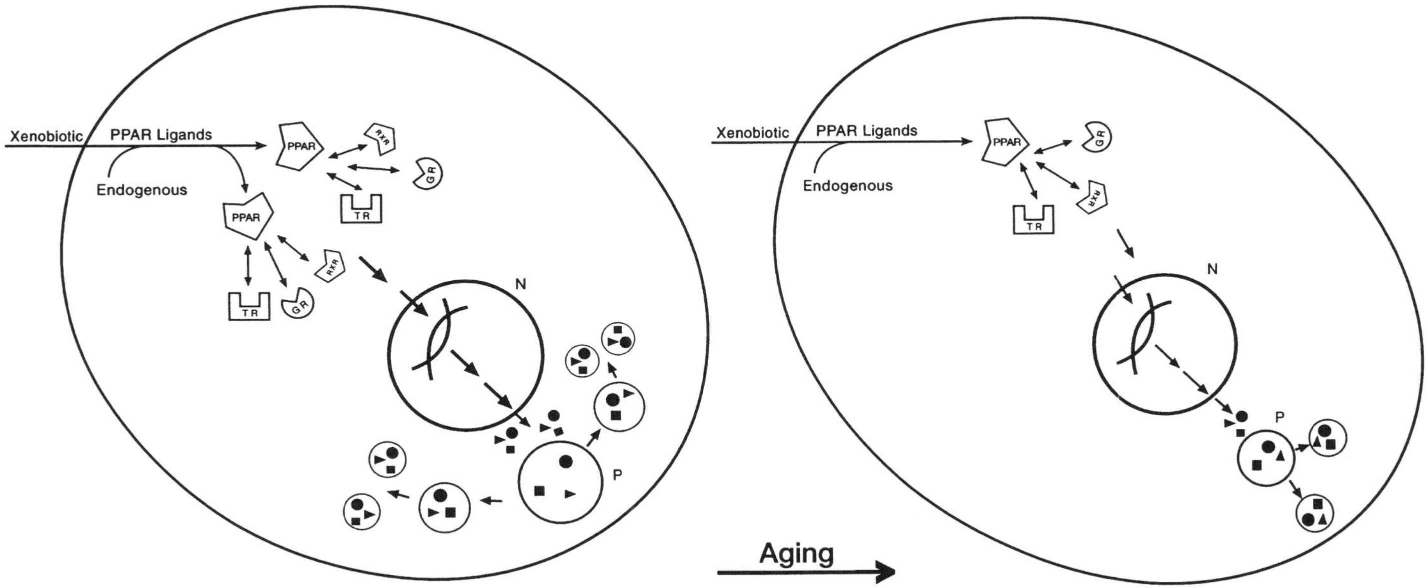

Figure 2. Schematic Depiction of the Effect of Aging on Proposed Signalling Pathways Involved in Maintaining and Proliferating Peroxisomes. GR, Glucocorticoid Receptor; PPAR, Peroxisome Proliferator-Activated Receptor; RXR, Retinoic Acid Receptor; TR, Thyroid Hormone Receptor; N, nucleus; P, peroxisome. Age-related down-regulation of members of the steroid receptor superfamily leads to insufficiency in signal transduction pathways necessary to maintain normal levels of basal peroxisomal enzyme activities. This, in turn, further contribute to the process of aging and to age-related diseases.

Furthermore, RXR ligands have recently been shown to produce effects similar to those produced by PPAR ligands (Mukherjee et al., 1998). A further example of the convergence of these receptors is manifested in the fact that peroxisome proliferators induce malic enzyme gene transcription, which is a known response to treatment with thyroid hormones (Pallet et al., 1997), through the action of heterodimers of $PPAR_{\alpha}$ and RXR (Castelein et al., 1994).

Recent findings show that expression of nuclear thyroid hormone receptor mRNA is reduced by 50% in the livers of male Wistar rats 24 months old compared to their 6 month old counterparts (Pallet et al., 1997). This decrease was accompanied by small changes in the binding characteristics of the receptors, both the B_{max} and K_d decreased by about 2 fold in the old rats. Similar changes were also observed for the mRNA of retinoic acid receptor (Pallet et al., 1997). The impact of aging on the farnesoid X receptor (FXR) is not known. However, the glucocorticoid receptor which is also involved in regulating PPAR-mediated pathways declines with aging (Sapolsky et al., 1983, 1984; Rigter et al., 1984; Lemberger et al., 1994). Synthesis of $PPAR_{\alpha}$ mRNA is stimulated by glucocorticoid hormones, an effect which was blocked by the glucocorticoid receptor antagonist RU 486 (Lemberger et al., 1994).

In addition to changes in receptor abundance and affinity, aging also causes modifications in cellular transduction pathways leading to altered gene expression which may be responsible for some important impairment in physiological functions (Roth, 1997). It remains to be investigated whether aging has an impact on the expression of the various peroxisome proliferator-activated receptor isoforms and/or binding of peroxisome proliferating chemicals to these receptors. This issue is the focus of current research endeavors in our laboratory.

Membrane structure, fluidity and permeability to ions undergo some age-dependent alterations (Zs-Nagy, 1979; Roth et al., 1995; Perichon and Bourre, 1996). These alterations have serious inhibitory effects on the rates of enzymatic catalysis involved in protein synthesis and may be responsible for the decline in the protein synthetic capacity of aged cells. They may also contribute to the onset of age-related diseases (Zs-Nagy, 1979; Roth et al., 1995; Perichon and Bourre, 1996). For example, membrane fatty acid composition, which is maintained through a balanced process of synthesis and degradation, affects membrane structure and function (Stubbs and Smith, 1984). While degradation of long-chain fatty acids occurs in both mitochondria and peroxisomes, very long-chain fatty acids are metabolized exclusively in peroxisomes (Jakobs and Wanders, 1991). Thus, peroxisomes appear to play an important role in maintenance of membrane structure and function by regulating fatty acid balance. Consequently, the decline in peroxisomal catabolism of fatty acids with aging may represent a potential mechanism by which these organelles are involved in the process of aging.

Peroxisomal PPAR Gene Products in Aged Animals

Studies have pointed to a general decrease in peroxisomal function with aging (Perichon and Bourre, 1995, 1996). However, this is not always the case as several studies failed to document measurable differences in peroxisomal enzyme activities between young and old animals (Cattley et al., 1991; Huber et al., 1991). Peroxisomes become smaller, but more numerous in older rats (Popper, 1986). In aged rats, peroxisomes show decreased catalase and acyl-CoA oxidase content, but increased content of thiolase and urate oxidase (Beier et al., 1993). These changes are thought to contribute to oxidative stress (Masters and Crane, 1995). Our laboratory, as well as others, previously showed that peroxisome proliferation was not accompanied by enhanced levels of oxidative damage in young, mature animals (Soliman et al., 1997; Youssef and Badr, 1998). Indeed, our data suggested that peroxisome proliferators may diminish hepatic levels of isoprostanes, a novel and sensitive marker of oxidative stress (Soliman et al., 1997). A recent report shows that induction of $PPAR_{\alpha}$ down-regulates plasma thiobarbituric acid-reactive substance, a marker of cellular oxidative stress (Inoue et al., 1997). These findings leave us with the possibility that potential down-regulation of $PPAR_{\alpha}$ with aging may lower the threshold for oxidative stress in senescent animals.

Basal Peroxisomal Enzyme Activities in Aged Animals

Apparent inconsistencies in the age-related effects on hepatic peroxisomal β-oxidation (Table 4) may have to do more with species differences and the age of animals at the time of sacrifice, than their age at the beginning of the experiment. For example, in some studies animals were kept for extended periods resulting in "younger" and "older" animals being senescent at the time of sacrifice (Table 4). Another potential reason for these apparent disparate results may reside in the ages of animal groups used, as liver peroxisomal enzyme activities appear to change dramatically and abruptly at certain age points during development and aging (Perichon and Bourre, 1995). Although not as abundant as in the liver or kidney, nervous tissues also contain peroxisomes (Imamura et al., 1994). Immunohistochemical studies on human brains suggest that peroxisomal proteins are closely related to neuronal maturation and gliogenesis of premyelination during development (Houdou et al., 1991, 1993; Imamura et al., 1994).

The decline in liver catalase activity, in male Fischer 344 rats, with aging has been uniformly observed in various studies (Table 5). In our laboratories, hepatic catalase activity declined by 40% between the ages of 4 and 100 weeks, with most of the decline (30%) observed between the ages of 50 and 100 weeks (Table 5). Similar findings were also observed by others (Perichon and Bourre, 1995) for both liver peroxisomal β-oxidation and catalase in both CSWV and OFi mice strains (Tables 4 & 5), and similar findings were reported for the enzymatic activity and mRNA coding for catalase in rat liver (Semsei et al., 1989; Xia et al., 1995). Transcription of the catalase gene decreased by approximately 60% between 6 and 29 months of age (Semsei et al., 1989). Thus, the age-related decline in catalase mRNA levels appears to arise primarily from a decrease in the transcription of the gene (Semsei et al., 1989). Quantitative analysis of kinetic parameters (rates of protein synthesis and degradation) revealed that the aged rat liver exhibited a decrease in rates of catalase synthesis without significant changes in degradation rates (Haining and Legan, 1973). In a recent study from our laboratory (Xu and Badr, 1999), hyperinsulinemia diminished hepatic peroxisomal catalase activity also by decreasing its rates of synthesis, without influencing its degradation. Since hyperinsulinemia predisposes individuals to develop several age-related diseases, it is possible that these diseases are mediated by the resulting imbalance between the capacity to produce and to decompose H_2O_2 in peroxisomes (Xu and Badr, 1999).

Table 4 Effect of age on hepatic peroxisomal β-oxidation

Species	Young	Old	Reference
Male Fischer 344 rats	100[a] (17)[b]	117 (28)	Huber et al., 1991
	100 (8)	115 (20)	Cattley et al., 1991
	100 (2)	92 (13)	Yamoto et al., 1995
Female CSWV mice	100 (11)	28 (20)	Perichon and Bourre, 1995
Female OFi mice	100 (11)	61 (24)	Perichon and Bourre, 1996

[a] Results are expressed as percent of activity in young animals within the same study.
[b] Numbers in parenthesis indicate age of animals, in months, at sacrifice.

Table 5 Liver catalase in animals of various ages

Species	Young	Old	Reference
Male Fischer 344 rats	100a (3)b	63 (24)	Badr and Birnbaum, 1991
	100 (5)	48 (26)	Fornos et al., 1990
	100 (6)	60 (26)	Xia et al., 1995
	100 (6)	61 (24)	Semsei et al., 1989
	100 (6)	48 (29)	Semsei et al., 1989
Female CSWV mice	100 (11)	71 (18)	Perichon and Bourre, 1995
Female OFi mice	100 (11)	66 (24)	Perichon and Bourre, 1996

[a] Results are expressed as percent of activity in young animals within the same study.
[b] Numbers in parenthesis are age of animals in months.

Data available show that myocardium catalase was significantly diminished in aged rats (Ji et al., 1991). The decline was more significant, however, in hearts of 26-month-old rats compared to their 31-month counterparts (Ji et al., 1991). Reasons for and/or significance of this apparent inconsistency are not readily discernable. Lung peroxisomal catalase activity also shows age-dependent patterns. In lungs from young rats, peroxisomal catalase activity is 3-4 times higher than in their aged counterparts (Amicarelli et al., 1997).

Peroxisome Proliferation in Aged Animals

Aging interferes with the ability of the liver to respond to peroxisome proliferating chemicals (Yamoto et al., 1995). While peroxisomal area relative to hepatocellular cytoplasmic area increased by about 7-fold in both 8 and 52 week old rats, following 14 days of receiving daily doses of 200 mg/kg clofibrate in diet, it increased only by 4 fold in 117 week old rats (Yamoto et al., 1995). In the same animals, increases in hepatic peroxisomal β-oxidation showed a similar pattern, with increases approaching 15 and 11 fold in 8 and 52 week old rats, but only 3 fold in the 117 week old rats (Yamoto et al., 1995). In our laboratory, perfluorooctanoic acid (PFOA) induced peroxisomal β-oxidation in male Fischer 344 rats of varying age groups to different age-related levels. Forty eight hours following 150 mg/kg PFOA, hepatic peroxisomal β-oxidation increased in 100 week old rats to levels equaling about 60% of those achieved in 10 week old rats (Table 6).

In contrast to the above findings, young and old male Fischer 344 rats given various concentrations of the peroxisome proliferator WY-14,643 in the diet exhibited comparable hepatic peroxisomal β-oxidation activity (Cattley et al., 1991; Table 6).

Table 6 Induction of liver Peroxisomal β-oxidation by various peroxisome proliferators in aged animals

Species	Young	Old	Reference
Male Fischer 344 rats	100a (3)b,c	44 (25)	Badr and Birnbaum, 1991
	100 (2)d	68 (13)	Yamoto et al., 1995
	100 (8)e	85 (20)	Cattley et al., 1991
Female OFi mice	100 (11)[f]	70 (24)	Perichon and Bourre, 1996

[a] Results are expressed as percent of peroxisome proliferator-induced activity in young animals within the same study.
[b] Numbers in parenthesis are age of animals in months.
[c] Animals were given perfluorooctanoic acid (150 mg/Kg, orally in corn oil 48 hrs prior to sacrifice).
[d] Animals were given clofibric acid in the diet (various concentrations to correspond to 200 mg/Kg/day for 14 day).
[e] Animals received WY-14,643 in the diet (0.1% for 22 weeks).
[f] Animals received clofibrate-supplemented diet (0.5% for 5 days).

Similarly, a study in male Wistar rats showed that age did not seem to influence either basal or nafenopin-induced induction of peroxisomal β-oxidation (Huber et al., 1991). In these long-term feeding studies (Cattley et al., 1991; Huber et al., 1991), animals were given diets containing peroxisome proliferators for 22-59 weeks rendering the age of young and old rats between 68-116 weeks at the conclusion of the experiment. Potentially, these are ages where responses to peroxisome proliferators are similar. In a more detailed study in our laboratories (unpublished data), striking results showing compound-specific effects were observed (Table 7). While aging-related differences appeared to exist in the response of the liver to certain peroxisome proliferators, aging did not seem to influence the response to other peroxisome proliferators studied (Table 7). Similar to results reported in the literature, WY-14,643 induced hepatic peroxisomal β-oxidation to comparable levels in young and old male Fischer rats (Table 7). Conversely, peroxisomal β-oxidation increased, in response to perfluorooctanoic acid, in livers of young rats by 40% above that observed older rats (Table 7).

These striking new data clearly show that the induction of peroxisomal β-oxidation in aged rats by peroxisome proliferating agents has a chemical-specific component.

Alterations in drug handling, including activation/deactivation, may be of critical importance to the final outcome on the level of peroxisomal β-oxidation.

Table 7 Age-related responses of hepatic peroxisomal β-oxidation to various peroxisome proliferators

Compound	Young 10 weeks old	Old 100 weeks old
PFOA	100 [a,b]	62*
WY-14,643	100 [c]	93
DEHP	100 [d]	137*

[a] Results are expressed as percent of hepatic enzyme activity in young rats.
[b] perfluorooctanoic acid (PFOA; 150 mg/Kg),
[c] WY-14,643 (250 mg/Kg),
[d] diethylhexylphthalate (DEHP; 1.2 g/Kg). Drugs in corn oil were given orally 48 hrs prior to sacrifice.
* Significantly different from corresponding young animal group.

Peroxisomal Metabolism in Age-Related Diseases

It has been shown that several diseases that are prevalent in the elderly population have the ability to alter peroxisomal metabolism in young, healthy animals. Among these conditions are diabetes (Horie et al., 1981; Thomas et al., 1989), changes in thyroid hormones levels (Fringes and Reith, 1982; Fringes et al., 1982), and an inflammatory response to endotoxin exposure (Levy et al., 1968). These conditions produce, in young animals, an effect similar to that initiated by the peroxisome proliferating xenobiotics, albeit not to the same extent. Whether there is a causal relationship between these diseases and the aging process is not clear. Furthermore, the peroxisomal response of senescent animals to these conditions is not known.

Nonperoxisomal PPAR Gene Products in Aged Animals

It has been postulated that mitochondria play a role in the aging process (Brierley et al., 1997). Mitochondrial respiratory chain function falls by half in humans between the ages of 20 and 80 years (Brierley et al., 1997). Mitochondria are the main source of energy in the cell, and they contain their own DNA (Mt DNA). Mt DNA is exclusively concerned with coding for 13 polypeptides, 22 tRNAs, 2 rRNAs and all components of the respiratory-chain/oxidative phosphorylation system (Brierley et al., 1997). There is good evidence that Mt DNA is more vulnerable to damage than nuclear DNA (Brierley et al., 1997). Interference with mitochondrial function may lead to cellular and tissue dysfunction or death and, thus, be causally related to aging (Brierley et al., 1997). Recent studies have revealed that PPAR is involved in the regulation of mitochondrial enzyme genes. Among those PPAR-regulated enzymes are 3-hydroxy-3-methylglutaryl-CoA synthase (Rodriguez et al., 1994) and medium-chain acyl-CoA dehydrogenase (Gulick et al., 1994). Furthermore, we, as well as others, have documented that peroxisome proliferators alter various mitochondrial functions, including oxidative phosphorylation (Winberg and Badr, 1995; Youssef and Badr, 1998).

Studies focusing on age-related changes in cytochrome P-450 demonstrated isozyme-selective decreases in CYP 2B1, CYP 2B2, CYP 2C11 and CYP 2E1 activity (Handler and Brian, 1997). Peroxisome proliferators are known to enhance the expression and activity of a specific cytochrome P-450 isozyme. This novel isozyme is designated as CYP 4A, which comprises several subforms (Youssef and Badr, 1998). Human liver microsomes have been shown to contain multiple members of CYP 4A, including CYP 4A9, CYP 4A11, CYP 4A12 and CYP 4F2 (Castle et al., 1995). In rat livers, aging was accompanied by a significant reduction in microsomal fatty acid ω-hydroxylation which is known to reflect the activity of CYP 4A 1 and other members of this gene family (Yamoto et al., 1995). The responsiveness of this activity to the inducing effect of the peroxisome proliferator clofibric acid was also significantly diminished in aged rats (Yamoto et al., 1995).

Studies show that the prevalence of hypothyroidism in the elderly is high (Robuschi et al., 1987), and that it is responsible for significant morbidity (Robuschi et al., 1987). Thyroid hormones play a role in regulating energy expenditure, and the basal metabolic rate decreases by about 50% between the ages of 3 and 85 years old (Robuschi et al., 1987). The thyroid hormone marker malic enzyme is a cytosolic enzyme which catalyzes the oxidative decarboxylation of malate to pyruvate, producing NADPH required for fatty acid biosynthesis (Castelein et al., 1994). The activity of malic enzyme and the level of expression of its gene are regulated by thyroid hormones (Badr, 1992; Castelein et al., 1994). Interestingly, virtually all known peroxisome proliferating chemicals have been shown to have a thyromimetic effect and to regulate the activity and gene expression of malic enzyme (Youssef and Badr, 1998). Recently, a cross talk between PPARs and the thyroid hormone receptor was proposed, and malic enzyme gene expression is reported to be regulated through a PPAR-mediated mechanism.

Aging, Peroxisomes and Health

The importance of the peroxisome to animal as well as to human life is evident by the fact that their absence, or the dysfunction of one or more peroxisomal enzymes, is invariably fatal (Schutgens et al., 1986; Xu et al., 1995; Singh, 1997). In addition, inhibiting peroxisomal enzyme activities resulted in the accelerated death of animals (Xu et al., 1995), suggesting the existence of a linkage between peroxisomal metabolism, aging and death.

Inherited Peroxisomal Disorders

Peroxisomal disorders are a group of inherited metabolic diseases which are classified into ten complementation groups (Shimozawa et al., 1998). These diseases involve multiple genes (Shimozawa et al., 1998) and are characterized by impairment of one or more functions of peroxisomes (Schutgens et al., 1986; Singh, 1997; Shimozawa et al., 1998). Recent studies have suggested that at least 16 different genes may be involved in the observed abnormalities in the assembly of peroxisomes (Moser et al., 1995; Singh, 1997), and the list of peroxisomal disorders continues to grow (Shimozawa et al., 1998). The combined incidence for the occurrence of these diseases is estimated to be 1 in 20,000 or greater (Singh, 1997). The cerebrohepatorenal syndrome, known as Zellweger Syndrome, was the first peroxisomal disorder identified (Goldfischer et al., 1973; Singh, 1997). This disease is characterized by the absence of morphologically distinguishable peroxisomes (Singh, 1997). The most common peroxisomal disease X-adrenoleukodystrophy results from the inability to metabolize very long chain fatty acids, a function which is located exclusively in peroxisomes (Singh et al., 1984; Singh, 1997). Patients suffering from this disease have peroxisomes of normal number and morphology, and the disease is caused by a single enzymatic abnormality, with lignoceroyl-CoA ligase being suggested as the missing enzyme (Singh, 1997). Other diseases, such as Rhizomelic Chondrodysplasia, are recognized as multiple peroxisomal enzymatic deficiencies but with intact peroxisome morphology, normal size and number (Singh et al., 1991; Singh, 1997).

Involvement of Peroxisome Proliferator-Activated Receptor in Age-Related Diseases

As the population of the U.S. and other industrial countries ages and average body weight increases, there is a more urgent need to understand the molecular basis of the aging process, as well as obesity and medical conditions clearly associated with obesity, such as type 2 diabetes and cardiovascular disease (Spiegelman, 1998). In this section, available data on potential roles of peroxisomes and/or various PPAR isoforms in age-related diseases are presented and discussed.

Obesity

There are two splice variants of PPARγ; γ1 and γ2 (Vidal-Puig, et al., 1997). Adipose expression of PPARγ2 mRNA has been shown to increase in obese male and female human subjects. In fact there is a strong positive correlation between the ratio of PPAR γ2/γ1 and the body mass index (Vidal-Puig, et al., 1997). Interestingly, differential expression of PPARγ in intra-abdominal and subcutaneous adipose tissues in humans (lower PPARγ expression in visceral adipose tissue in lean individuals) may underscore the association of visceral, but not subcutaneous fat with obesity and obesity-related cardiovascular and metabolic problems (Lefebvre et al., 1998a). Ligand activation of PPARγ in cultured fibroblasts or preadipocytes stimulates adipose cell differentiation, including the cessation of cell growth (Spiegelman and Flier, 1996). It has also been shown that when stimulated by its strongest activators, PPARα also induced adipogenesis, though to a lesser extent and more slowly than that stimulated by PPARγ (Spiegelman and Flier, 1996). PPARδ could not stimulate detectable fat cell formation (Brun et al., 1996; Spiegelman and Flier, 1996). Future development of PPARγ antagonists as potential treatment for obesity presents an exciting hope for the multitudes of overweight individuals.

Diabetes

Recognition that PPARγ is the receptor for the anti-diabetic thiazolidinedione drugs, such as troglitazone, linked this receptor to glucose homeostasis in rodents and humans (Forman et al., 1995; Kliewer at al., 1995). *In vitro* experiments revealed that insulin induces the expression of PPARγ mRNA in human adipocytes (Vidal-Puig, et al., 1997). Furthermore, there appears to be an inverse relationship between the level of PPARγ mRNA and that of nonfunctional insulin receptors in human visceral adipose tissue (Lefebvre et al., 1998a). PPARγ ligands improve insulin sensitivity *in vivo* (Spiegelman and Flier, 1996). Whether this effect occurs primarily through these receptors in adipose tissue and/or through their low levels in non-adipose tissues (muscle and liver), is not known (Spiegelman and Flier, 1996). It is estimated that 730,000 Americans currently take $PPAR_{\gamma}$ agonists to improve insulin action in diabetes and advanced clinical trials are underway to assess their activity in other insulin-resistant states such as polycystic ovarian syndrome (Spiegelman, 1998). In addition, clinical trials are underway to assess the ability of $PPAR_{\gamma}$ agonists to prevent diabetes and to regulate cell growth in cancers of the adipogenic lineage, liposarcoma (Spiegelman, 1998). Recent reports link troglitazone with liver failure, tempering the excitement for this drug and possibly for this entire class of once promising anti-diabetogenic agents.

Cardiovascular Diseases

One of the earliest steps in the formation of atherosclerotic plaques in the walls of large blood vessels is the conversion of macrophages to cholesterol-engorged foam cells (Steinberg et al., 1989). These are cells that have internalized lipids and stored them as cytoplasmic droplets, which gives the cells their characteristic foamy appearance under light microscopy. It has been postulated that the PPARγ receptor is involved in the conversion of monocytes to foam cells (Steinberg et al., 1989). PPARγ is induced in human monocytes following exposure to oxidized lipoprotein and is expressed at high levels in the foam cells of atherosclerotic lesions (Tontonoz et al., 1998). Changes in macrophage lipid metabolism, leading to the formation of foam cells, include increased free and esterified cholesterol content, increased content of lipoprotein oxidation products and increase in acyl CoA: cholesterol acyltransferase and acid cholesterol-ester hydrolase activities (Tontonoz et al., 1998). Hence, modulating PPARγ expression/activity would have an obvious potential pharmacological value in preventing and/or treating atherosclerosis. Accumulation of LDL cholesterol by monocytes is an early stage of atherogenesis (Tontonoz et al., 1998).

Aging-associated decrease in myocardial antioxidant activity, including the peroxisomal catalase (Ji et al., 1991), is expected to contribute to enhanced oxidative stress in the aging heart. This is a particularly important consideration in light of the fact that the aging heart produces more oxygen free radicals compared to younger hearts (Ji et al., 1991). Furthermore, a recent report shows that PPARα may be involved in reducing oxidative conditions in blood (Inoue et al., 1997). Potential down-regulation of this receptor with aging may result in enhanced oxidation of LDL, followed by its accumulation in monocytes and foam cell formation.

Inflammation and Immunity

Inflammation is a local response to exogenous molecules, infection or injury (Devchand et al., 1996). Chronic inflammation from infectious and noninfectious sources contribute to about one-third of the world's cancer (Ames et al., 1993). While the immune response in the elderly decreases against exogenous antigens, autoimmune reactivity increases (Wick et al., 1991). These effects are blamed on changes in the lipid metabolism of cells of the immune system (Wick et al., 1991). Diminished *in vitro* responsiveness of lymphocytes from the elderly correlates with decreased plasma membrane fluidity (Rivnany et al., 1979, 1980), which may influence the accessibility and function of surface receptors (Wick et al., 1991). Furthermore, monocytes and macrophages express high levels of $PPAR_\gamma$ mRNA (Jiang et al., 1998) and a very wide range of $PPAR_\gamma$ ligands and certain nonsteroidal antiinflammatory drugs have been shown to strongly inhibit inflammatory cytokine production by stimulated human monocytes (Jiang et al., 1998). Therefore, changes in immune cell membrane lipid composition, which may result from age-related depression of peroxisomal lipid metabolism or altered regulation of PPARγ with age, may explain the altered immune reactivity in old age.

In addition to the potential involvement of PPARγ in age-related altered immunity, PPARα may also be involved in this immune modification. Leukotriene B_4, a potent chemotactic agent involved in the initiation, coordination and amplification of the inflammatory response, has recently been shown to activate $PPAR_\alpha$ (Devchand et al., 1996). Consequently, it was postulated that activation of $PPAR_\alpha$ by leukotriene B_4 would enhance the catabolism of this mediator to its less potent metabolites (Devchand et al., 1996).

Cancer

The incidence of various cancers increases with about the fifth power of age in both short-lived species, such as rats, and long-lived species, such as humans (Ames et al., 1993), clustering in the terminal part of the life span (Herman, 1981). Oxidative damage to DNA, among other macromolecules, has been postulated to be a major source of damage, leading to aging and cancer (Ames et al., 1993). Among sources of endogenous oxidants are by-products of mitochondrial and microsomal metabolism, oxidative bursts, which accompany phagocytic activity, and peroxisomes (Ames et al., 1993). Sequesteration of H_2O_2-producing enzymes in peroxisomes may act as a natural defense mechanism against oxidative damage in young animals (Ames et al., 1993). Therefore, changes in peroxisomal membrane structure with aging (Masters and Crane, 1995) may lead to the leakage of these enzymes into the cytosol with subsequent enhanced potential for oxidative damage and cancer in old animals.

The potential role of $PPAR_\gamma$ in cancer formation in humans is controversial. While two recent studies implicate $PPAR_\gamma$ in promotion and development of colon cancer, a third report indicates a possible protective role for PPARγ agonists against colon cancer in humans. Lefebvre et al., (1998b) and Saez et al., 1998 present evidence that mice genetically predisposed to develop polps in the colon show increased numbers of polyps when subjected to $PPAR_\gamma$ agonists orally. Conversely, Sarraf et al., 1998 demonstrate that human colon tumor cell lines both in culture and in nude mice respond to $PPAR_\gamma$ agonists with a reduced rate of growth and an increased degree of differentiation. Obviously, further studies are needed to clarify the exact potential outcome from exposure to $PPAR_{\gamma}$ agonists in humans. These drugs have been reported as useful for the differentiation in liposarcomas and breast tumors (Seed, 1998).

Another connection between peroxisomes and cancer arises from the fact that aged animals are more susceptible to the hepatocarcinogenic effect of peroxisome proliferators compared to their young counterparts (Kraupp-Grasl et al., 1991; Huber et al., 1991). Long-term treatment with either of the peroxisome proliferating chemicals nafenopin or WY-14,643 produced numerous hepatocellular adenomas and carcinomas in old male Wistar and Fischer 344 rats, while producing only fewer changes in young animals (Kraupp-Grasl et al., 1991; Huber et al., 1991). Since the enhanced susceptibility of old rats did not appear to result from enhanced hepatic oxidative damage, it was hypothesized that peroxisome proliferators may promote spontaneously occurring preneoplastic foci. These foci are more abundant in the livers of aged animals as a natural consequence of aging (Kraupp-Grasl et al., 1991). Hypolipidemic drugs are preferentially prescribed to older individuals who may have high levels of preneoplastic hepatocytes. Promotion of these cells by peroxisome proliferators has the potential to result in liver cancer. Furthermore, $PPAR_{\gamma}$ has also been reported to be involved in some common cancers (Spiegelman, 1998).

Alternatively, several studies have shown that peroxisome proliferators suppress apoptosis in the liver in a process mediated by PPARα (Roberts et al., 1998). Apoptosis appears to be a safeguard to prevent cells with DNA damage from progressing to a tumor (Ellis et al., 1991). In primary cultures of rat hepatocytes and in a FaO rat hepatoma cell line, apoptosis was induced by TGFβ1, an effect which was significantly reduced by the co-administration of the peroxisome proliferator nafenopin (Bayly et al., 1994). Furthermore, nafenopin suppressed hepatic apoptosis *in vivo* and its withdrawal resulted in a 100- to 200-fold elevation in apoptosis (Roberts et al., 1995). Hepatocytes generated during nafenopin-induced hyperplasia were not the same that underwent apoptosis upon nafenopin withdrawal (Roberts et al., 1995). Hepatocytes resistant to apoptotic death may represent preferential targets for promotion by peroxisome proliferators, and hence suppression of apoptosis may play a role in the hepatocarcinogenecity of this class of non-genotoxic compounds. Senescent cells are resistant to apoptotic stimuli (Warner, 1997). This characteristic may contribute to the enhanced vulnerability of the aged liver to the hepatocarcinogenic effect of peroxisome proliferators.

Summary

Despite the myriad of peroxisomal enzymes involved in essential cellular metabolic pathways, little attention has been focused on the relationship between changes in the activities of these enzymes and the process of cellular aging. A few reports in the literature suggest the presence of an aging-associated decline in hepatic peroxisomal enzyme activities and in their response to peroxisome proliferating chemicals.

The exact basis for the relationship between peroxisomal deficit, aging and death is not clear. However, an aging-related decline in peroxisomal protein synthesis, among other proteins, is expected to diminish peroxisomal capacity to metabolize very long-chain fatty acids. This effect may be manifested in an altered composition of the structure of cellular and organellar membranes. For example, it has been reported that aging causes an increase in the cholesterol/phospholipid ratio in rat liver microsomal, mitochondrial and cellular membranes (Grinna, 1976; Zs-Nagy, 1979). Phospholipids are essential for the activation of various signal transduction pathways. This includes the modulation of protein kinase C activity by membrane phospholipids (Takai et al., 1984). Accordingly, changes in membrane phospholipids as a result of age-related diminished peroxisomal metabolism may trigger a cascade of events and/or interrupt normal important cellular functions. Since PPARs regulate gene expression of enzymes associated with lipid homeostasis (Devchand et al., 1996), it is feasible that age-related changes in the expression and/or activity of these receptors may be involved in aging and age-related diseases. Also, it becomes clear that altered expression and/or responsiveness of PPARs in aged subjects could potentially result in the modulation of inflammatory intensity and duration, obesity, diabetes and cancer.

Another connection between peroxisomal senescence and cellular aging is exemplified by the fact that aging-associated changes in membrane phospholipid composition is also accompanied by an increase in Na, K-ATPase activity and intracellular potassium content (Zs-Nagy, 1979). Increased intracellular ion concentrations are postulated to reduce the activity of the whole translational machinery, increase aggregation of macromolecules, retarding all enzymatic catalysis (Hey et al., 1976; Zs-Nagy, 1979), and to interfere with mitotic activity (Cone, 1971). This, in turn, may initiate a vicious cycle leading to further diminution of peroxisomal metabolism. Peroxisomes grow by the posttranslational incorporation of new content and membrane proteins into preexisting peroxisomes which then divide to form daughter peroxisomes in a dynamic "fission" process (Lazarow and Fujiki, 1985).

The cloning and identification of PPARγ as a nuclear receptor that can regulate important facets of energy balance has offered new opportunities to understand several key metabolic processes. This form of PPAR is considered a promising candidate gene for involvement in several human disorders including obesity and type 2 diabetes mellitus (Yen, et al., 1997). Although adipose tissue has been recognized as a principal site of expression of PPARγ, this receptor is also expressed, albeit at lower levels, in many other tissues and cell types (Mueller et al., 1998). The apparent contradiction between the need to develop PPARγ agonists to treat insulin resistance and the need to develop PPARγ antagonists to treat obesity will require further understanding of the potential multiple roles of these receptors and the intricate interrelationships between various metabolic pathways.

Although the existence of a causal relationship between peroxisomal senescence and cellular aging has not been established, the importance of these organelles to normal cell functions and cellular survival makes the investigation of their role in aging and aging-related diseases an urgent task. The impact of aging on extracellular matrix, growth factors, cytokines and nonparenchymal cells, among other factors which are known to influence peroxisomal metabolism, will be of utmost importance to our understanding of the relationship between aging and peroxisomal metabolism. Alterations in these factors, with aging, may explain aging-related effects on peroxisomal population and enzyme activities and their role in aging-related disease and death.

Acknowledgement

Aging studies in our laboratories are supported by grants from the NIH, CA/OD 74384 and AG 18479.

References

Ames, B.N., Shigenaga, M.K. & Hagen, T.M. (1993). Oxidants, antioxidants and the degenerative diseases of aging. Proc. Natl. Acad. Sci. U. S.A. 90, 7915-7922.

Amicarelli, F., Di Ilio, C., Mascicco, L., Bonfigli, A., Zarivi, O., D'Andrea, M.R., Di Giulio, C. & Miranda, M. (1997). Aging and detoxifying responses to hypoxic or hyperoxic treatment. Mech. Agein Dev. 97, 215-226.

Badr, M. & Birnbaum, L. (1991). Selective induction of peroxisomal enzyme activities by perfluorooctanoic acid in aged rats. The Toxicologist 187, 11.

Badr, M. (1992). Induction of peroxisomal enzyme activities by diethylhexylphthalate in thyroidectomized rats with thyroid replants. J. Pharmacol. Exp. Ther. 263, 1105-1110.

Bayly, A., Roberts, R. & Dive, C. (1994). Suppression of liver cell apoptosis *in vitro* by the non-genotoxic hepatocarcinogen and peroxisome proliferator nafenopin. J. Cell Biol. 125, 197-203.

Beier, K., Volkl, A. & Fahimi, H. (1993). The impact of aging on enzyme proteins of rat liver peroxisomes-quantitative analysis by immunoblotting and immunoelectron microscopy. Virchows Arch. B. 63, 139-146.

Bentley, P., Elcombe, C., Grasso, P., Stringer, D. & Wiegand, H. (1993). Hepatic peroxisome proliferation in rodents and its significance for humans. Fd. Chem. Toxicol. 31, 857-907.

Birnbaum, L.S. (1980). Hepatic drug metabolism in senescent mice. Exp. Geront. 15, 259-267.

Brierley, E., Johnson, M., James, O. & Turnbull, D. (1997). Mitochondrial involvement in the aging process. Facts and controversies. Mol. Cell. Biochem. 174, 325-328.

Brun, R.P., Tontonoz, P., Forman, B.M., Ellis, R., Chen, J., Evans, R.M. & Spiegelman, B.M. (1996). Differential activation of adipogenesis by multiple PPAR isoforms. Genes Dev. 10, 974-984.

Bucher, N.L.R. (1963). Regeneration of mammalian liver. Int. Rev. Cytol. 15, 245-300.

Castelein, H., Gulick, T., Declercq, P., Mannaerts, G., Moore, D. & Baes, M. (1994). The peroxisome proliferator activated receptor regulates malic enzyme gene expression. J. Biol. Chem. 269, 26754-26758.

Castle, P., Merdink, J., Okita, J., Wrighton, S. & Okita, R. (1995). Human liver lauric acid hydroxylase activities. Drug Metab. Dispos. 23, 1037-1043.

Cattley, R.C., Marsman, D.S. & Popp, J.A. (1991). Age-related susceptibility to the carcinogenic effect of the peroxisome proliferator WY-14'643 in rat liver. Carcinogenesis 12, 469-473.

Cone, C.D. (1971). Unified theory on the basic mechanism of normal mitotic control and oncogenesis. J. Theor. Biol. 30, 151-181.

Cutler, R.G. (1982). Longevity is determined by specific genes, in: Testing the theories of aging (Adelman, R.C. and Roth, G.S., eds.) pp 25-114, CRC Press, Boca Raton, FL.

De Craemer, D., Vamecq, J., Roels, F., Vallee, L., Pauwels, M. & Van den Branden, C. (1994). Peroxisomes in liver, heart, and kidney of mice fed a commercial fish oil preparation: original data and review on peroxisomal changes induced by high-fat diets. J. Lipid Res. 35, 1241-1250.

De la Iglesia, F., Lewis, J., Buchanan, R., Marcus, E. & Mc Mahon, G. (1982). Light and electron microscopy of liver in hyperlipoproteinemic patients under long-term gemfibrozil treatment. Atherosclerosis 43, 19-37.

Del Rio, L.A., Sandalio, L.M. & Palma, J.M. (1990). A new cellular function for peroxisomes related to oxygen free radicals. Experientia 46, 989-992.

Devchand, P., Keller, H., Peters, J., Vazquez, M., Gonzalez, F. & Wahli, W. (1996). The $PPAR_{\alpha}$ - leukotriene B_4 pathway to inflammation control. Nature 384, 39-43.

Dice, J.F. (1985). Cellular theories of aging as related to the liver. Hepatology 5, 508-513.

Dice, J.F. (1993). Cellular and molecular mechanisms of aging. Physiol. Rev. 73, 149-159.

Dice, J.F. & Goff, S.A. (1988). Aging and the liver. In: The Liver: Biology and Pathobiology. (Arias, I.M., Jackoby, W.B., Poppe,r H., Schachter, D. and Shafritz, D.A., eds.), Raven Press Ltd., New York, pp. 1245-1258.

Eacho, P., Foxworthy, P., Johnson, W., Hoover, D. & White, S. (1986). Hepatic peroxisomal changes induced by a tetrazole-substituted alkoxyacetophenone in rats and comparison with other species. Toxicol. Appl. Pharmacol. 83, 430-437.

Ellis, R, Yuan, J. & Horvitz H (1991) Mechanisms and functions of cell death. Annu. Rev. Cell Biol. 7, 663-689.

Forman, B.M., Tontonoz, P., Chen, J., Brun, R.P., Spiegelman, B.M. & Evans, R.M. (1995).15-deoxy - $\Delta^{12,14}$- PGJ_2 is a ligand for the adipocyte determination factor $PPAR_{\gamma}$. Cell 83, 803-812.

Forman, B., Chen, J. & Evans, R. (1997). Hypolipidemic drugs, polyunsaturated fatty acids and eicosanoids are ligands for peroxisome proliferator-activated receptors α and δ. Proc. Natl. Acad. Sci. USA 94, 4312-4317.

Fornos, P., Duncan, C., Lawrence, R. & Jenjinson, S. (1990). Enhancement of H_2O_2-induced hepatic oxidant injury in aged rats. FASEB. J. 4, A302.

Fringes, B. & Reith, A. (1982). Time-course of peroxisome biogenesis during adaptation to mild hyperthyroidism in rat liver. Lab. Invest. 47, 19-26.

Fringes, B., Reith, A. & Riede, U. (1982). Peroxisome biogenesis in rat liver during adaptation to mild hyperthyroidism: A morphometric/stereologic study by electron microscopy. Ann. N. Y. Acad. Sci. 386, 514-518.

Genova, M., Castelluccio, C., Fato, R., Parenti Castelli, G., Merlo Pich, M., Formiggini, G., Bovina, C., Marchetti, M. & Lenaz, G. (1995). Major changes in complex I activity in mitochondria from aged rats may not be detected by direct assay of NADH:coenzyme Q reductase. Biochem. J. 311, 105-109.

Goldfischer, S., Moore, C.L., Johnson, A.B., Spiro, A.J., Valsamis, M.P., Wisnieski, H.K., Ritch, R.H., Norton, W.T., Rapin, I. & Gartner, L.M. (1973). Peroxisomal and mitochondrial defects in the cerebro-hepato-renal syndrome. Science 182, 62-64.

Gottlicher, M., Widmark, E., Li, Q. & Gustafsson, J-A.(1992). Fatty acids activate a chimera of the clofibric acid-activated receptor and the glucocorticoid receptor. Proc. Natl. Acad.Sci. USA 89, 4653-4657.

Greene, M., Blumberg, B., McBride, O., Yi, H., Kronquist, K., Kwan, K., Hsieh, L., Greene, G. & Nimer, S. (1995). Isolation of the human peroxisome proliferator activated receptor gamma cDNA: expression in hematopoitic cells and chromosomal mapping. Gene. Exp. 4, 281-200.

Grinna, L.S. & Barber, A.A. (1976). Lipid changes in the microsomal and mitochondrial membranes of rat liver during aging. Fed. Proc. 35, 1425.

Gulick, T., Cresci, S., Caira, T., Moore, D. & Kelly, D. (1994). The peroxisome proliferator-activated receptor regulates mitochondrial fatty acid oxidative enzyme gene expression. Proc. Natl. Acad. Sci. USA 91, 11012-11016.

Gustafsson, J-A. (1998). Fatty acids in control of gene expression. Nutr. Rev. 56, S20-S21

Haining, J.L. & Legan, J.S. (1973). Catalase turnover in rat liver and kidney as a function of age. Expt. Geront. 8, 85-91.

Handler, J. & Brian, W. (1997). Effect of aging on mixed-function oxidation and conjugation by isolated perfused rat livers. Biochem. Pharmacol. 54, 159-164.

Hansford, R.G. & Castro, F. (1985). Age-linked changes in the activity of enzymes of the tricarboxylic cycle and lipid oxidation and carnitine content in muscles of rat. Mech. Ageing Dev. 19, 191-200.

Herbener, G.H. (1976). A morphometric study of age dependent changes in mitochondrial population of mouse liver and heat. J. Gerontol. 31, 8-12.

Herman, D. (1981). The aging process. Proc. Natl. Acad. Sci. USA 78, 7124-7128.

Hey, M.J., Clough, M.J. & Taylor, D.J. (1976). Ion effects on macromolecules in aqueous solution. Nature 262, 807-809.

Higgintham, M.B., Morris, K.G., Williams, R.S., Coleman, R.E. & Cobb, F.R. (1985). Physiological basis for the age-related decline in aerobic work capacity. Am. J. Cardiol. 57, 1374-1379.

Horie, S., Ishii, H. & Suga, T. (1981). Changes in peroxisomal fatty acid oxidation in the diabetic rat liver. J. Biochem. 90, 1691-1696.

Hossack, K.F. & Bruce, R.A. (1982). Maximal cardiac function in sedentary normal men and women. Comparison of age-related changes. J. Appl. Physiol. 53, 799-804.

Houdou, S., Kuruta, H., Hasegawa, M., Konomi, H., Takashima, S., Suzuki, Y. & Hashimoto, T. (1991). Developmental immunohistochemistry of catalase in human brain. Brain Res. 556, 267-270.

Houdou, S., Takashima, S. & Suzuki, Y. (1993). Immunohistochemical expression of peroxisomal enzymes in developing human brains. Mol. Chem. Neuropath. 19, 235-248.

Huber, W., Kraupp-Grasl, B., Esterbauer, H. & Schulte-Hermann, R. (1991). Role of oxidative stress in age-dependent hepatocarcinogenesis by the peroxisome proliferator nafenopin in the rat. Cancer Res. 51, 1789-1792.

Imamura, A., Kamei, A., Suzuki, Y., Orii, T. & Takashima, S. (1994). Developmental immunohistochemistry of bifunctional protein in human brain. Brain Res. 640, 236-239.

Inoue, I., Noji, S., Shen, M-z., Takahashi, K. & Katayama, S. (1997). The peroxisome proliferator-activated receptor α (PPARα) regulates the plasma thiobarbituric acid-reactive substance (TBARS) level. Biochem. Biophys. Commun. 237, 606-610.

Issemann, I. & Green, S. (1990). Activation of a member of the steroid hormone receptor superfamily by peroxisome proliferators. Nature 347, 645-650.

Jakobs, B. & Wanders, R. (1991). Conclusive evidence that very long chain fatty acids are oxidized exclusively in peroxisomes in human skin fibroblasts. Biochem. Biophys. Res. Commun. 178, 842-847.

Jiang, C., Ting, A. & Seed, B. (1998). $PPAR_{\gamma}$ agonists inhibit production of monocyte inflammatory cytokines. Nature 391, 82-86.

Ji, L., Dillon, D. & Wu, E. (1991). Myocardial aging: antioxidant enzyme systems and related biochemical properties. Am. J. Physiol. 261, R386-R392.

Kliewer, S., Forman, B., Blumberg, B., Ong, E., Borgmeyer, U., Mangelsdorf, D., Umesono, K. & Evans, R. (1994). Differential expression and activation of a family of murine peroxisome proliferator-activated receptors. Proc. Natl. Acad. Sci. USA 91, 7344-7349.

Kliewer, S.A., Lenhard, J.M., Wilson, T.M., Patel, I., Morris, D.C. & Lenhard, J.M. (1995). A prostaglandin J_2 metabolite binds peroxisome proliferator-activated receptor γ and promotes adipocyte differentiation. Cell 83, 813-819.

Kliewer, S., Sundseth, S., Jones, S., Brown, P., Wisely, G.B., Koble, C., Devchand, P., Wahli, W., Wilson,T., Lenhard, J. & Lehman, J. (1997). Fatty acids and eicosanoids regulate gene expression through direct interactions with peroxisome proliferator-activated receptors α and γ. Proc. Natl. Acad. Sci. (USA). 94, 4318-4323.

Kraupp-Grasl, B., Huber, W., Taper, H. & Schulte-Hermann, R. (1991). Increased susceptibility of aged rats to hepatocarcinogenesis by the peroxisome proliferator nafenopin and the possible involvement of altered liver foci occurring spontaneously. Cancer Res. 51, 666-671.

Lake, B., Evans, J., Gray, T., Korosi, S. & North, C. (1989). Comparative studies on nafenopin-induced hepatic peroxisome proliferation in the rat, syrian hamster, guinea pig and marmoset. Toxicol. Appl. Pharmacol. 99, 148-160.

Lazarow, P. & Fujiki, Y. (1985). Biogenesis of peroxisomes. Ann. Rev. Cell Biol. 1, 489-530.

Lefebvre, A-M., Laville, M., Vega, N., Riou, J., Gaal, L., Auwerx, J. & Vidal, H. (1998a). Depot-specific differences in adipose tissue gene expression in lean and obese subjects. Diabetes 47, 98-103.

Lefebvre, A-M., Chen, I., Desreumaux, P., Najib, J., Fruchart, J-C., Geboes, K., Briggs, M., Heyman, R. & Auwerx, J. (1998b). Activation of the peroxisome proliferator-activated receptor γ promotes the development of colon tumors in C57BL/6J-APC^{Min}/+ mice. Nat. Med. 4, 1053-1057.

Lemberger, T., Staels, B., Saladin, R., Desvergne, B., Auwerx, J. & Wahli, W. (1994). Regulation of the peroxisome proliferator-activated receptor α gene by glucocorticoids. J. Biol. Chem. 269, 24527-24530.

Levy, E., Slusser, R. & Ruebner, B. (1968). Hepatic changes produced by a single dose of endotoxin in the mouse. Am. J. Path. 52, 477-502.

Makarides, S. (1983). Protein synthesis and degradation during aging and senescence. Biol. Rev. 58, 343-422.

Martin, G.M. (1992). Biological mechanisms of ageing, in: Oxford textbook in geriatric medicine (Evans, G.J. and Williams, J.F. eds.), Oxford Medical Publications, Oxford.

Masters, C. & Crane, D. (1995). On the role of the peroxisome in ontogeny, aging and degenerative disease. Mech. Aging Dev. 80, 69-83.

Masters, C. (1998). On the role of the peroxisome in the metabolism of drugs and xenobiotics. Biochem. Pharmacol. 56, 667-673.

McMillin, J.B., Taffet, G.E., Taegtmeyer, H., Hudson, E.K. & Tate, C.A. (1993). Mitochondrial metabolism and substrate competition in the aging Fischer rat heart. Cardiovascu. Res. 27, 2222-2228.

Mehlhorn, R. & Cole, G. (1985). The free radical theory of aging: a critical review. Adv. Free Radical. Biol. & Med. 1, 165-223.

Moser, A.E., Rasmussen, M., Naidu, S., Watkins, P., McGuinness, M., Hajra, A.K., Chen, G., Raymond, G., Liu, A., Gordon. D., Garnaas, K., Walton, D., Skjeldal, O., Guggenheim, M.A., Jackson, L.G., Elias, E.R. & Moser, H.W. (1995). Phenotype of patients with peroxisomal disorders subdivided into 16 complimentation groups. J. Pediatr. 127, 13-22.

Mueller, E., Sarraf, P., Tontonoz, P., Evans, R.M., Martin, K.J., Zhang, M., Fletcher, C., Singer, S. & Spiegelman, B.M. (1998). Terminal differentiation of human breast cancer through $PPAR_{\gamma}$. Mol. Cell 1, 465-470.

Mukherjee, R., Jow, L., Noonan, D. & McDonnell, D. (1994). Human and rat peroxisome proliferator activated receptor (PPARs) demonstrate similar tissue distribution but different responsiveness to PPAr activators. J. Steroid Mol. Biol. 51, 157-166.

Mukherjee, R., Strasser, J., Jow, L., Hoener, P., Paterniti, J. & Heyman, R. (1998). RXR agonists activate $PPAR_{\alpha}$-inducible genes, lower triglycerides and raise HDL levels *in vivo* Arterioscler Thromb. Vasc. Biol. 18, 272-276.

O'Brien, M.L.O., Rangwala, S.M., Henry, K.W., Weinberger, C., Crick, D.C., Waecher, C.J., Feller, D.R. & Noonan, D.J. (1996) Convergence of three steroid receptor pathways in the mediation of nongenotoxic hepatocarcinogenesis. Carcinogenesis 17, 185-190.

Oesch, F., Schladt, L., Steinberg, P. & Thomas, H. (1988). Concomitant induction of cytosolic epoxide hydrolase and peroxisomal β-oxidation by hypolipidemic compounds in rat and guinea pig liver. Arch.

Toxicol. 12(suppl), 248-255.

Orr, W. & Sohal, R. (1994). Extension of life span by overexpression of superoxide dismutase and catalase in Drosophila melanogaster. Science 263, 1128-1130.

Paget, G. (1963). Experimental studies of the toxicity of atromid with particular reference to fine structural changes in the livers of rodents. J. Atheroscler. Res. 3, 729-736.

Pallet, V., Azais-Braesco, V., Enderlin, V., Grolier, P., Noel-Suberville, C., Garcin, H. & Higueret, P. (1997). Aging decreases retinoic acid and triiodothyronine nuclear expression in rat liver: exogenous retinol and retinoic acid differentially modulates this decreased expression. Mech. Aging Dev. 99, 123-136.

Perez-Campo, R., Lopez-Torres, M., Cadenes, S., Rojas, C. & Barja, G. (1998). The rate of free radical production as a determinant of the rate of aging: evidence from comparative approach. J. Comp. Physiol. B 168, 149-158.

Perichon, R. & Bourre, J.M. (1995). Peroxisomal β-oxidation activity and catalase activity during development and aging in mouse liver. Biochimie 77, 288-293.

Perichon, R. & Bourre, J.M. (1996). Age-related decrease in liver peroxisomal fatty acid oxidation in control and clofibrate-treated mice. A biochemical study and mechanistic approach. Mech. Ageing Dev. 87, 115-126.

Popper, H. (1986). Aging and the liver, in Progress in Liver Diseases (Popper, H. and Schaffner, F., eds.) vol. 8, pp 659-683, Grune and Stratton, Inc., New York.

Rigter, H., Veldhuis, H. & de Kloet, E. (1984). Spatial orientation and the hippocampal corticosterone receptor system of old rats: effect of $ACTH_{4-9}$ analogue ORG2766. Brain Res. 309, 393-398.

Rikans, L.E. & Moore, D.R. (1987). Effect of age and sex on allyl alcohol hepatotoxicity in rats: role of liver alcohol and aldehyde dehydrogenase activities. J. Pharmacol. Exp. Ther. 243, 20-26.

Rivany, B., Globerson, A. & Shinitzky, M. (1979). Viscosity of lymphocyte plasma membrane in aging mice and its possible relation to serum cholesterol. Mech. Ageing Dev. 10, 71-79.

Rivnay, B., Bergman, S., Shinitzky, M. & Globerson, A. (1980). Correlation between viscosity, serum cholesterol, lymphocyte activation and aging in man. Mech. Ageing Dev. 12, 119-126.

Roberts, R., Soames, A., Gill, J., James, N. & Wheeldon, E. (1995). Non-genotoxic hepatocarcinogens stimulate DNA synthesis and their withdrawal induces apoptosis but in different hepatocyte populations. Carcinogenesis 16, 1693-1698.

Roberts, R., James, N., Woodyatt, N., Macdonald, N. & Tugwood, J. (1998). Evidence for the suppression of apoptosis by the peroxisome proliferator activated receptor alpha (PPARα). Carcinogenesis 19, 43-48.

Robuschi, G., Safran, M., Braverman, L., Gnudi, A. & Roti, E. (1987). Hypothyroidism in the elderly. Endocrine Rev. 8, 142-153.

Rodeheffer, R.J., Gerstenblith, G., Becker, I.C., Fleg, J.L., Weisfelder, M.L. & Lakatta, J.G. (1984). Excercise cardiac output is maintained with advancing age in healthy human subjects: cardiac dilatation and increased stroke volume compensate for diminished heart rate. Circulation 69, 203-213.

Rodriguez, J., Gil-Gomes, G., Hegardt, F. & Haro, D. (1994). Peroxisome proliferator activated receptor mediates induction of the mitochondrial 3-hydroxy-3-methylglutaryl-CoA synthase gene by fatty acids. J. Biol. Chem. 269, 18769-18772.

Roffe, C. (1991). Ageing of the heart. Br. J. Biomed Sci. 55, 136-148.

Roth, G., Joseph, J. & Mason, R. (1995). Membrane alterations as causes of impaired signal transduction in Alzheimer's disease and aging. T.I.N.S. 18, 203-206.

Roth, G.S. (1997). Age changes in signal transduction and gene expression. Mech. Ageing Dev. 98, 231-238.

Rubin, H. (1997). Cell aging *in vivo* and *in vitro*. Mech. Ageing Dev. 98, 1-35.

Saez, E., Tontonoz, P., Nelson, M., Alvarez, J., U, T., Baird, S. Thomazy, V. & Evans, R. (1998). Activation of the nuclear receptor $PPAR_\gamma$ enhances colon polyps formation. Nature Medicine 4, 1058-1061.

Sakata, N., Meng, J., Jimi, S., Segawa, M. & Takebayashi, S. (1995). Aging of the aorta and atherosclerosis. Role of non-enzymatic glycation of collagen. Jp. J. Geriatrics 32, 336-343.

Sakuma, M., Yamada, J. & Suga, T. (1992). Comparison of the inducing effect of dehydroepiandrosterone on hepatic peroxisome proliferation-associated enzymes in several rodent species. Biochem. Pharmacol. 43, 1269-1273.

Sapolsky, R., Krey, L. & McEwen, B. (1983). The adrenocortical stress response in the aged male rat: impairment of recovery from stress. Exp. Gerontol. 18, 55-64.

Sapolsky, R., Krey, L., McEwen, B. & Rainbow, T. (1984). Do vasopressin-related peptides induced hippocampal corticosterone receptors? Implications for aging. J. Neurosci. 4, 1479-1485.

Sarraf, P., Mueller, E., Jones, D., King, F., DeAngelo, D., Parrtridge, J., Holden, S., Chen, L., Singer, S., Fletcher, C. & Spiegelman, B. (1998). Differentiation and reversal of malignant changes in colon cancer through $PPAR_\gamma$. Nature Medicine 4, 1046-1052.

Schapiro, H., Hotta, S.S., Outten, W.E. & Klein, A.W. (1982). The effect of aging on rat liver regeneration. Experentia 38, 1075-1076.

Schmucker, D.L. & Wang, M.M. (1980). Age-related changes in liver drug metabolism: Structure vs. function. Proc. Soc. Exp. Biol. Med. 165, 178-187.

Schutgens, R., Heyman, H., Wanders, R., Vandenbosch, H. & Tager, J. (1986). Peroxisomal disorders: A newly recognized group of genetic diseases. Eur. J. Pediatr. 144, 430-440.

Seed, B. (1998). $PPAR_\gamma$ and colorectal carcinoma: conflicts in a nuclear family. Nat. Med.4, 1004-1005.

Semsei, I., Rao, G. & Richardson, A. (1989). Changes in the expression of superoxide dismutase and catalase as a function of age and dietary restriction. Biochem. Biophys. Res. Commun. 164, 620-625.

Sher, T., Yi, H., McBride, O. & Gonzalez, F. (1993). cDNA cloning, chromosomal mapping and functional characterization of the human peroxisome proliferator activated receptor. Biochemistry 32, 5598-5604.

Shimozawa, N., Suzuki, Y., Zhang, Z., Imamura, A., Tsukamoto, T., Osumi, T., Tateishi, K., Okumoto, K., Fujiki, Y., Orii, T., Barth, P., Wanders, R. & Kondo, N. (1998). Peroxisome biogenesis disorders: identification of a new complementation group distinct from peroxisome-deficient CHO mutants and not complemented by human PEX 13. Biochem. Biophys. Res. Commun. 243, 368-371.

Singh, S.N. (1973). Effect of age on the activity and citrate inhibition of malate dehydrogenase of the brain and heart of rats. Experientia 29, 42-43.

Singh, I., Moser, A.E., Goldfischer, S. & Moser, H.W. (1984). Lignoceric acid is oxidized in the peroxisome: Implications for the Zellweger cerebro-hepato-renal syndrome and adrenoleukodystrophy. Proc. Natl. Acad. Sci. U.S/A. 81, 4203-4207.

Singh, I., Lazo, O., Contreras, M., Stanley, W. & Hashimoto, T. (1991). Rhizomelic chondrodysplasia punctata: biochemical studies of peroxisomes isolated from cultured skin fibroblasts. Arch. Biochem. Biophys. 286, 277-283.

Singh, I. (1997). Biochemistry of peroxisomes in health and disease. Mol. Cell. Biochem. 167, 1-29.

Soliman, M., Cunningham, M., Morrow, J., Roberts, J. & Badr, M. (1997). Evidence against peroxisome proliferation-induced hepatic oxidative damage: levels of esterified isoprostanes in livers of mice fed a diet containing [4-chloro-6(2,3-xylidino)-2-pyrimidinylthio] acetic acid (WY-14,643). Biochem. Pharmacol. 53, 1369-1374.

Spiegelman, B.M. & Flier, J.S. (1996). Adipogenesis and obesity: rounding out the big picture Cell 87, 377-389.

Spiegelman, B. (1998). $PPAR_\gamma$ in monocytes: less pain, any gain? Cell 93, 153-155.

Staubli, R., Hess, R. & Weibel, E. (1969). Correlated morphometric and biochemical studies on the rat liver cell. II. Effects of phenobarbital on rat hepatocytes. J. Cell Biol. 42, 92-112.

Steinberg, D. Parthasarathy, S., Carew, T.E., Khoo, J.C. & Witzum, J.L. (1989). Beyond cholesterol. Modification of low-density lipoprotein that increase its atherogenicity. N. Engl. J. Med. 320, 915-924.

Stubbs, C. & Smith, A. (1984). The modification of mammalian membrane polyunsaturated fatty acid composition in relation to membrane fluidity and function. Biochim. Biophs. Acta 779, 89-137.

Su, J-L., Simmons, C., Wisely, B., Ellis, B. & Winegar, D. (1998). Monitoring of PPAR alpha protein expression in human tissue by the use of PPAR alpha-specific mab. Hybridoma 17, 47-53.

Takai, Y., Kikkawam, U., Kaibuchi, K. & Nishizuka, Y. (1984) Membrane phospholipid metabolism and signal transduction for protein phosphorylation, in Adv. Cyclic Nucleotides and Protein Phosphorylation Res. (Greengard P and Robison GA eds), 18, 119-158, Raven Press, New York.

Tauchi, H. & Sato, T. (1978). Hepatic cells of the aged, in Liver and Aging (Kitani K ed) pp 3-19, Elsevier, Amsterdam.

Thomas, H., Schladt, L., Knehr, M. & Oesch, F. (1989). Effect of diabetes and starvation on the activity of rat liver epoxide hydrolase, glutathione-S transferase and peroxisomal β–oxidation. Biochem. Pharmacol. 38, 4291-4297.

Tolbert, N. (1981). Metabolic pathways in peroxisomes and glyoxysomes. Annu. Rev. Biochem. 50, 133-157.

Tontonoz, P., Singer, S., Forman, B., Sarraf, P., Fletcher, J., Fletcher, C., Brun, R., Mueller, E., Altoik, S., Oppenheim, H., Evans, R. & Spiegelman, B. (1997). Terminal differentiation of human liposarcoma cells induced by ligands for peroxisome proliferator-activated receptors γ and the retinoid X receptor. Proc. Natl. Acad. Sci. USA 94, 237-241.

Tontonoz, P., Nagy, L., Alvarez, J.G.A., Thomazy, V.A. & Evans, R.M. (1998). $PPAR_{\gamma}$ promotes monocyte/macrophage differentiation and uptake of oxidized LDL. Cell 93, 241-252.

Tsai, M. & O'Malley, B. (1994). Molecular mechanisms of action of steroid/thyroid receptor superfamily members. Annu. Rev. Biochem. 63,451-486.

Van den Bosch, H., Schutgens, R.B.H., Wanders, R.J.A. & Tager, J.M. (1992). Biochemistry of peroxisomes. Annu. Rev. Biochem. 61, 157-197.

Vidal-Puig, A., Considine, R., Jimenez-Linan, M., Werman, A., Pories, W., Caro, J. & Flier, J. (1997). Peroxisome proliferator-activated receptor gene expression in human tissues. Effects of obesity, weight loss, and regulation by insulin and glucocorticoids. J. Clin. Invet. 99, 2416-2422.

Warner, H. (1997). Aging and regulation of apoptosis, in Curr Topics in Cell Reg (Stadtman, E. and Chock, P., eds.), 35, 107-121, Academic Press, San Diego.

Watanabe, T., Horie, S., Yamada, J., Isaji, M., Nishigaki, T., Naito, J. & Suga, T. (1989). Species differences in the effect of bezafibrate, a hypolipidemic agent, on hepatic peroxisome-associated enzymes. Biochem. Pharmacol. 38, 367-371.

Weibel, E., Staubli, W., Gangi, H. & Hess, F. (1969). Correlated morphometric and biochemical studies on the liver cell. I. Morphometric model, stereologic methods, and normal morphometric data for rat liver. J. Cell Biol. 42, 68-91.

Weindruch, R.H., Cheung, M.K., Verity, M.A. & Walford, R.L. (1980). Modification of mitochondrial respiration by aging and dietary restriction. Mech. Ageing Dev. 12, 375-392.

Wick, G., Huber, L., Qing-Bo, X., Jarosch, E., Schonitzer, D. & Jurgens, G. (1991). Ann. N.Y. Acad. Sci. 621, 277-290.

Wilson, P. & Franke, L.M. (1975). The effect of age on mitochondrial ultrastructure. Gerontologia 21, 81-94.

Winberg, L. & Badr, M. (1995). Mechanism of phthalate-induced inhibition of hepatic mitochondrial β-oxidation. Toxicol. Lett. 76, 63-69.

Wolf, G. (1998). Fatty acids bind to and activate peroxisome proliferator-activated receptors α and γ.

Nutr. Rev. 56, 61-63.

Xia, E., Rao G, Van Remmen, H., Heydari, A. & Richardson, A. (1995). Activities of antioxidant enzymes in various tissues of male Fischer 344 rats are altered by food restriction. J. Nutr. 125, 195-201.

Xiao, R.P. & Lakatta, E. (1991). Mechanism of altered beta- adrenergic modulation of the cardiovascular system with aging. Rev. Bio. Gerontol. 1, 309-322.

Xu, L., Ash, M., Abdel-aleem, S., Lowe, J. & Badr, M. (1995). Hyperinsulinemia inhibits hepatic peroxisomal β-oxidation in rats. Horm. Metab. Res. 27, 76-78.

Xu, L. & Badr, M. (1999). Enhanced potential for oxidative stress in hyperinsulinemic rats. Imbalance between hepatic peroxisomal hydrogen peroxide production and decomposition due to hyperinsulinemia. Horm. Metab. Res. 31, 278-282.

Yamoto, T., Ohashi, Y., Teranishi, M., Takaoka, M., Manabe, S., Matsunuma, N. & Makita, T. (1995). Age-related changes in the susceptibility to clofibric acid, a hypolipidemic agent, of male rat liver. Toxicol. Lett. 78, 141-145.

Yen, C-J., Beamer, B., Negri, C., Silver, K., Brown, K., Yarnall, D., Burns, D., Roth, J. & Shuldiner, A. (1997). Molecular scanning of the human peroxisome proliferator activated receptor γ (hPPARγ) gene in diabetic caucasians: identification of a Pro 12Ala hPPARγ2 missense mutation. Biochem. Biophys. Commun. 241, 270-274.

Youssef, J. & Badr, M. (1998). Extraperoxisomal targets of peroxisome proliferators. Mitochondrial, microsomal and cytosolic effects. Implication for health and disease CRC Crit. Rev. Toxicol. 28, 1-33.

Youssef, J. & Badr, M. (1999). Biology of senescent liver peroxisomes: Role of hepatocellular aging and disease. Environ. Health Perspec. 107, 791-797, 1999.

Zs-Nagy, I. (1979). The role of membrane structure and function in cellular aging: a review. Mech. Ageing Dev. 9, 237-246.

PERTURBED CALCIUM HOMEOSTASIS IN NEURODEGENERATIVE DISORDERS

MARK P. MATTSON

Chief, Laboratory of Neurosciences, National Institute on Aging
Professor, Department of Neuroscience, Johns Hopkins University

Table of Contents

Introduction

The calcium ion serves as a signal for a variety of responses of neurons to environmental stimuli (see Mattson, 1992; Berridge, 1998 for review). For example, Ca^{2+} influx stimulates neurotransmitter release, induces changes in the neuronal cytoskeleton (e.g., regulation of neurite elongation and growth cone motility), regulates the expression of a variety of genes, modulates the activity of an array of kinases and phosphatases, and (of particular relevance to the present chapter) plays a major role in neuronal degeneration and death in various neurological disorders. The types of signals that exert their effects on neurons by changing intracellular calcium levels range from membrane depolarization, to various neurotransmitters (e.g., glutamate and acetylcholine), to neurotrophic factors such as brain-derived neurotrophic factor, to cytokines such as tumor necrosis factor. Neurons possess a complex array of mechanisms for regulating intracellular free calcium levels (Figure 1; and see Mattson and Mark, 1996 for review). Under resting conditions the free calcium concentration ($[Ca^{2+}]$) in the cytoplasm of most mammalian neurons is in the range of 70-150 nM, while the extracellular $[Ca^{2+}]$ is approximately 1-2 mM. This 10,000-fold $[Ca^{2+}]$ gradient is maintained largely by

Interorganellar Signaling in Age-Related Disease, Edited by Mark P. Mattson, 29—54

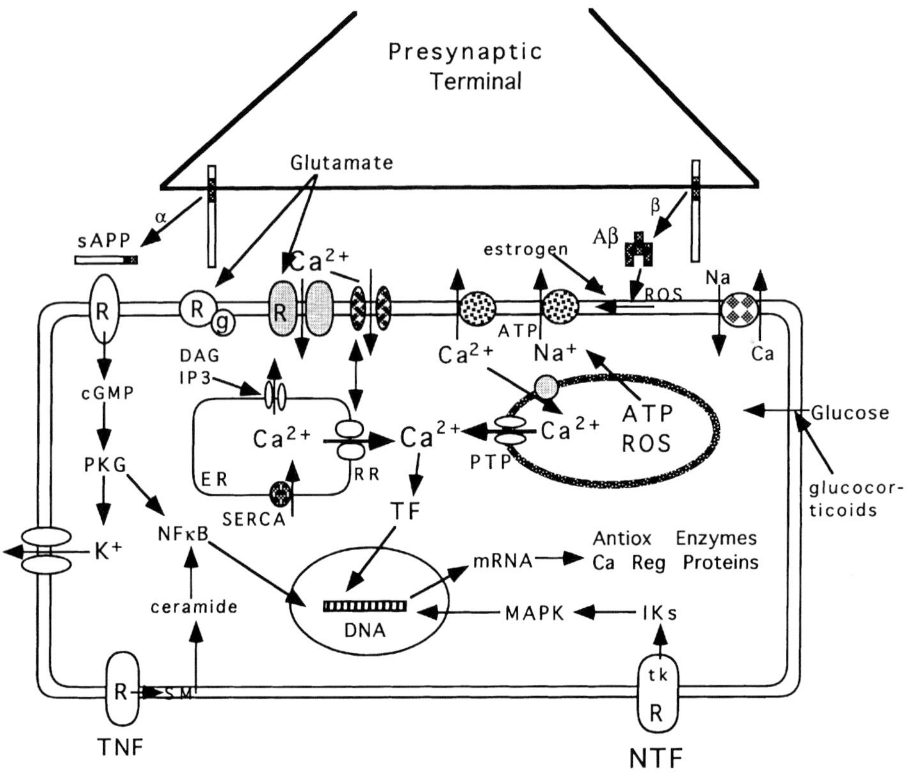

Figure 1. Mechanisms involved in the regulation of intracellular calcium levels in neurons. See text for discussion. Aβ, amyloid β-peptide; DAG, diacylglycerol; ER, endoplasmic reticulum; g, GTP-binding protein; IK, intermediate kinases; MAPK, mitogen activated protein kinase; NTF, neurotrophic factor; PKG, cyclic GMP-dependent protein kinase; R, PTP, permeability transition pore; receptor; ROS, reactive oxygen species; RR, ryanodine receptor; sAPP, secreted form of amyloid precursor protein; SERCA, smooth endoplasmic reticulum calcium ATPase; SM, sphingomyelinase; tk, tyrosine kinase; TNF, tumor necrosis factor.

the activity of the plasma membrane Ca^{2+}-ATPase. The $[Ca^{2+}]$ inside mitochondria and endoplasmic reticulum is also higher than in the cytoplasm, and these organelles play an important role in calcium signaling in neurons. Mitochondria sequester calcium via uptake by a calcium uniporter, and endoplasmic reticulum (ER) contains a Ca^{2+}-ATPase that is distinct from, but analagous to, the plasma membrane Ca^{2+}-ATPase.

Elevation of cytoplasmic $[Ca^{2+}]$ occurs mainly via influx through plasma membrane channels, and release from the ER and mitochondria. The major types of plasma membrane calcium channels are voltage-dependent calcium channels and ligand-gated calcium channels. Voltage-dependent calcium channels are divided into subtypes according to their ion conductance properties and their pharmacological profiles

(Racay et al., 1996). For example, L-type calcium channels flux high levels of calcium, remain open for relatively long time periods, and are blocked by dihydropyridines such as nifedipine and nimodipine. Other types of cloned voltage-dependent calcium channels include T, N and P. Prominent among the ligand-gated calcium channels are the ionotropic glutamate receptors. The latter receptors are divided into two classes, N-methyl-D-aspartate (NMDA) receptors and AMPA/kainate receptors (O'Brien et al., 1998). NMDA receptors will open only when the membrane is depolarized and flux very high levels of calcium. AMPA/kainate receptors are activated by glutamate at resting membrane potentials, and flux primarily sodium, although certain subtypes of these receptors also flux considerable calcium. Both voltage-dependent and ligand-gated calcium channels are modulated by phosphorylation, which is regulated by various kinases and phosphatases (Swope et al., 1999).

The endoplasmic reticulum and mitochondria play very important roles in local regulation of intracellular calcium levels in dendrites and axons (Spacek and Harris, 1997; Finch and Augustine, 1998; Mattson and Partin, 1999; Takechi et al., 1998). Calcium is released from the ER in response to agonists that activate cell surface receptors coupled to GTP-binding proteins that activate phospholipase C. Phospholipase C catalyzes the hydrolysis phosphatidylinositol bis-phosphate in the membrane resulting in liberation of diacylglycerol and ionositol triphosphate (IP_3). Diacylglycerol activates protein kinase C and IP_3 activates IP_3 receptors located in the ER membrane. IP_3 receptors are calcium channels that, when activated, provide a conduit for calcium release from the ER. In addition to IP_3-sensitive stores, neurons also possess ER that contain ryanodine receptors that are calcium channels activated by caffeine and ryanodine. Mitochondria contain millimolar concentrations of calcium, and the uptake of calcium by mitochondria is increasingly recognized as playing important roles in both physiological and pathological settings (Budd, 1998; Simpson and Russell, 1998). The mechanisms that control release of calcium are not yet well understood, but formation of reversible and irreversible permeability transition pores that span both mitochondrial membranes appears to be a major control mechanism.

In addition to movement into and out of the cell, and into and out of organelles, neurons often express one or more calcium-binding proteins that serve important roles in regulating cytoplasmic [Ca^{2+}]. Examples include calbindin D28k, parvalbumin, calretinin and calreticulin (Baimbridge et al., 1992). Calbindin D28k is expressed in subpopulations of neurons in the brain (e.g., dentate granule neurons in the hippocampus) that appear to be relatively resistant to neuronal degeneration in various human neurodegenerative disorders and animal models (Sloviter et al., 1991). Cell culture studies have shown that hippocampal neurons expressing calbindin (Mattson et al., 1991) or calretinin (Pike and Cotman, 1995) are less vulnerable to excitotoxic and metabolic insults compared to neurons lacking these calcium-binding proteins.

The cell cytoskeleton, particularly the actin-based system of microfilaments, is emerging as an important modulator of neuronal calcium homeostasis. Depolymerization of actin with cytochalasin D results in reduced currents through voltage-dependent calcium channels and NMDA receptors (Johnson and Byerly, 1993; Rosenmund and Westbrook, 1993). Pretreatment of hippocampal neurons with cytochalasin D results in reduced calcium responses to glutamate, and increased resistance of the neurons to excitotoxicity in cell culture and *in vivo* (Furukawa et al., 1995). Studies of mice lacking the calcium-activated actin-severing protein gelsolin have shown that actin depolymerization in response to calcium influx serves to reduce further calcium influx through membrane channels and thereby protects neurons against excitotoxicity and ischemia (Furukawa et al., 1997; Endres et al., 1999). The mechanism whereby actin filaments modulate calcium influx is not yet clear. However, roles for actin-binding proteins and interactions with proteins that comprise focal adhesion complexes seem likely. Indeed, emerging data suggest roles for components of focal adhesion complexes, such as integrins and cadherins, in the regulation of neuronal calcium homeostasis (Bixby and Bookman, 1998). Such mechanisms may modulate neuronal vulnerability in various pathophysiological settings because experimental manipulation of cell adhesion molecule interactions in cultured neurons can greatly change their vulnerability to excitotoxic and metabolic insults (Gary and Mattson, 2001).

The systems that regulate neuronal calcium homeostasis can be modulated by environmental signals. One class of such signals that has striking effects on neuronal calcium regulating systems is neurotrophic factors. It was discovered that several different neurotrophic factors including basic fibroblast growth factor (bFGF), insulin-like growth factors (IGFs), brain-derived neurotrophic factor (BDNF) and nerve growth factor (NGF) can protect neurons in culture and *in vivo* against excitotoxic and ischemic insults that are known to involve disruption of calcium homeostasis (see Mattson and Furukawa, 1996; Mattson and Lindvall, 1997 for review). Analyses of cellular calcium homeostasis in cultured hippocampal neurons showed that neurons pretreated with these growth factors exhibit attenuated increases in intracellular calcium levels following exposure to glutamate or metabolic insults (Mattson et al., 1989; Cheng and Mattson, 1991, 1992, 1994). The specific mechanisms whereby neurotrophic factors stabilize neuronal calcium homeostasis may include modulation of expression of glutamate receptor subunits (Mattson et al., 1993a; Cheng et al., 1995), induction of calcium-binding proteins (Collazo et al., 1992; Cheng et al., 1994a), and increased production of antioxidant enzymes (Mattson et al., 1995). Cytokines such as tumor necrosis factor (Cheng et al., 1994a; Barger et al., 1995), and novel trophic factors such as the secreted form of β-amyloid precursor protein (Mattson et al., 1993b; Furukawa et al., 1996), are examples of other environmental signals that modulate neuronal calcium homeostasis.

Shared Mechanisms of Perturbed Calcium Homeostasis

Whereas the genetic and environmental factors that initiate neurodegenerative processes are quite different in different neurodegenerative disorders, the neurodegenerative cascades in each disorder seem to converge on a few shared pathways. The convergence points include perturbed synaptic signaling (particularly overactivation of glutamate receptors), increased oxyradical production, and impaired energy metabolism (Mattson, 1998a). As will be described below, each of these alterations promotes sustained elevations of intracellular calcium levels. Such sustained increases in $[Ca^{2+}]_i$ cause cell dysfunction and death by activating proteases such as calpains and caspases (see Chan and Mattson,

1999 for review), and by inducing oxyradical production (Lafon-Cazal et al., 1993; Mattson et al., 1995).

Altered Synaptic Signaling: Excitotoxicity and Apoptosis

Overactivation of glutamate receptors has been implicated in the neurodegenerative process in essentially every neurological disorder that involves degeneration of neurons in the central nervous system. These disorders include: stroke (Choi, 1994), traumatic injury (Lynch and Dawson, 1994), temporal lobe epilepsy (Bernard et al., 1997), Alzheimer's disease (Mattson et al., 1993c), Huntington's disease (Perry and Hansen, 1990), Parkinson's disease (Beal, 1998a), and amyotrophic lateral sclerosis (Ince et al., 1998). The vulnerable populations of neurons in each disorder (e.g., hippocampal CA1 neurons in stroke and Alzheimer's disease, substantia nigra dopaminergic neurons in Parkinson's disease, and spinal cord motor neurons in amyotrophic lateral sclerosis) express high levels of ionotropic glutamate receptors. Treatment of animals (and/or cultured neurons) with glutamate receptor antagonists results in reduced neuronal degeneration in experimental models of these disorders. Glutamate receptors are concentrated in postsynaptic spines wherein they are activated by glutamate released from the presynaptic terminal. In various experimental models, degeneration of dendrites appears to occur prior to degeneration of the cell body or axon (Olney, 1985; Rothman and Olney, 1986; Mattson et al., 1989), and observation consistent with the "excitotoxicity" hypothesis. Levels of $[Ca^{2+}]_i$ rise to extremely high levels (1-100 μM) in postsynaptic regions of dendrites during normal synaptic activity, but normally recover towards baseline levels very rapidly. However, studies of experimental models of epilepsy, stroke and Alzheimer's disease demonstrate that levels of dendritic $[Ca^{2+}]_i$ do not recover following glutamate receptor activation, resulting in dendritic degeneration (Mattson et al., 1989; Cheng and Mattson, 1991; Mattson et al., 1992).

Activation of glutamate receptors under conditions of oxidative stress and/or reduced energy availability is particularly deleterious to neuronal calcium regulation and survival (see next two sections). The following sequence of events appears to lead to dendritic degeneration and cell death resulting from glutamate receptor activation. Activation of AMPA/kainate receptors induces membrane depolarization and activation of NMDA receptors and voltage-dependent calcium channels. Calcium influx through NMDA receptors and voltage-dependent channels results in a marked elevation of $[Ca^{2+}]_i$ which, in turn, promotes mitochondrial membrane depolarization and increased oxyradical production. These events occur in both excitotoxic necrosis and apoptosis (Ankarcrona et al., 1995; Kruman et al., 1999a). In necrosis the neuron swells, the integrity of the plasma membrane is compromised, and both mitochondria and the endoplasmic reticulum become swollen and dysfunctional. In apoptosis the neuron shrinks, organellar integrity is maintained, and one or more cysteine proteases called caspases are activated (Chan and Mattson, 1999). Caspases appear to orchestrate the cleavage of structural and enzymatic substrates that lead to nuclear DNA condensation and fragmentation. Studies of a preparation called the synaptosome (which consists of intact synaptic terminals that have been isolated from their parent neurons) and of cultured hippocampal neurons have shown that apoptotic biochemical cascades can be activated locally in dendritic terminals by excitotoxic and oxidative insults (Figure 2; Mattson et al., 1998a, b).

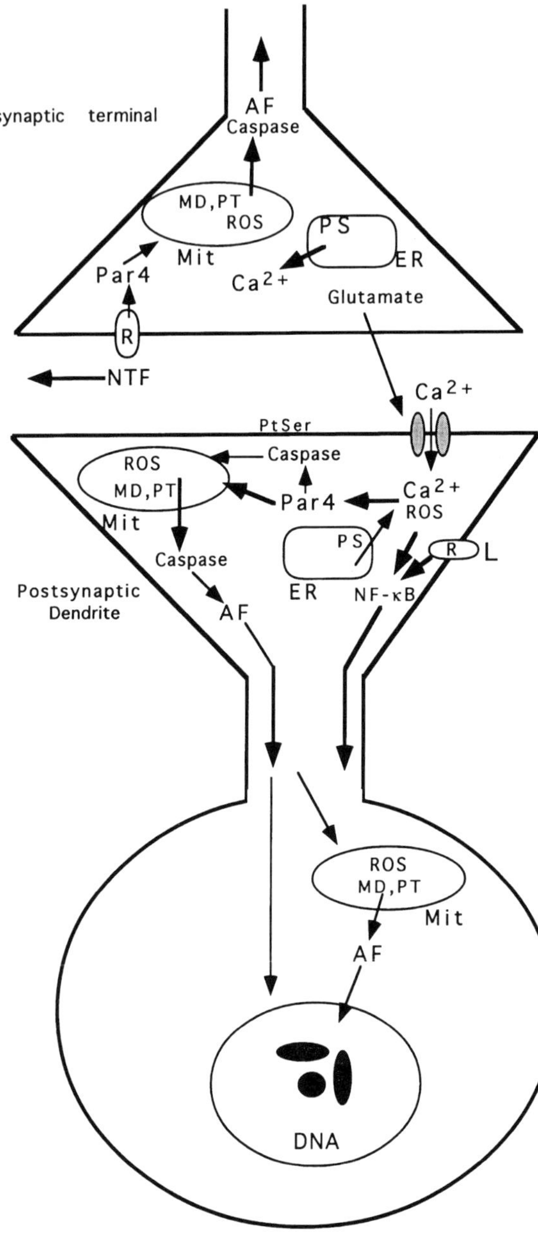

Figure 2. Synaptic activation of apoptotic biochemical cascades. See text for discussion. AF, apoptotic factor; ER, endoplasmic reticulum; MD, membrane depolarization; Mit, mitochondria; PT, permeability transition; ROS, reactive oxygen species. Modified from Mattson et al., 1998a.

During development of the nervous system, many neurons die during the period of synaptogenesis (Boonman and Isacson, 1999). This natural neuronal apoptosis (also called programmed cell death) can be prevented by administration of neurotrophic factors, and exacerbated by treatment with antibodies against neurotrophic factors, suggesting that neuronal death results from lack of sufficient target-derived neurotrophic factor. Thus, insufficient activation of neurotrophic factor receptors located in presynaptic terminals leads to axonal degeneration and subsequent neuronal apoptosis. Increasing evidence suggests that reduced neurotrophic factor signaling may also contribute to the neurodegenerative process in both acute and chronic neurodegenerative disorders.

Oxidative Stress

Increased levels of protein oxidation, membrane lipid peroxidation and/or oxidative damage to DNA have been documented in association with the neurodegenerative process in disorders ranging from Alzheimer's, Parkinson's and Huntington's diseases, to amyotrophic lateral sclerosis, and ischemic brain injury (Ferrante et al., 1997; Beal, 1998b; Jenner and Olanow, 1998; Pedersen et al., 1998). Reactive oxygen species (ROS) that contribute to neuronal damage include superoxide anion radical, hydrogen peroxide, hydroxyl radical and peroxynitrite (Figure 3). Major sources of ROS include the mitochondrial electron transport chain, which produces large quantities of superoxide. Superoxide is normally converted to hydrogen peroxide via activity of superoxide dismutases (SODs). Hydrogen peroxide is converted to water via activity of glutathione peroxidase and catalase. However, in the presence of ferrous iron (Fe^{2+}) hydroxyl radical is produced from hydrogen peroxide. Hydroxyl radical is a

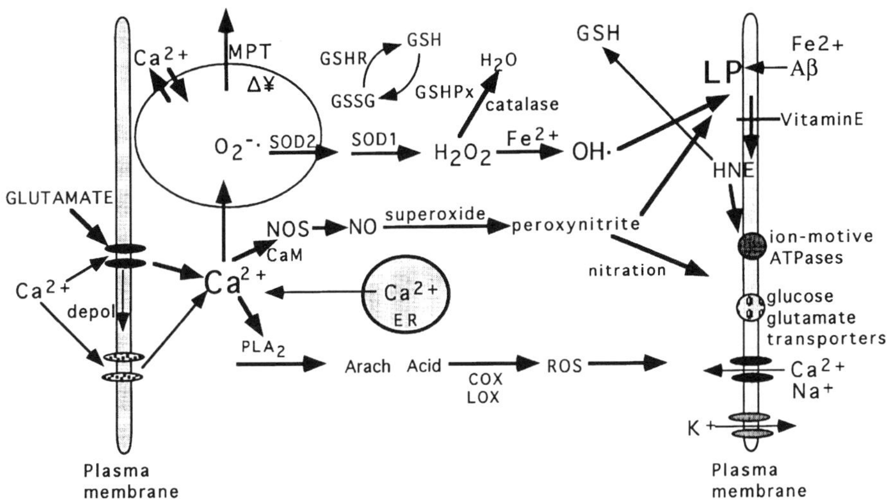

Figure 3. Cellular mechanisms for the generation and removal of oxyradicals. See text for discussion. Aβ, amyloid β-peptide; COX, cyclooxygenase; GSH, glutathione; HNE, 4-hydroxynonenal; LOX, lipoxygenase; LP, lipid peroxidation; MPT, membrane permeability transition; NOS, nitric oxide synthase; SOD, superoxide dismutase. Modified from Mattson, 1998b.

potent inducer of membrane lipid peroxidation. Another source of ROS is the activity of the calcium-activated enzyme nitric oxide synthase, which generates nitric oxide. Nitric oxide can interact with superoxide to produce peroxynitrite, which is a strong inducer of membrane lipid peroxidation and can also induce nitration of proteins on tyrosine residues.

There are several mechanisms whereby oxidative stress disrupts neuronal calcium homeostasis. Increased lipid peroxidation can impair the function of several membrane transporters that play important roles in the maintenance of $[Ca^{2+}]_i$. Impairment of ion-motive ATPases (Na^+/K^+-ATPase and Ca^{2+}-ATPase) results in membrane depolarization and impaired calcium extrusion (Mark et al., 1995, 1997a). Impairment of the glucose transporter (GLUT-3) results in depletion of ATP levels because neurons rely on a constant supply of glucose (Mark et al., 1997b). Oxidative stress in astrocytes may also indirectly perturb neuronal calcium homeostasis. For example, Fe^{2+} and other conditions that induce lipid peroxidation can impair glutamate transport in astrocytes (Keller et al., 1997; Blanc et al., 1998). Recent findings have elucidated the mechanism whereby lipid peroxidation impairs the function of membrane transporter proteins. Lipid peroxidation generates the aldehyde 4-hydoxynonenal, which can covalently modify proteins on cysteine, lysine and histidine residues. Such covalent modification by 4-hydroxynonenal impairs protein function and promotes protein crosslinking (Mark et al., 1997a, b; Keller et al., 1997; Blanc et al., 1998).

Other major mechanisms whereby oxidative stress disrupts neuronal calcium homeostasis include disruption of mitochondrial and ER function. Increased oxidative stress in mitochondria and ER disrupts the ability of these organelles to take up and retain calcium (Marks, 1997; Budd, 1998; Mattson et al., 1998c; Kruman et al., 1999a). Calcium imaging studies have shown that mitochondria and ER play important roles in buffering physiological calcium loads (Berridge, 1998; Budd, 1998). Evidence that disruption of mitochondrial and ER calcium regulation contributes to excitotoxic and apoptotic neuronal degeneration comes from studies showing that agents that stablize mitochondrial calcium homeostasis (e.g., cyclosporine) or suppress calcium release from ER (e.g., dantrolene) protect cultured neurons against excitotoxic and oxidative insults (Keller et al., 1998; Guo et al., 1997).

Metabolic Compromise

Perhaps the best-studied paradigm of neuronal injury resulting from metabolic impairment is ischemic stroke (Kristian and Siesjo, 1998). When the blood supply to neurons is disrupted, the supply of glucose and oxygen is cut off and cellular ATP levels plummet rapidly. ATP depletion leads to failure of ion-motive ATPases with resulting membrane depolarization and calcium influx and release from ER (Figure 1). ATP is also required for various phosphorylation reactions that regulate a variety of cellular processes including various signaling cascades and metabolic pathways. Thus, in stroke it is very clear that metabolic compromise is a critical event that leads to disruption of neuronal calcium homeostasis. Emerging data also support a role for decreased energy availability to the degeneration of neurons in age-related disorders such as Alzheimer's and Parkinson's disease. Positron emission topography studies have shown that glucose uptake is reduced in vulnerable brain regions of Alzheimer's patients, and the decreased glucose uptake may precede neuronal degeneration (Hoyer et al., 1991; Jagust et al., 1991). Further evidence that energy production is reduced in neurons in Alzheimer's disease comes from studies showing reduced levels of α-ketoglutarate dehydrogenase complex activity in brain tissue (Gibson et al., 1998). Interestingly, activity of this enzyme complex is also decreased in fibroblasts of Alzheimer's disease suggesting a widespread metabolic alteration. Parkinson's disease may also involve impaired energy metabolism, as levels of mitochondrial complex I activity are reduced in brain cells of Parkinson's patients (Schapira et al., 1990).

Disorder-Specific Mechanisms of Altered Calcium Homeostasis

What are the upstream events that ultimately lead to perturbed neuronal calcium homeostasis in neurodegenerative disorders? Considerable insight into the primary alterations in several of the major neurodegenerative disorders have arisen from application of knowledge obtained from neurochemical, cell biological and molecular genetic analyses of human patients to the development of experimental models of the disorders. The identification of mutations responsible for inherited forms of neurodegenerative disorders are leading rapidly to a clearer understanding of disease pathogenesis. As will be described below, studies in which these mutations are expressed in cultured cells and transgenic mice are revealing an array of pathogenic mechanisms that can lead to perturbed neuronal calcium homeostasis.

Alzheimer's Disease

Alzheimer's disease is a progressive age-related disorder characterized by degeneration of neurons in brain regions, such as the hippocampus and certain regions of cerebral cortex, that are involved in learning and memory processes (Mattson and Geddes, 1997). Histological examination of brains of Alzheimer's patients reveals three prominent abnormalities: (1) Extracellular accumulation of insoluble diffuse and fibrillar deposits (plaques) consisting primarily of amyloid β-peptide (Aβ), a 40-42 amino acid peptide; (2) Intraneuronal accumulation of straight and paired-helical filaments (i.e., neurofibrillary tangles) comprised of the microtubule-associated protein tau; and (3) Degeneration of synapses. Degenerated neurites are often associated with fibrillar amyloid plaques, suggesting a role for Aβ in the neurodegenerative process. Cell culture studies have shown that Aβ can induce degeneration of hippocampal neurons (Yankner, 1996). A particularly striking effect of Aβ is to increase neuronal vulnerability to excitotoxicity by a mechanism involving disruption of cellular calcium homeostasis (Mattson et al., 1992, 1993d). When Aβ interacts with the plasma membrane of neurons it induces membrane lipid peroxidation (Butterfield et al., 1994; Goodman and Mattson, 1994). The membrane lipid peroxidation leads to impairment of ion-motive ATPases and glucose transporters, and both of these events render neurons vulnerable to excitotoxicity and apoptosis (Mark et al., 1995, 1997a, b; Kruman et al., 1997). Further evidence that Aβ and perturbed calcium homeostasis contribute to the neurodegenerative process in Alzheimer's disease are data showing that excessive activation of glutamate receptors and voltage-dependent calcium channels can induce antigenic alterations in the cytoskeleton similar to those seen in neurofibrillary tangles (Mattson, 1990; Elliot et al., 1993; Stein-Behrens et al., 1994).

The deposition of Aβ in the brains of Alzheimer's victims appears to result from altered proteolytic processing of the amyloid precursor protein (APP) in a manner that increases production of Aβ (Figure 4; Mattson, 1997). APP is a transmembrane protein with a single membrane-spanning domain and a large extracellular N-terminal domain; the Aβ sequence is located at the cell surface. In addition to increasing Aβ production, altered proteolytic processing of APP also decreases production of a secreted form of APP called sAPPα. sAPPα is a very potent neurotrophic factor that appears to act by stabilizing calcium homeostasis. sAPPα protects hippocampal neurons in culture and *in vivo* against excitotoxic and metabolic insults (Mattson et al., 1993b). Calcium imaging analyses have shown that sAPPα reduces basal levels of $[Ca^{2+}]_i$ and suppresses calcium responses to glutamate and membrane depolarization in hippocampal neurons (Mattson et al., 1993b; Furukawa et al., 1996). Studies of the signaling pathway for sAPPα implicate the involvement of cyclic GMP production, activation of potassium channels, and activation of the anti-apoptotic transcription factor NF-κB (Furukawa et al., 1996; Barger and Mattson, 1996). Moreover, data from electrophysiological studies in the hippocampal slice preparation (Ishida et al., 1997), and *in vivo* studies with APP antibodies (Doyle et al., 1990), suggest that sAPPα plays an important role in the learning and memory process. Consistent with a role for modulation of synaptic calcium homeostasis in the actions of sAPPα on neuronal plasticity and survival are data showing that sAPPα can enhance glutamate and glucose transport locally in synaptic terminals (Mattson et al., 1999). Because levels of sAPPα may be reduced in Alzheimer's disease, the experimental data suggest a role for this reduction in synaptic dysfunction and neuronal degeneration (Mattson, 1997).

Additional findings suggest that insufficient neurotrophic factor signaling may contribute to perturbed cellular calcium homeostasis and the neurodegenerative process in Alzheimer's disease. Levels of BDNF (Phillips et al., 1991) and NGF (Scott and Crutcher, 1994) were reported to be decreased in certain regions of cerebral cortex of Alzheimer's patients. Basic FGF (Mattson et al., 1993d) and activity-dependent neurotrophic factor (ADNF; Brenneman et al., 1998) protected cultured hippocampal neurons against Aβ-induced mitochondrial dysfunction and cell degeneration. In the case of bFGF it was shown that pretreatment of neurons with this neurotrophic factor resulted in increased resistance to the calcium-elevating and excitotoxicity-promoting effects of Aβ (Mattson et al., 1993a; Mark et al., 1997c); the mechanism may involve modulation of the expression glutamate receptor subunits (Mattson et al., 1993a; Cheng et al., 1995) and/or antioxidant enzymes (Mattson et al., 1995). In addition to transcription-mediated mechanisms, neurotrophic factors may also exert local transcription-independent effects at the level of the synapse. For example, we have found that bFGF and ADNF can protect synaptic terminals against oxidative impairment of glucose and glutamate transporter functions, and can stabilize mitochondrial function in synaptic terminals (Guo and Mattson, 1999). These findings suggest that such neurotrophic factors may

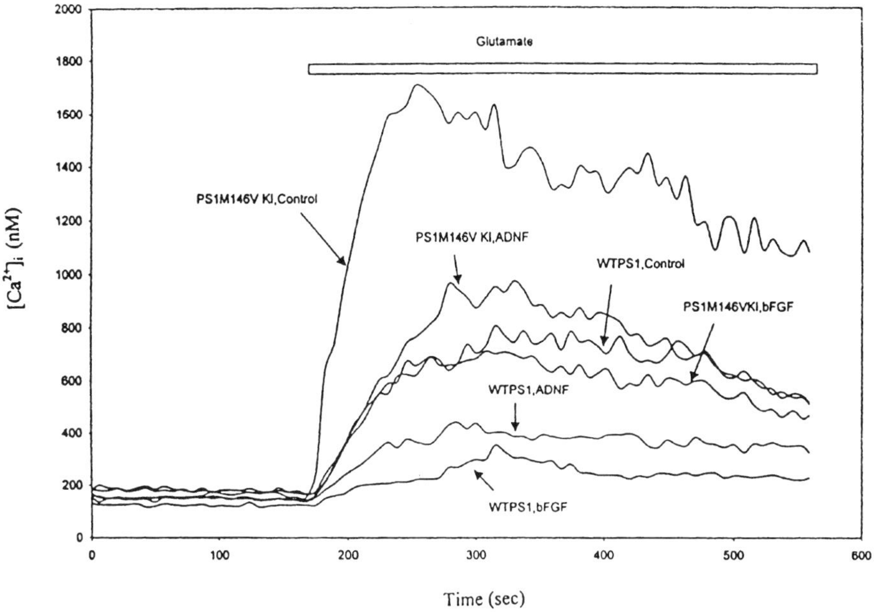

Figure 4. Hippocampal neurons expressing an Alzheimer's presenilin-1 mutation exhibit enhanced calcium responses to glutamate, an effect counteracted by neurotrophic factors. Hippocampal cultures from wild-type mice (WTPS1) and presenilin-1 mutant knockin mice (PS1M146VKI) were pretreated for 24 h with saline (Con), basic fibroblast growth factor (bFGF) or activity-dependent neurotrophic factor peptide (ADNF9). The $[Ca^{2+}]_i$ was then monitored prior to and following exposure to 50 μM glutamate.

exert a local beneficial affect on synaptic calcium homeostasis at the level of the synapse. Interestingly, recent electrophysiological studies suggest that neurotrophic factors can modulate synaptic plasticity in a manner consistent with roles in learning and memory (Figurov et al., 1996; Albensi and Mattson, 1999).

Postmenopausal women who take estrogen replacement therapy have a reduced risk for AD compared to women who do not take estrogens. Experimental data provide evidence that estrogen (17β-estradiol) can exert direct protective actions on neurons. For example, treatment of cultured hippocampal neurons with estrogen results in increased resistance of the neuron to oxidative insults (e.g., exposure to Aβ or Fe^{2+}) and excitotoxicity (Goodman et al., 1996). Estrogen possesses inherent antioxidant properties, and its ability to suppress membrane lipid peroxidation appears to account for its neuroprotective actions. In this way, estrogen can stabilize neuronal calcium homeostasis and preserve mitochondrial function (Goodman and Mattson, 1996; Keller and Mattson, 1997; Mattson et al., 1997a). Another type of steroid hormone that can modulate neuronal calcium homeostasis is glucocorticoids, which are produced by the adrenal cortex in response to various physiological and psychological stressors.

Experimental studies have shown that glucocorticoids can increase neuronal vulnerability to excitotoxic and oxidative insults by a mechanism involving disruption of calcium homeostasis (Elliot et al., 1993; Stein-Behrens et al., 1994; Goodman et al., 1996). There is evidence for alterations in the neuroendocrine system controlling glucocorticoid production in Alzheimer's patients suggesting a contribution of glucocorticoids to the neurodegenerative process in this disease.

Mutations in the presenilin-1 (PS1) gene located on chromosome 14 have been linked to many cases of early-onset autosomal dominant inherited Alzheimer's disease (see Mattson and Guo, 1999 for review). PS1 is an integral membrane protein that is widely expressed in neurons throughout the brain, wherein it is localized mainly in the ER. When mutant PS1 is overexpressed in cultured cells and transgenic mice (Scheuner et al., 1996; Duff et al., 1996), or expressed at normal levels in knockin mice (Guo et al., 1999a), levels of the particularly neurotoxic long form of Aβ ($A\beta_{1-42}$) are increased. The altered APP processing resulting from PS1 mutations appears to be secondary to a primary effect of the mutations on ER calcium homeostasis. Thus, cultured neural cells expressing mutant PS1 exhibit enhanced calcium release from IP_3- and ryanodine-sensitive stores (Guo et al., 1996, 1997). Expression of PS1 mutations in cultured cells (Guo et al., 1996, 1997, 1998a, b, c) and knockin mice (Guo et al., 1999a, b, c) results in increased vulnerability of the cells to apoptosis induced by various insults including exposure to Aβ, metabolic insults and trophic factor withdrawal. PS1 mutations also increase the vulnerability of hippocampal neurons to excitotoxicity induced by glutamate in cell culture and *in vivo* (Guo et al., 1999b). The latter study showed that calcium responses to glutamate were greatly enhanced in neurons from PS1 mutant knockin mice (Figure 4). Perturbed ER calcium homeostasis plays a central role in the endangering actions of PS1 mutations, because agents that suppress calcium release from ER can counteract the effects of the PS1 mutations (Guo et al., 1997).

Parkinson's Disease

The symptoms of Parkinson's disease include motor abnormalities (akinesia, tremor and rigidity) that result from progressive degeneration of dopaminergic neurons in the substantia nigra. Increased levels of oxidative stress (membrane lipid peroxidation and protein nitration) and mitochondrial dysfunction (reduced complex I activity) are associated with the neurodegenerative process (Jenner and Olanow, 1998). Both genetic and environmental factors likely contribute to the increased oxidative stress and mitochondrial dysfunction. Mutations in α-synuclein, a major component of the insoluble aggregates that comprise the characteristic brain lesions (Lewy bodies) in PD, appear to be responsible for a small percentage of PD cases. If and how alterations in α-synuclein might lead to perturbed calcium homeostasis in dopaminergic neurons remains to be determined. Environmental toxins are suspected because the toxin 1-methyl-4-phenyl-1,2,3,6-tetrahydropyridine (MPTP) can induce Parkinson-like pathology and symptomology in mice, monkeys and humans (Langston, 1998). The specific events implicated in inducing oxidative stress and disruption of neuronal calcium homeostasis include aberrant dopamine metabolism (Figure 5; Fu et al., 1998), increased accumulation of iron and resulting lipid peroxidation in the substantia nigra (Dexter et al., 1989), and excessive nitric oxide production (Good et al., 1998). As described above, oxidative stress promotes neuronal degeneration by impairing the function of various membrane transporters. Mitochondrial calcium homeostasis would also be compromised by increased oxidative stress and impairment of complex I activity. Indeed, introduction of mitochondria from Parkinson's patients into cultured cells results in altered cellular calcium homeostasis (Sheehan et al., 1997).

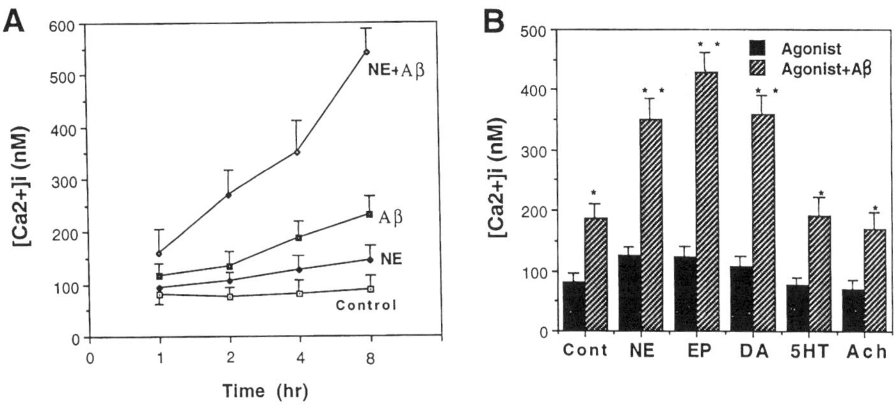

Figure 5. Catecholamines exacerbate Aβ-induced elevation of intracellular calcium levels in cultured neurons. (**A**) Hippocampal cultures were exposed for the indicated time periods to vehicle (Control), 10 μM norepinephrine (NE), 5 μM Aβ, or 5 μM Aβ plus 10 μM NE. The $[Ca^{2+}]_i$ in individual neurons was then measured and values are the mean and SD of determinations made in 4 separate cultures (measurements made in 20-30 neurons/culture). (**B**) Cultures were exposed for 4 hr to 5 μM Aβ alone, to the indicated neurotransmitters alone (10 μM each), or to combinations of Aβ (5 μM) plus a neurotransmitter (10 μM). The $[Ca^{2+}]_i$ was then measured and values are the mean and SD of determinations made in 4 separate cultures. *$p<0.01$ compared to corresponding control value. **$p<0.01$ compared to value in Control cultures exposed to Aβ alone. EP, epinephrine; DA, dopamine; 5HT, serotonin; Ach, acetylcholine. Modified from Fu et al., 1998.

Studies of postmortem brain tissue from Parkinson's patients have documented increased levels of calcium-dependent proteases in substantia nigra neurons, suggesting increased calcium-mediated proteolysis (Mouatt-Prigent et al., 1996). Dopaminergic neurons that express the calcium-binding protein calretinen are resistant to levodopa-induced neurotoxicity consistent with a role for differential cellular calcium regulation in the selective vulnerability of a subpopulation of dopamine-producing neurons (Isaacs et al., 1997). Interestingly, calcium imaging studies have shown that exposure of mesencephalic neurons to MPP^+ disrupts calcium homeostasis only in dopaminergic neurons and not in other neurons of that cell population (Chen et al., 1995).

Overactivation of glutamate receptors and underactivation of neurotrophic factor signaling pathways may contribute to the neurodegenerative process in Parkinson's disease. Dopaminergic neurons in the substantia nigra express high levels of AMPA receptors and, in animal models, are vulnerable to excitotoxicity. Moreover, AMPA receptor antagonists have been shown to reduce damage and improve behavioral outcome following MPTP administration in monkeys (Klockgether et al., 1991). As in other neurodegenerative disorders, there is considerable evidence that neurotrophic factors can protect neurons against the aberrant conditions that promote neurodegeneration in Parkinson's disease. Several neurotrophic factors including glial cell line-derived neurotrophic factor (GDNF) and brain-derived neurotrophic factor (BDNF) can protect dopaminergic neurons in culture and *in vivo* against insults relevant to Parkinson's disease including exposure to Fe^{2+} and MPTP or MPP^+ (Spina et al., 1992; Gash et al., 1998). GDNF may even be able to reverse the symptoms of Parkinson's disease by a mechanism involving recovery of function of damaged dopaminergic neurons (Gash et al., 1998). Although the mechanisms whereby GDNF and BDNF benefit dopaminergic neurons are not clear, it has been shown that BDNF can stabilize calcium homeostasis in other neuronal populations (Cheng and Mattson, 1994).

Amyotrophic Lateral Sclerosis

Dysfunction and degeneration of motor neurons in the spinal cord resulting in progressive paralysis are the diagnostic features of amyotrophic lateral sclerosis (ALS). Both sporadic and inherited forms of ALS manifest similar pathological and clinical phenotypes, suggesting that different initiating causes lead to a mechanistically similar neurodegenerative cascade involving oxidative stress and aberrant calcium regulation (Siddique, 1991; Pedersen et al., 1998). Analyses of spinal cord tissue from ALS patients has provided evidence that levels of intracellular calcium are increased in motor neurons (Siklos et al., 1996). The motor neuron populations which are most vulnerable (spinal cord) contain low levels of calcium-binding proteins whereas less vulnerable motor neurons, such as those of cranial nerve nuclei, contain higher levels of calcium-binding proteins (Alexianu et al., 1994). Considerable insights into the pathogenic mechanisms of ALS have followed from the discovery that some cases of inherited ALS are caused by mutations in Cu/Zn-superoxide dismutase (Cu/Zn-SOD) (Rosen, 1993). Cu/Zn-SOD converts superoxide radical anions ($O_2^{\cdot-}$) to hydrogen peroxide, and also possesses peroxidase activity which can generate hydroxyl radical upon interaction of the enzyme-associated Cu^+ with hydrogen peroxide (Wiedau-Pazos et al., 1996). When expressed in transgenic mice ALS-linked mutations in Cu/Zn-SOD result in progressive degeneration of spinal cord motor neurons and a clinical phenotype remarkably similar to that of ALS patients (Gurney et al., 1994; Wong et al., 1995). Analyses of spinal cords of ALS patients and "ALS mice" have revealed evidence for increased membrane lipid peroxidation and protein nitration (Ferrante et al., 1997; Pedersen et al., 1998). As appears to be the case in Alzheimer's disease, 4-hydroxynonenal may be a key mediator of neuronal degeneration in ALS. As evidence, levels of 4-hydroxynonenal are increased in the CSF and spinal cord tissue of ALS patients (Figure 6; Pedersen et al., 1998; Smith et al., 1998) and exposure of cultured motor neurons to 4-hydroxynonenal results in impairment of glucose and glutamate transport and apoptosis (Pedersen et al., 1999). Presumably, the increased oxidative stress and dysfunction of membrane transporters lead to disruption of calcium homeostasis and increased vulnerability of neurons to excitotoxicity. Indeed, recent studies of calcium homeostasis in motor neurons from transgenic mice expressing mutant Cu/Zn-SOD suggest that the mutant enzyme promotes oxidative stress, mitochondrial dysfunction and perturbed calcium homeostasis resulting in increased vulnerability to excitotoxicity (Kruman et al., 1999b).

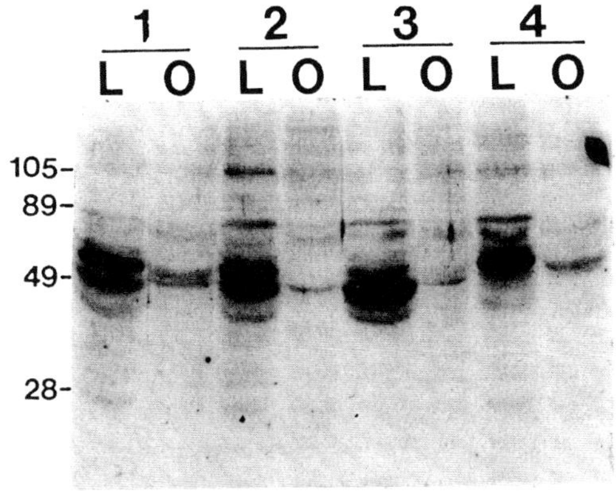

Figure 6. Levels of proteins covalently modified by the lipid peroxidation product 4-hydroxynonenal are increased in lumbar spinal cord tissue from patients with amyotrophic lateral sclerosis. Western blot of protein samples from lumbar spinal cord tissue (L) and occipital cortex tissue (O) from four different ALS patients (50 μg protein/lane) probed with an antibody against 4-hydroxynonenal-modified proteins. Note much greater extent of immunoreactive proteins in the spinal cord tissue (the region of the nervous system most affected in ALS) as compared to occipital cortex tissue (a region of the brain unaffected in ALS patients). Modified from Pedersen et al., 1998.

Among the evidence implicating overactivation of glutamate receptors in the pathogenesis of ALS are data showing that: spinal cord motor neurons are vulnerable to excitotoxicity resulting from overactivation of glutamate receptors (particularly AMPA/kainate receptors); levels of glutamate are increased in the cerebrospinal fluid of ALS patients; and glutamate transport is reduced in the brain and spinal cord of ALS patients (see Rothstein, 1995 for review). Moreover, motor neurons from Cu/Zn-SOD mutant mice exhibit enhanced calcium responses to glutamate and increased vulnerability to excitotoxicity compared to motor neurons from wild-type mice (Roy et al., 1998; Kruman et al., 1999b). Further evidence for perturbed calcium homeostasis and an excitotoxic mechanism in ALS comes from studies of an intriguing syndrome localized to a few islands in Southeast Asia near Guam (Garruto and Yase, 1986). The affected individuals develop progressive motor neuron degeneration and paralysis similar to ALS, and also develop Parkinson-like symptoms and dementia. Increased levels of calcium are present in affected neurons from these patients. This disorder appears to have an environmental cause as there is no clear pattern of inheritance, and when the islanders move to industrialized countries the disease disappears. There is considerable evidence that the cause of the disorder may be an excitotoxin or other neurotoxic agent in the seed of the cycad plant (Spencer et al., 1987). The islanders use the cycad seeds to make flour, which is a major component of their diet. Experimental data suggest that substances in the cycad seed can induce excitotoxic neuronal injury (Spencer et al., 1987).

Stroke

Stroke is a major cause of disability and death in older persons. From the affected neuron's perspective, stroke involves loss of blood supply (due to blood vessel occlusion or rupture) resulting in oxygen and glucose deprivation. There are excellent animal and cell culture models that have proven valuable in elucidating the mechanisms that lead to neuronal degeneration following stroke. Cell culture models include glucose deprivation, hypoxia, and combined glucose/oxygen deprivation. The animal model that most closely resembles stroke in humans is the middle cerebral artery occlusion — reperfusion paradigm in rats and mice (Endres et al., 1999). Overwhelming data supports a major role for perturbed neuronal calcium homeostasis in the pathogenesis of stroke. In each cell culture and animal model, the insult results in increases of $[Ca^{2+}]_i$, and pharmacological manipulations that suppress the increases of $[Ca^{2+}]_i$ (e.g., glutamate receptor antagonists, calcium channel blockers, calcium chelators) protect the neurons (see Mattson and Mark, 1996 for review). Increased levels of oxidative stress, which may occur in response to excessive calcium influx, contribute to the neurodegenerative process in stroke models (Mattson et al., 1995). Additional downstream effectors of calcium-induced neuronal injury in stroke models include proteases of the calpain (Lee et al., 1997) and caspase (Schulz et al., 1999) families.

In a transient global forebrain ischemia model (which mimics cardiac arrest) CA1 neurons in hippocampus are selectively vulnerable. These neurons express high levels of NMDA receptors, but have relatively low levels of calcium-binding proteins such as calbindin D28k. In contrast, hippocampal neurons that are less vulnerable express lower levels of NMDA receptors (CA3 neurons) or high levels of calbindin D28k (Sloviter et al., 1991; Monyer et al., 1994). Studies of cultured hippocampal neurons overexpressing or underexpressing calbindin and NMDA receptor subunits support neuroprotective and neuroendangering roles, respectively, for these proteins in ischemic and excitotoxic brain injury (Mattson et al., 1993a; Cheng et al., 1994a). Studies of mice lacking calbindin or NMDA receptor subunits are consistent with pivotal roles of these calcium-regulating proteins in selective neuronal vulnerability, although their roles appear more complicated than originally thought. For example, CA1 neurons in calbindin heterozygous knockout mice are less vulnerable to seizure-induced injury compared to wild-type or calbindin homozygous knockout mice (Gary et al., 1998). The latter study also showed that lack of calbindin does not render dentate granule neurons vulnerable to seizure-induced injury, demonstrating that calbindin is not the major factor responsible for resistance of dentate neurons in this model.

Several neurotrophic factors have been shown to protect cultured neurons against hypoxia and glucose deprivation, and several such neurotrophic factors have also proven beneficial in animal models of stroke (see Mattson, 1996 for review). Examples of factors that protect against ischemia-relevant insults in culture include bFGF (Cheng and Mattson, 1991), NGF (Cheng and Mattson, 1991), IGF-1 and IGF-2 (Cheng and Mattson,

1992), BDNF (Cheng and Mattson, 1994), neurotrophin-4/5 (Cheng et al., 1994b), transforming growth factor-β (Prehn et al., 1994), PDGF (Cheng and Mattson, 1995), TNFα (Cheng et al., 1994a; Mattson et al., 1997b) and sAPPα (Mattson et al., 1993b). The latter studies provided evidence that each of these neurotrophic factors stabilizes cellular calcium homeostasis. Examples of *in vivo* data supporting anti-ischemic roles for neurotrophic factors include: reduced infarct size and improved behavioral outcome following focal cerebral ischemia in rats administered bFGF (Koketsu et al., 1994); reduced damage to CA1 hippocampal neurons following transient global forebrain ischemia in rats administered NGF (Shigeno et al., 1991) or aAPPα (Smith-Swintosky et al., 1994); and increased infarct size following middle cerebral artery occlusion in mice lacking TNF receptors (Bruce et al., 1996).

As might be expected, based upon their general roles in regulating neuronal calcium homeostasis, mitochondria and ER play important roles in ischemic brain injury. Dysregulation of mitochondrial calcium handling has been demonstrated in cell culture paradigms relevant to ischemic neuronal injury (Murphy et al., 1996; Yu et al., 1998), and agents that block mitochondrial membrane permeability transition (e.g., cyclosporin A) can reduce the extent of ischemic brain injury *in vivo* (Uchino et al., 1998). Ischemic conditions may also promote release of ER calcium (Paschen and Doutheil, 1999), and agents that block calcium release from ER (e.g., dantrolene) reduced ischemic brain injury in animal models (Zhang et al., 1993).

Conclusions

Perturbed neuronal calcium homeostasis occurs in many different neurodegenerative disorders. The causes of the perturbed calcium homeostasis include increased levels of oxidative and metabolic stress, and mutations affecting proteins that modulate specific aspects of cellular calcium homeostasis. Overactivation of glutamate receptors under conditions of increased oxidative and metabolic stress appears to contribute greatly to the increased vulnerability of neurons to age-related degeneration. Mitochondria and the endoplasmic reticulum play pivotal roles in regulating neuronal calcium homeostasis, and alterations in the calcium-regulating functions of these organelles contribute to the pathogenesis of several acute and chronic neurodegenerative disorders. The specific molecular and biochemical alterations responsible for disruption of neuronal calcium homeostasis are being elucidated in disorders ranging from Alzheimer's and Parkinson's diseases to stroke. The nervous system possesses a remarkable array of neuroprotective signal transduction pathways including those activated by neurotrophic factors and cytokines. By modulating the expression and/or function of various calcium-regulating signaling pathways, such signaling pathways stabilize neuronal calcium homeostasis. In addition, antioxidants such as vitamin E and uric acid, and hormones such as estrogen, stabilize calcium homeostasis by preserving the function of membrane ion-motive ATPases and glucose and glutamate transporters. A better understanding of systems that regulated neuronal calcium homeostasis, and how they are altered by environmental and genetic aberrancies, will lead to novel approaches for the prevention and treatment of neurodegenerative disorders.

References

Albensi, B.C. & Mattson, M.P. (1999). Evidence for the involvement of TNF and NF-κB in hippocampal synaptic plasticity. Synapse 35, 151-159.

Alexianu, M.E., Ho, B.K., Mohamed, A.H., La Bella, V., Smith, R.G. & Appel, S.H. (1994). The role of calcium-binding proteins in selective motoneuron vulnerability in amyotrophic lateral sclerosis. Ann. Neurol. 36, 846-858

Ankarcrona, M., Dypbukt, J.M., Bonfoco, E., Zhivotovsky, B., Orrenius, S., Lipton, S.A. & Nicotera, P. (1995). Glutamate-induced neuronal death: a succession of necrosis or apoptosis depending on mitochondrial function. Neuron 15, 961-973.

Baimbridge, K.G., Celio, M.R. & Rogers, J.H. (1992). Calcium-binding proteins in the nervous system. Trends Neurosci. 15, 303-308.

Barger, S.W., Horster, D., Furukawa, K., Goodman, Y., Krieglestein, J. & Mattson, M.P. (1995): Tumor necrosis factors α and β protect neurons against amyloid β-peptide toxicity: evidence for involvement of a κB-binding factor and attenuation of peroxide and Ca^{2+} accumulation. Proc. Natl. Acad.Sci. USA 92, 9328-9332.

Barger, S.W. & Mattson, M.P. (1996). Induction of neuroprotective κB-dependent transcription by secreted forms of the Alzheimer's β-amyloid precursor. Mol. Brain Res. 40, 116-126.

Beal, M.F. (1998a). Excitotoxicity and nitric oxide in Parkinson's disease pathogenesis. Ann Neurol. 44, S110-S114.

Beal, M.F. (1998b). Mitochondrial dysfunction in neurodegenerative diseases. Biochim. Biophys. Acta 1366, 211-223.

Bernard, C., Hirsch, J. & Ben-Ari, Y. (1997). Glutamatergic excitatory receptors and temporal lobe epilepsy. Rev. Neurol. (Paris) 153 Suppl 1, S14-24.

Berridge, M.J. (1998). Neuronal calcium signaling. Neuron 21, 13-26.

Bixby, J.L. & Bookman, R.J. (1998). Intracellular mechanisms of axon growth induction by CAMs and integrins: some unresolved issues. Perspect. Dev. Neurobiol. 4, 147-156.

Blanc, E.M., Keller, J.N., Fernandez, S. & Mattson, M.P. (1998). 4-hydroxynonenal, a lipid peroxidation product, inhibits glutamate transport in astrocytes. Glia 22, 149-160.

Boonman, Z. & Isacson, O. (1999). Apoptosis in neuronal development and transplantation: role of caspases and trophic factors. Exp. Neurol. 156, 1-15.

Brenneman, D.E., Hauser, J., Neale, E., Rubinraut, S., Fridkin, M., Davidson, A. & Gozes, I. (1998). Activity-dependent neurotrophic factor: structure-activity relationships of femtomolar-acting peptides. J. Pharmacol. Exp. Ther. 285, 619-627.

Bruce, A.J., Boling, W., Kindy, M.S., Peschon, J., Kraemer, P.J., Carpenter, M.K., Holtsberg, F.W. & Mattson, M.P. (1996). Altered neuronal and microglial responses to brain injury in mice lacking TNF receptors. Nature Med. 2, 788-794.

Budd, S.L. (1998). Mechanisms of neuronal damage in brain hypoxia/ischemia: focus on the role of mitochondrial calcium accumulation. Pharmacol. Ther. 80, 203-229.

Butterfield, D.A., Hensley, K., Harris, M., Mattson, M.P. & Carney, J. (1994). β-amyloid peptide free radical fragments initiate synaptosomal lipoperoxidation in a sequence-specific fashion: implications to Alzheimer's disease. Biochem. Biophys. Res. Commun. 200, 710-715.

Chan, S.L. & Mattson, M.P. (1999). Caspase and calpain substrates: roles in synaptic plasticity and cell death. J. Neurosci. Res. 58, 167-190.

Koutsilieri, E. & Rausch, W.D. (1995). MPP+ selectively affects calcium homeostasis in mesencephalic cell cultures from embryonal C57/Bl6 mice. J. Neural. Transm. Gen. Sect. 100, 153-163.

Cheng, B. & Mattson, M.P. (1991). NGF and bFGF protect rat and human central neurons against hypoglycemic damage by stabilizing calcium homeostasis. Neuron 7, 1031-1041.

Cheng, B. & M.P. Mattson (1992). IGF-I and IGF-II protect cultured hippocampal and septal neurons against calcium-mediated hypoglycemic damage. J. Neurosci. 12, 1558-1566.

Cheng, B. & M.P. Mattson (1994). NT-3 and BDNF protect CNS neurons against metabolic/excitotoxic insults. Brain Res. 640, 56-67.

Cheng, B., Christakos, S. & Mattson, M.P. (1994a): Tumor necrosis factors protect neurons against excitotoxic/metabolic insults and promote maintenance of calcium homeostasis. Neuron 12, 139-153.

Cheng, B., Goodman, Y., Begley, J.G. & Mattson, M.P. (1994b). Neurotrophin-4/5 protects hippocampal and cortical neurons against energy deprivation- and excitatory amino acid-induced injury. Brain Res. 650, 331-335.

Cheng, B. & Mattson, M.P. (1995). PDGFs protect hippocampal neurons against energy deprivation and oxidative injury: evidence for induction of antioxidant pathways. J. Neurosci. 15, 7095-7104.

Cheng, B., Furukawa, K., O'Keefe, J.A., Goodman, Y., Kihiko, M., Fabian, T. & Mattson, M.P. (1995). Basic fibroblast growth factor selectively increases AMPA-receptor subunit GluR1 protein level and differentially modulates Ca^{2+} responses to AMPA and NMDA in hippocampal neurons. J. Neurochem. 65, 2525-2536.

Choi, D.W. (1994). Calcium and excitotoxic neuronal injury. Ann. N.Y. Acad. Sci. 747, 162-171.

Collazo, D., Takahashi, H. & McKay, R.D. (1992). Cellular targets and trophic functions of neurotrophin-3 in the developing rat hippocampus. Neuron 9, 643-656.

Dexter, D.T., Carter, C.J., Wells, F.R., Javoy-Agid, F., Agid, Y., Lees, A., Jenner, P. & Marsden, C.D. (1989). Basal lipid peroxidation in substantia nigra is increased in Parkinson's disease. J. Neurochem. 52, 381-389.

Doyle, E., Bruce, M.T., Breen, K.C., Smith, D.C., Anderton, B. & Regan, C.M. (1990), Intraventricular infusions of antibodies to amyloid β-protein precursor impair the acquisition of a passive avoidance response in the rat. Neurosci. Lett. 115, 97-102.

Duff, K., Eckman, C., Zehr, C., Yu, X., Prada, C.-M., Perez-Tur, J., Hutton, M., Buee, L., Harigaya, Y., Yager, D., Morgan, D., Gordon, M. N., Holcomb, L., Refolo, L., Zenk, B., Hardy, J. & Younkin, S. (1996). Increased amyloid-β42(43) in brains of mice expressing mutant presenilin 1. Nature 383, 710-713

Elliott, E., Mattson, M.P., Vanderklish, P., Lynch, G., Chang, I. & Sapolsky, R.M. (1993). Corticosterone exacerbates kainate-induced alterations in hippocampal tau immunoreactivity and spectrin proteolysis *in vivo*. J. Neurochem. 61, 57-67.

Endres, M., Fink, K., Zhu, J., Stagliano, N.E., Bondada, V., Geddes, J.W., Azuma, T., Mattson,. M.P., Kwiatkowski, D.J. & Moskowitz, M.A. (1999). Neuroprotective effects of gelsolin during murine stroke. J. Clin. Invest. 103, 347-354.

Ferrante, R.J., Browne, S.E., Shinobu, L.A., Bowling, A C., Baik, M.J., MacGarvey, U., Kowall, N.W., Brown Jr., R.H. &. Beal, M.F. (1997). Evidence of increased oxidative damage in both sporadic and familial amyotrophic lateral sclerosis. J. Neurochem. 69, 2064-2074.

Figurov, A., Pozzo-Miller, L.D., Olafsson, P., Wang, T. & Lu, B. (1996). Regulation of synaptic responses to high-frequency stimulation and LTP by neurotrophins in the hippocampus. Nature 381, 706-709.

Finch, E.A. & Augustine, G.J. (1998). Local calcium signaling by inositol-1,4,5-trisphosphate in Purkinje cell dendrites. Nature 396, 753-756.

Fu, W., Luo, H., Parthasarathy, S. & Mattson, M.P. (1998). Catecholamines potentiate amyloid β-peptide neurotoxicity: involvement of oxidative stress, mitochondrial dysfunction, and perturbed calcium homeostasis. Neurobiol. Dis. 5, 229-243.

Furukawa, K., Smith-Swintosky, V.L. & Mattson, M.P. (1995). Evidence that actin depolymerization protects hippocampal neurons against excitotoxicity by stabilizing $[Ca^{2+}]_i$. Exp. Neurol. 133, 153-163.

Furukawa, K., Barger, S.W., Blalock, E. & Mattson, M.P. (1996). Activation of K^+ channels and suppression of neuronal activity by secreted β-amyloid precursor protein. Nature 379, 74-78.

Furukawa, K., Fu, W., Witke, W., Kwiatkowski, D.J. & Mattson, M.P. (1997). The actin-severing protein gelsolin modulates calcium channel and NMDA receptor activities and vulnerability to excitotoxicity in hippocampal neurons. J. Neurosci. 17, 8178-8186.

Garruto, R. & Yase, Y. (1986). Neurodegenerative disorders of the western Pacific: the search for mechanisms for pathogenesis. TINS 368-374.

Gary, D.S. & Mattson, M.P. (2000). Integrin signaling via the PI3 kinase ñ Akt pathway increases neuronal resistance to glutamate-induced apoptosis. J. Neurochem. 76, in press.

Gary, D.S., Bruce-Keller, A.J., Kindy, M.S. & Mattson, M.P. (1998). Ischemic and excitotoxic brain injury is enhanced in mice lacking the p55 tumor necrosis factor receptor. J. Cereb. Blood Flow Metab. 18, 12831297.

Gash, D.M., Zhang, Z. & Gefhardt, G. (1998). Neuroprotective and neurorestorative properties of GDNF. Ann. Neurol. 44, S121-125.

Gibson, G.E.,. Sheu, K.F & Blass, J.P. (1998). Abnormalities of mitochondrial enzymes in Alzheimer disease. J. Neural. Transm. 105, 855-870.

Good, P.F., Hsu, A., Werner, P., Perl, D.P. & Olanow, C.W. (1998). Protein nitration in Parkinson's disease. J. Neuropathol. Exp. Neurol. 57, 338-342.

Goodman, Y. & Mattson, M.P. (1994). Secreted forms of β-amyloid precursor protein protect hippocampal neurons against amyloid β-peptide-induced oxidative injury. Exp. Neurol. 128, 1-12.

Goodman, Y., Bruce, A.J. Cheng, B. & Mattson, M.P. (1996). Estrogens attenuate and corticosterone exacerbates excitotoxicity, oxidative injury and amyloid β-peptide toxicity in hippocampal neurons. J. Neurochem. 66, 1836-1844.

Guo, Q., Furukawa, K., Sopher, B.L., Pham, D.G., Robinson, N., Martin, G.M. & Mattson, M.P. (1996). Alzheimer's PS-1 mutation perturbs calcium homeostasis and sensitizes PC12 cells to death induced by amyloid β-peptide. NeuroReport 8, 379-383.

Guo, Q., Sopher, B.L., Pham, D.G., Furukawa, K., Robinson, N., Martin, G.M. & Mattson, M.P. (1997). Alzheimer's presenilin mutation sensitizes neural cells to apoptosis induced by trophic factor withdrawal and amyloid β-peptide: involvement of calcium and oxyradicals. J. Neurosci. 17, 4212-4222.

Guo, Q., Christakos, S., Robinson, N. & Mattson, M.P. (1998a). Calbindin blocks the pro-apoptotic actions of mutant presenilin-1: reduced oxidative stress and preserved mitochondrial function. Proc. Natl. Acad. Sci. USA 95, 3227-3232.

Guo, Q., Robinson, N. & Mattson, M.P. (1998b). Secreted APPα counteracts the pro-apoptotic action of mutant presenilin-1 by activation of NF-κB and stabilization of calcium homeostasis J. Biol. Chem. 273, 12341-12351.

Guo, Q., Fu, W., Xie, J., Luo, H., Sells, S.F., Geddes, J.W., Bondada, V., Rangnekar, V. & Mattson, M.P. (1998c). Par-4 is a novel mediator of neuronal degeneration associated with the pathogenesis of Alzheimer's disease. Nature Med. 4:957-962.

Guo, Z.H. & Mattson, M.P. (2000). Neurotrophic factors protect cortical synaptic terminals against amyloid- and oxidative stress-induced impairment of glucose transport, glutamate transport and mitochondrial function. Cereb. Cortex. 10, 50-57.

Guo, Q., Sebastian, L., Sopher, B.L., Miller, M.W., Glazner, G.W., Ware, C.B., Martin, G.M. & Mattson, M.P. (1999a). Neurotrophic factors [activity-dependent neurotrophic factor (ADNF) and basic fibroblast growth factor (bFGF)] interrupt excitotoxic neurodegenerative cascades promoted by a presenilin-1 mutation. Proc. Natl. Acad. Sci. USA 96, 4125-4130.

Guo, Q., Fu, W., Sopher, B.L., Miller, M.W., Ware, C.B., Martin, G.M. & Mattson, M.P. (1999b). Increased vulnerability of hippocampal neurons to excitotoxic necrosis in presenilin-1 mutant knockin mice. Nature Med. 5, 101-107.

Guo, Q., Sebastian, L., Sopher, B.L., Miller, M.W., Ware, C.B., Martin, G.M. & Mattson, M.P. (1999c). Increased vulnerability of hippocampal neurons from presenilin-1 mutant knock-in mice to amyloid β-peptide toxicity: central roles of superoxide production and caspase activation. J. Neurochem. 72, 1019-1029.

Gurney, M.E., Pu, H., Chiu, A.Y., Dal Canto, M.C., Polchow, C.Y., Alexander, D.D., Caliendo,. J., Hentati, A., Kwon, Y.W. & Deng, H.X. (1994). Motor neuron degeneration in mice that express a human CuZn superoxide dismutase mutation. Science 264, 1772-1775.

Hoyer, S., Nitsch, R. & Oesterreich, K. (1991). Predominant abnormality in cerebral glucose utilization in late-onset dementia of the Alzheimer type: a cross-sectional comparison against advanced late-onset and incipient early-onset cases. J. Neural Transm. 3, 1-14.

Ince, P.G., Lowe, J. & Shaw, P.J. (1998). Amyotrophic lateral sclerosis: current issues in classification, pathogenesis and molecular pathology. Neuropathol. Appl. Neurobiol. 24, 104-117.

Isaacs, K.R., Wolpoe, M.E. & Jacobowitz, D.M. (1997). Calretinin-immunoreactive dopaminergic neurons from embryonic rat mesencephalon are resistant to levodopa-induced neurotoxicity. Exp Neurol 146, 25-32

Ishida, A., Furukawa, K., Keller, J.N. & Mattson, M.P. (1997). Secreted form of β-amyloid precursor protein shifts the frequency dependence for induction of LTD, and enhances LTP in hippocampal slices. NeuroReport 8, 2133-2137.

Jagust, W.J., Seab, J.P. Huesman, R.H. Valk, P.E., Mathis, C.A., Reed, B.R., Coxson, P.G. & Budinger, T.F. (1991). Diminished glucose transport in Alzheimer's disease: Dynamic PET studies. J. Cereb. Blood Flow Metab. 11, 323-330.

Jenner, P. & Olanow, C.W. (1998). Understanding cell death in Parkinson's disease. Ann. Neurol. 44, S72-84.

Johnson, B. & Byerly, L. (1993). A cytoskeletal mechanism for Ca^{2+} channel metabolic dependence and inactivation by intracellular Ca^{2+}. Neuron 10, 797-804.

Keller, J.N. & Mattson, M.P. (1997). 17β-estradiol attenuates oxidative impairment of synaptic Na^{+}/K^{+}-ATPase activity, glucose transport and glutamate transport induced by amyloid β-peptide and iron. J. Neurosci. Res. 50, 522-530.

Keller, J.N., Pang, Z., Geddes, J.W., Begley, J.G., Germeyer, A., Waeg, G. & Mattson, M.P. (1997). Impairment of glucose and glutamate transport and induction of mitochondrial oxidative stress and dysfunction in synaptosomes by amyloid β-peptide: role of the lipid peroxidation product 4-hydroxynonenal. J. Neurochem. 69, 273-284.

Keller, J.N., Kindy, M.S., Holtsberg, F.W., St. Clair, D.K., Yen, H.C., Germeyer, A., Steiner, S.M., Bruce-Keller, A.J., Hutchins, J.B. & Mattson, M.P. (1998). Mitochondrial MnSOD prevents neural apoptosis and reduces ischemic brain injury: suppression of peroxynitrite production, lipid peroxidation and mitochondrial dysfunction. J. Neurosci. 18, 687-697.

Klockgether, T., Turski, L., Honore, T., Zhang, Z.M., Gash, D.M., Kurlan, R. & Greenamyre, J.T. (1991). The AMPA receptor antagonist NBQX has antiparkinsonian effects in monoamine-depleted rats and MPTP-treated monkeys. Ann. Neurol. 30, 717-723.

Koketsu, N., Berlove, D.J., Moskowitz, M.A., Kowall, N.W., Caday, C.G. & Finklestein, S.P. (1994). Pretreatment with intraventricular basic fibroblast growth factor decreases infarct size following focal cerebral ischemia in rats. Ann. Neurol. 35, 451-457.

Kristian, T. & Siesjo, B.K. (1998). Calcium in ischemic cell death. Stroke 29, 705-718.

Kruman, I., Bruce-Keller, A.J., Bredesen, D.E., Waeg, G. & Mattson, M.P. (1997). Evidence that 4-hydroxynonenal mediates oxidative stress-induced neuronal apoptosis. J. Neurosci. 17, 5089-5100.

Kruman, I., Pang, Z., Geddes, J.W. & Mattson, M.P. (1999a). Pivotal role of mitochondrial calcium uptake in neural cell apoptosis and necrosis. J. Neurochem. 72, 529-540.

Kruman, I., Pedersen W.A. & Mattson, M.P. (1999b). ALS-linked Cu/Zn-SOD mutation increases vulnerability of motor neurons to excitotoxicity by a mechanism involving increased oxidative stress and perturbed calcium homeostasis. Exp. Neurol. In press.

Lafon-Cazal, M., Pietri, S., Culcasi, M. & Bockaert, J. (1993). NMDA-dependent superoxide production and neurotoxicity. Nature 364, 535-537.

Langston, J.W. (1998). Epidemiology versus genetics in Parkinson's disease: progress in resolving an age-old debate. Ann. Neurol. 44, S45-S52.

Lee, K.S., Yanamoto, H., Fergus, A., Hong, S.C., Kang, S.D., Cappelletto, B., Toyoda, T., Kassell, N.F., Bavbek, M. & Kwan A.L. (1997). Calcium-activated proteolysis as a therapeutic target in cerebrovascular disease. Ann. N. Y. Acad. Sci. 825, 95-103.

Lynch, D.R. & Dawson, T.M. (1994). Secondary mechanisms in neuronal trauma. Curr. Opin. Neurol. 7, 510-516.

Mark, R.J., Hensley, K., Butterfield, D.A. & Mattson, M.P. (1995). Amyloid β-peptide impairs ion-motive ATPase activities: evidence for a role in loss of neuronal Ca^{2+} homeostasis and cell death. J. Neurosci. 15, 6239-6249.

Mark, R.J., Lovell, M.A., Markesbery, W.R., Uchida, K. & Mattson, M.P. (1997a). A role for 4-hydroxynonenal in disruption of ion homeostasis and neuronal death induced by amyloid β-peptide. J. Neurochem. 68, 255-264.

Mark, R.J., Pang, Z., Geddes, J.W., Uchida, K. & Mattson, M.P. (1997b). Amyloid β-peptide impairs glucose uptake in hippocampal and cortical neurons: involvement of membrane lipid peroxidation. J. Neurosci. 17, 1046-1054.

Mark, R.J., Keller, J.N., Kruman, I. & Mattson, M.P. (1997c). Basic FGF attenuates amyloid β-peptide-induced oxidative stress, mitochondrial dysfunction, and impairment of Na^+/K^+-ATPase activity in hippocampal neurons. Brain Res. 756:205-214.

Marks, A.R. (1997). Intracellular calcium-release channels: regulators of cell life and death. Am. J. Physiol. 272, H597-605.

Mattson, M.P., Murrain, M., Guthrie, P.B. & Kater, S.B. (1989). Fibroblast growth factor and glutamate: opposing roles in the generation and degeneration of hippocampal neuroarchitecture. J. Neurosci. 9, 3728-3740.

Mattson, M.P. (1990). Antigenic changes similar to those seen in neurofibrillary tangles are elicited by glutamate and calcium influx in cultured hippocampal neurons. Neuron 4, 105-117.

Mattson, M.P., Rychlik, B., Chu, C. &. Christakos, S. (1991). Evidence for calcium-reducing and excitoprotective roles for the calcium binding protein (calbindin-D28k) in cultured hippocampal neurons. Neuron 6, 41-51.

Mattson, M.P. (1992). Calcium as sculptor and destroyer of neural circuitry. Exp. Gerontology 27, 29-49.

Mattson, M.P., Cheng, B., Davis, D. Bryant, K., Lieberburg, I. & Rydel, R.E. (1992). β-amyloid peptides destabilize calcium homeostasis and render human cortical neurons vulnerable to excitotoxicity. J. Neurosci. 12, 376-389.

Mattson, M.P., Kumar, K., Cheng, B., Wang, H., Michaelis, E.K. (1993a). Basic FGF regulates the expression of a functional 71 kDa NMDA receptor protein that mediates calcium influx and neurotoxicity in cultured hippocampal neurons. J. Neurosci. 13, 4575-4588.

Mattson, M.P., Cheng, B., Culwell, A. Esch, F., Lieberburg, I. & Rydel, R.E. (1993b). Evidence for excitoprotective and intraneuronal calcium-regulating roles for secreted forms of β-amyloid precursor protein. Neuron 10, 243-254.

Mattson, M.P., Barger, S.W., Cheng, B., Lieberburg, I., Smith-Swintosky, V.L. & Rydel, R.E. (1993c). β-amyloid precursor protein metabolites and loss of neuronal calcium homeostasis in Alzheimer's disease. Trends Neurosci. 16, 409-415.

Mattson, M.P., Tomaselli, K. & Rydel, R.E. (1993d). Calcium-destabilizing and neurodegenerative effects of aggregated β-amyloid peptide are attenuated by basic FGF. Brain Res. 621:35-49.

Mattson, M.P., Lovell, M.A., Furukawa, K. & Markesbery, W.R. (1995). Neurotrophic factors attenuate glutamate-induced accumulation of peroxides, elevation of $[Ca^{2+}]_i$ and neurotoxicity, and increase antioxidant enzyme activities in hippocampal neurons. J. Neurochem. 65:1740-1751.

Mattson, M.P. (1996). Neuroprotective signal transduction: relevance to stroke. Neurosci. Biobehav. Rev. 21: 193-206.

Mattson, M.P. & Furukawa, K. (1996). Programmed cell life: anti-apoptotic signaling and therapeutic strategies for neurodegenerative disorders. Restorative Neurol. Neurosci. 9, 191-205.

Mattson, M.P. & Mark, R.J. (1996). Excitotoxicity and excitoprotection *in vitro*. In: B. Siesjo and T. Wielock (eds) Cellular and Molecular Mechanisms of Ischemic Brain Damage. Adv. Neurol., Lippincott-Raven, New York, 71, 1-37.

Mattson, M.P. (1997). Cellular actions of β-amyloid precursor protein, and its soluble and fibrillogenic peptide derivatives. Physiol. Rev. 77, 1081-1132.

Mattson, M.P. & Geddes, J.W. (1997). The Aging Brain Adv. Cell Aging Gerontology, JAI Press, Greenwich, CT, 354 pp.

Mattson, M.P. & Lindvall, O. (1997). Neurotrophic factor and cytokine signaling in the aging brain. In (M.P. Mattson and J.W. Geddes, eds) The Aging Brain, Adv. Cell Aging Gerontol. 2, 299-345.

Mattson, M.P., Robinson, N. & Guo, Q. (1997a). Estrogens stabilize mitochondrial function and protect neural cells against the pro-apoptotic action of mutant presenilin-1. NeuroReport 8: 3817-3821.

Mattson, M.P., Goodman, Y., Luo, H., Fu, W. & Furukawa, K. (1997b). Activation of NF-κB protects hippocampal neurons against oxidative stress-induced apoptosis: evidence for induction of Mn-SOD and suppression of peroxynitrite production and protein tyrosine nitration. J. Neurosci. Res. 49, 681-697.

Mattson, M.P. (1998a). Free radicals, calcium, and the synaptic plasticity - cell death continuum: emerging roles of the transcription factor NF-κB. Int. Rev. Neurobiol. 42, 103-168.

Mattson, M.P. (1998b). Modification of ion homeostasis by lipid peroxidation: roles in neuronal degeneration and adaptive plasticity. Trends Neurosci. 21, 53-57.

Mattson, M.P., Guo, Q., Furukawa, K. & Pedersen, W.A. (1998). Presenilins, the endoplasmic reticulum, and neuronal apoptosis in Alzheimer's disease. J. Neurochem. 70, 1-14.

Mattson, M.P., Keller, J.N. & Begley, J.G. (1998a). Evidence for synaptic apoptosis. Exp. Neurol. 153, 35-48.

Mattson, M.P., Partin, J. & Begley, J.G. (1998b). Amyloid β-peptide induces apoptosis-related events in synapses and dendrites. Brain Res. 807, 167-176.

Mattson, M.P. & Guo, Q. (1999). The Presenilins. The Neuroscientist 5, 112-124.

Mattson, M.P. & Partin, J. (1999). Evidence for mitochondrial control of neuronal polarity. J. Neurosci. Res. 56, 8-20.

Mattson, M.P., Guo, Z.H. & Geiger, J.D. (1999). Secreted form of amyloid precursor protein enhances basal glucose and glutamate transport, and protects against oxidative impairment of glucose and glutamate transport in synaptosomes by a cyclic GMP-mediated mechanism. J. Neurochem. 73:532-537.

Monyer, H., Burnashev, N., Laurie, D.J., Sakmann, B. & Seeburg, P.H. (1994) Developmental and regional expression in the rat brain and functional properties of four NMDA receptors. Neuron 12, 529-540.

Mouatt-Prigent, A., Karlsson, J.O., Agid, Y. & Hirsch, E.C. (1996). Increased M-calpain expression in the mesencephalon of patients with Parkinson's disease but not in other neurodegenerative disorders involving the mesencephalon: a role in nerve cell death? Neurosci. 73, 979-987.

Murphy, A.N., Bredesen, D.E., Cortopassi, G., Wang, E. & Fiskum, G. (1996) Bcl-2 potentiates the maximal calcium uptake capacity of neural cell mitochondria. Proc. Natl. Acad. Sci. U. S.A. 93, 9893-9898.

O'Brien, R.J., Lau, L.F. & Huganir, R.L. (1998). Molecular mechanisms of glutamate receptor clustering at excitatory synapses. Curr. Opin. Neurobiol. 8:364-369.

Olney, J.W. (1985). Excitatory transmitters and epilepsy-related brain damage. Int. Rev. Neurobiol. 27, 337-362.

Paschen, W. & Doutheil, J. (1999). Disturbances of the functioning of endoplasmic reticulum: a key mechanism underlying neuronal cell injury? J. Cereb. Blood Flow Metab. 19, 1-18.

Pedersen, W.A., Cashman, N. & Mattson, M.P. (1999). The lipid peroxidation product 4-hydroxynonenal impairs glutamate and glucose transport and choline acetyltransferase activity in NSC-19 motor neuron cells. Exp. Neurol. 155:1-10.

Pedersen, W.A., Fu, W., Keller, J.N., Markesbery, W.R., Appel, S.H., Smith, R., G., Kasarskis, E. & Mattson, M.P. (1998). Protein modification by the lipid peroxidation product 4-hydroxynonenal in spinal cord tissue of ALS patients. Ann. Neurol. 44, 819-824.

Perry, T.L. & Hansen, S. (1990). What excitotoxin kills striatal neurons in Huntington's disease? Clues from neurochemical studies. Neurology 40, 20-24.

Phillips, H.S., Hains, J.M., Armanini, M., Laramee, G., Johnsos, S.A. & Winslow, J.W. (1991). BDNF mRNA is decreased in the hippocampus of individuals with Alzheimer's disease. Neuron 7, 695-702.

Pike, C.J. & Cotman, C.W. (1995). Calretinin-immunoreactive neurons are resistant to beta-amyloid toxicity *in vitro*. Brain Res. 671, 293-298.

Prehn, J.H., Bindokas, V.P., Marcuccilli, C.J., Krajewski, S., Reed, J.C. & Miller, R.J. (1994). Regulation of neuronal Bcl2 protein expression and calcium homeostasis by transforming growth factor type beta confers wide-ranging protection on rat hippocampal neurons. Proc. Natl. Acad. Sci. U. S.A. 91, 12599-12603.

Racay, P., Kaplan, P. & Lehotsky, J. (1996). Control of Ca^{2+} homeostasis in neuronal cells. Gen. Physiol. Biophys. 15, 193-210.

Rosen, D.R. (1993). Mutations in Cu/Zn superoxide dismutase gene are associated with familial amyotrophic lateral sclerosis. Nature 362, 59-62

Rosenmund, C. & Westbrook, G. (1993). Calcium-induced actin depolymerization reduces NMDA channel activity. Neuron 10, 805-814.

Rothman, S.M. & Olney, J.W. (1986). Glutamate and the pathophysiology of hypoxic-ischemic brain damage. Ann. Neurol. 19, 105-111.

Rothstein, J.D. (1995). Excitotoxicity and neurodegeneration in amyotrophic lateral sclerosis. Clin. Neurosci. 3, 348-359.

Roy, J., Minotti, S., Dong, L., Figlewicz, D.A. & Durham, H.D. (1998). Glutamate potentiates the toxicity of mutant Cu/Zn-superoxide dismutase in motor neurons by postsynaptic calcium-dependent mechanisms. J. Neurosci. 18, 9673-9684.

Schapira, A.H.V., Cooper, J.M., Dexter, D., Clark, J.B., Jenner, P. & Marsden, C.D. (1990). Mitochondrial complex I deficiency in Parkinson's disease. J Neurochem. 54, 823-827.

Scheuner, D. Eckman, C., Jensen, M., Song, X., Citron, M., Suzuki, N., Bird, T.D., Hardy, J., Hutton, M.,

Kukull, W., Larson, E., Levy-Lahad, E., Viitanen, M., Peskind, E., Poorkaj, P., Schellenberg, G., Tanzi, R., Wasco, W., Lannfelt, L., Selkoe, D. & Younkin, S. (1996). The amyloid β protein deposited in the senile plaques of Alzheimer's disease is increased *in vivo* by the presenilin 1 and 2 and APP mutations linked to familial Alzheimer's disease. Nature Med. 2, 864-870

Schulz, J.B., Weller, M. & Moskowitz, M.A. (1999). Caspases as treatment targets in stroke and neurodegenerative diseases. Ann. Neurol. 45, 421-429.

Scott, S.A. & Crutcher, K.A. (1994). Nerve growth factor and Alzheimer's disease. Rev. Neurosci. 5, 179-211.

Sheehan, J.P., Swerdlow, R.H., Parker, W.D., Miller, S.W., Davis, R.E. & Tuttle, J.B. (1997). Altered calcium homeostasis in cells transformed by mitochondria from individuals with Parkinson's disease. J. Neurochem. 68, 1221-1233.

Shigeno, T., Mima, T., Takakura, K., Graham, D.I., Kato, G., Hashimoto, Y. & Furukawa, S. (1991). Amelioration of delayed neuronal death in the hippocampus by nerve growth factor. J. Neurosci. 11, 2914-2919.

Siddique, T. (1991). Molecular genetics of familial amyotrophic lateral sclerosis. Adv. Neurol. 56, 227-231

Siklos, L., Engelhardt, J., Harati, Y.,Smith, R.G., Joo, F. & Appel, S.H. (1996). Ultrastructural evidence for altered calcium in motor nerve terminals in amyotropic lateral sclerosis. Ann. Neurol. 39, 203-216.

Simpson, P.B. & Russell, J.T. (1998). Role of mitochondrial Ca^{2+} regulation in neuronal and glial cell signalling. Brain. Res. Rev. 26, 72-81.

Sloviter, R.S., Sollas, A.L., Barbaro, N.M. & Laxer, K.D. (1991). Calcium-binding protein (calbindin-D28K) and parvalbumin immunocytochemistry in the normal and epileptic human hippocampus. J. Comp. Neurol. 308, 381-396.

Smith, R.G., Henry, Y.K., Mattson, M.P. & Appel, S.H. (1998). Presence of 4-hydroxynonenal in cerebrospinal fluid of patients with sporadic amyotrophic lateral sclerosis. Ann. Neurol. 44, 696-699.

Smith-Swintosky, V.L., Pettigrew, L.C., Craddock, S.D., Culwell, A.R., Rydel, R.E. & Mattson, M.P. (1994). Secreted forms of beta-amyloid precursor protein protect against ischemic brain injury. J. Neurochem. 63, 781-784.

Spacek, J. & Harris, K.M. (1997). Three-dimensional organization of smooth endoplasmic reticulum in hippocampal CA1 dendrites and dendritic spines of the immature and mature rat. J. Neurosci. 17, 190-203.

Spencer, P.S., Nunn, P.B., Hugon, J., Ludolph, A.C.,Ross, S.M., Roy, D.N. & Robertson, R.C. (1987). Guam amyotrophic lateral sclerosis-Parkinsonism-dementia linked to a plant excitant neurotoxin. Science 237, 517-522

Spina, M.B., Squinto, S.P., Miller, J., Lindsay, R.M. & Hyman, C. (1992). Brain-derived neurotropic factor protects dopamine neurons against 6-hydroxydopamine and N-methyl-4-phenylpyridinium ion toxicity: involvement of the glutathione system. J. Neurochem. 59, 99-106.

Stein-Behrens, B., Mattson, M.P., Chang, I., Yeh, M. & Sapolsky, R.M. (1994). Stress excacerbates neuron loss and cytoskeletal pathology in the hippocampus. J. Neurosci. 14, 5373-5380.

Swope, S.L., Moss, S.I., Raymond, L.A. & Huganir, R.L. (1999). Regulation of ligand-gated ion channels by protein phosphorylation. Adv. Second Messenger Phosphoprotein Res. 33, 49-78.

Takechi, H., Eilers, J. & Konnerth, A. (1998). A new class of synaptic response involving calcium release in dendritic spines. Nature 396, 757-760.

Uchino, H., Elmer, E., Uchino, K., Li, P.A., He, Q.P., Smith, M.L. & Siesjo, B.K. (1998). Amelioration by cyclosporin A of brain damage in transient forebrain ischemia in the rat. Brain Res. 812, 216-26.

Wiedau-Pazos, M., Goto, J.J., Rabizadeh, S., Gralla, E B., Roe, J.A., Lee, M.K., Valentine, J.S. & Bredesen, D.E. (1996). Altered reactivity of superoxide dismutase in familial amyotrophic lateral sclerosis. Science 271, 515-518.

Wong, P.C., Pardo, C.A., Borchelt, D.R., Lee, M.K., Copeland, N.G., Jenkins, N.A., Sisodia, S.S,. Cleveland, D.W. & Price, D.L. (1995). An adverse property of a familial ALS-linked SOD1 mutation causes motor neuron disease characterized by vacuolar degeneration of mitochondria. Neuron 14, 1105-1116.

Yankner, B.A. (1996). Mechanisms of neuronal degeneration in Alzheimer's disease. Neuron 16, 921-932.

Yu, Z.F., Bruce-Keller, A.J., Goodman, Y. & Mattson, M.P. (1998). Uric acid protects neurons against excitotoxic and metabolic insults in cell culture, and against focal ischemic brain injury *in vivo*. J. Neurosci. Res. 53, 613-625.

Yu, Z.F., Luo, H., Fu, W. & Mattson, M.P. (1999a). The endoplasmic reticulum stress-responsive protein GRP78 protects neurons against excitotoxicity and apoptosis: suppression of oxidative stress and stabilization of calcium homeostasis. Exp. Neurol. 155, 302-314.

Yu, Z.F. & Mattson, M.P. (1999b). Dietary restriction and 2-deoxyglucose administration reduce focal ischemic brain damage and improve behavioral outcome: evidence for a preconditioning mechanism. J. Neurosci. Res. 57, 830-839.

Zhang, L., Andou, Y., Masuda, S., Mitani, A. & Kataoka, K. (1993). Dantrolene protects against ischemic, delayed neuronal death in gerbil brain. Neurosci. Lett. 158, 105-108.

MITOCHONDRIAL DYSFUNCTION IN DIABETES MELLITUS

MATTHIAS F. BAUER,
SABINE HOFMANN & KLAUS-DIETER GERBITZ

Institute of Clinical Chemistry, Molecular Diagnostics and Mitochondrial Genetics & Diabetes Research Group, Academic Hospital Munich-Schwabing, Koelner Platz 1, 80804 Munich, Germany

Table of Contents

Introduction

Over the past decade, mitochondrial dysfunction has been implicated in a broad variety of degenerative diseases, cancer and aging. Functional investigations in cell lines and model organisms, as well as studies in patients suffering from these disorders, have revealed much about the complex situation of mitochondrial genetics, which involves

Interorganellar Signaling in Age-Related Disease, Edited by Mark P. Mattson, 55—101

an interplay between the mitochondrial and the nuclear genome. The essential role of mitochondrial oxidative phosphorylation (OXPHOS) in cellular energy production and signaling, the generation of reactive oxygen radicals and the proposed role of mitochondria in the initiation of apoptosis have suggested a number of novel mechanisms for the pathogenesis of mitochondrial disorders. In many complex genetic human diseases, the pathogenetical role of mitochondria has gained a large experimental support over time, indicating the decline of mitochondrial functionality as one of the driving forces for the development and persistence of these disease processes.

In the following we will provide an update on the current knowledge about the relationship between mitochondrial dysfunction and the common mechanisms which act during the development of diabetes.

For a better understanding of the complexity of mitochondrial disorders one has to consider the unique genetic, structural and metabolic features of these organelles. The functions of mitochondria are confined to a relatively small number of metabolic pathways, which are closely interwoven with each other. These processes are in particular the carnitine-dependent transport of fatty acids and their β-oxidation, pyruvate carboxylation and decarboxylation, the Krebs-cycle, iron metabolism, the urea cycle, parts of steroid and nitric oxide (NO) synthesis, and the RC-OXPHOS (summarized in Figure 1). Many of the biochemical pathways located within the mitochondria feed into in the energy-generating system, thereby combining substrate oxidation with the generation of a transmembrane proton motif force to produce ATP. ATP constitutes an universally usable form of energy in all cellular compartments and is released from the mitochondrion via specialized carrier proteins to supply most of the energy consuming biosynthetic reactions of the cell. Moreover, mitochondria were shown to influence a variety of cellular mechanisms and pathways (e.g. apoptotic cell death or insulin secretion) (Figure 1).

Depending on their energy demand, different cell types contain different numbers of mitochondria ranging from only a few in the spermatozoon to several thousand in cells of skeletal muscle, heart and brain (Taylor and Piko, 1987). Tissues and organ systems with a high number of mitochondria and a high-energy demand are known to be predilection points for mitochondrial dysfunction. In particular, the cells of the central nervous system, skeletal and cardiac muscle, liver, kidney and pancreatic islets are reliant on oxidative energy metabolism. Mitochondrial dysfunction therefore causes a heterogeneous group of disorders presenting with a broad spectrum of clinical phenotypes ranging in their severity from systemic, lethal pediatric disease to late-onset, tissue-specific neurodegenerative disorders (Wallace, 1994b).

During the past decade several lines of evidence focussed on the conclusion that mitochondrial dysfunction may play an important role in the development of diabetes mellitus due to impaired insulin production and secretion (Gerbitz et al., 1996). The following aspects support this hypothesis: (i), diabetes mellitus is frequently observed as one symptom in patients with complex mitochondrial disorders, in particular in those carrying mitochondrial DNA (mtDNA) mutations (Gerbitz et al., 1995; Maassen and Kadowaki, 1996); (ii), diabetes mellitus, or at least the most frequent non-insulin dependent form (NIDDM), is more often transmitted from the mother than from the father suggesting the involvement of maternally transmitted susceptibility

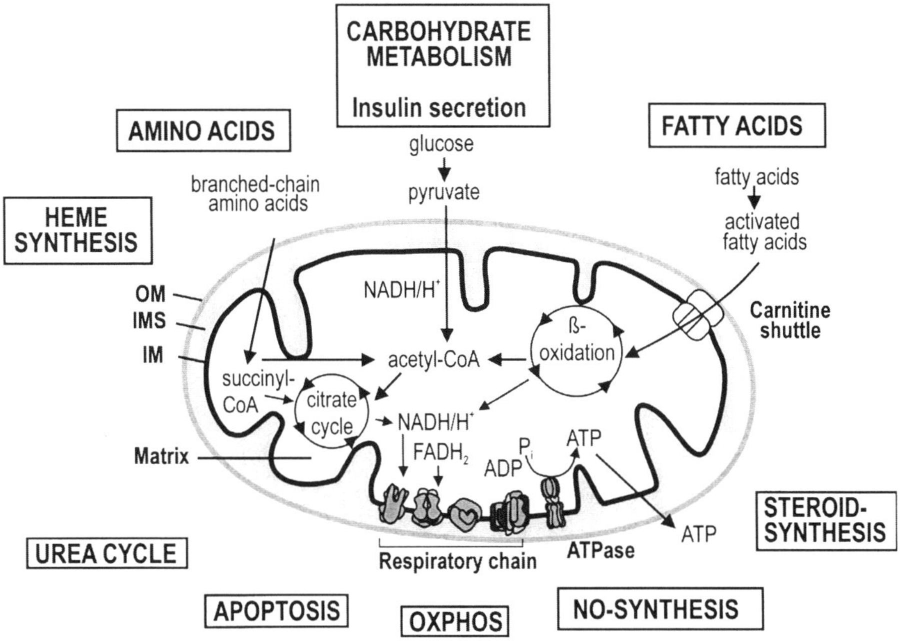

Figure 1. Mitochondrial metabolic pathways.

factors (like the mitochondrial DNA) and/or maternal environmental conditions in the development of diabetes mellitus (Alcolado and Alcolado, 1991; Thomas et al., 1994); (iii) the generation of oxygen radicals along the mitochondrial respiratory chain has been suggested to be involved in the progress of neurodegenerative disorders and normal aging - a mechanism which is also used to explain selective damage of pancreatic β-cells in certain forms of diabetes (Esposito et al., 1999); (iv) apoptosis has been identified as a mechanism of pancreatic β-cell death in autoimmune diabetes. As mitochondria play an important role in the initial steps of the programmed cell death, it was concluded that mitochondrial dysfunction is at least indirectly involved in β-cell loss (Rabinovitch et al., 1999); (v) moreover, various animal models have been established which indicate mitochondrial dysfunction as a possible contributor to the development of diabetes (Newgard and McGarry, 1995).

Mitochondria: Structure and Function

Mitochondrial Structure and Biogenesis

The mitochondria are made up of two highly specialized membrane systems, the outer and inner membrane, and two aqueous compartments, the matrix and the intermembrane space.

The membranes contain carrier proteins and channels that regulate the exchange of substrates between the compartments. The inner membrane is richer in proteins than the outer membrane and contains the high molecular weight complexes constituting the respiratory chain. The total number of different mitochondrial proteins can only be estimated to date to be in the range between 1000 and 2000. Most of these proteins are encoded by the nuclear genome and synthesized as precursor proteins at cytosolic ribosomes. During the biogenesis of mitochondria, these proteins have to be taken up from the cytosol in order to form new mitochondria based on the growth and subsequent division of preexisting organelles. The precursor proteins are picked up by the mitochondria in a multi-step process which is facilitated by the coordinated action of independent translocation systems in the mitochondrial outer and inner membranes (Figure 2) (for review see Neupert, 1997).

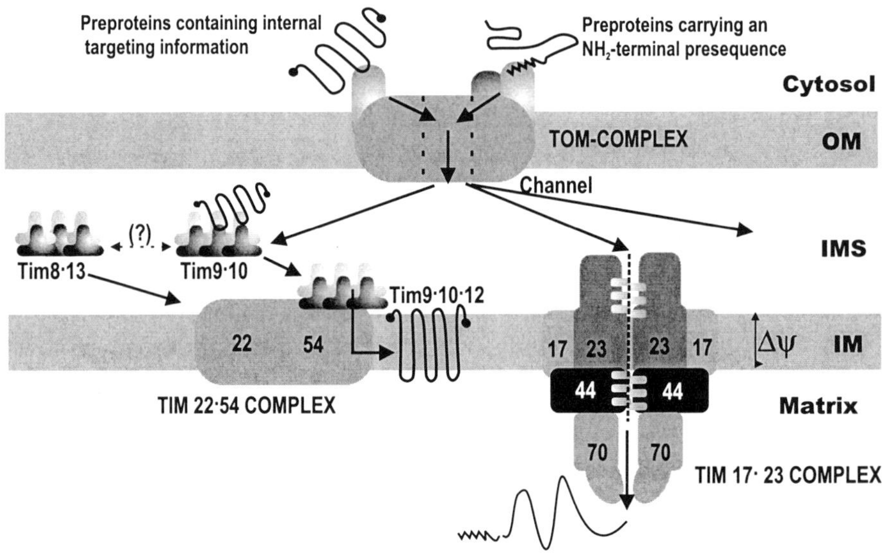

Figure 2. Working model of the protein translocation machineries in the outer and inner membranes involved in the biogenesis of mitochondria.

During the past years a number of components and a series of divergent pathways involved in import, sorting, membrane insertion and assembly of nuclear DNA-encoded mitochondrial proteins have been identified. In the yeast *Saccharomyces cerevisiae*, three distinct translocases have been described to mediate mitochondrial protein import. One translocase, the TOM (Translocase of the Outer Membrane) complex, is located in the outer membrane and is used by all nuclear encoded precursors (Figure 2). For import of preproteins into the matrix and into the inner membrane the TOM complex cooperates with two distinct translocation machineries in the inner membrane, named TIM23•17 Complex and TIM22•54 Complex (Bauer et al., 2000; Neupert, 1997; Rassow et al., 1999). The TIM23•17 Complex mediates the import of precursor proteins into mitochondrial matrix in an ATP- and membrane-potential dependent fashion (Bauer et al., 1996). For import and membrane insertion of a subset of hydrophobic carrier proteins such as the ANT (=ADP/ATP) carrier into the inner membrane the TOM complex cooperates with the TIM22•54 Complex via additional soluble components located in the intermembrane space. Membrane integrated and soluble components together constitute the second translocation system of the inner membrane (Kerscher et al., 1997; Koehler et al., 1998; Nelson et al., 1998; Sirrenberg et al., 1996, 1997; 1998). Identification of mitochondrial protein import systems in different species has revealed that TOM and TIM complexes of yeast and mammals are conserved and apparently composed of homologous components (Bauer et al., 1999).

Mitochondrial Pathways Particularly Important for Normal β-Cell Function

Carnitine Shuttle and β-Oxidation of Fatty Acids

The pathways of carnitine-mediated transport and β-oxidation of fatty acids are summarized in Figure 3. All of the fatty acids to be utilized for energy production by intramitochondrion β-oxidation come from outside the mitochondrium, either through biosynthesis or through transport of triacylglycerol esters from fat depots. Since the enzymes of β-oxidation can only act on CoA esters, fatty acids located in the cytosol have to be esterified by a set of acyl-CoA synthases (for review see Eaton et al., 1996). Translocation of "activated" fatty acids (acyl-CoA) across the inner mitochondrial membrane does not occur spontaneously but is mediated by a carnitinedependent shuttle system (Figure 3). Prior to translocation the CoA esters are transformed to carnitine esters by the action of several different carnitine-acyl transferases. Two distinct forms of carnitine-palmitoyl transferase (CPT I and CPT II) are known to work in cooperation with a carnitine-acylcarnitine translocase. CPT II also carries free carnitine from the matrix to the cytosol thereby regenerating the shuttle system (Huizing et al., 1997; Indiveri et al., 1997a, b Palmieri et al., 1996).

The "activated" fatty acids are sequentially degraded during repeated cycles of β-oxidation. This process is characterized by the cleavage of two carbon atoms between position alpha(-2) and beta(-3) of the acyl-CoA ester which results in the formation of acetyl-CoA (Figure 3). During the first step, acyl-CoA dehydrogenation is performed by three acyl-CoA dehydrogenases differing in their substrate specificity which depends on the length of the respective acyl-chain (short-chain acyl-CoA dehydrogenase, SCAD, active on C_4 to C_6; medium-chain acyl-CoA dehydrogenase, MCAD, active on C_4 to C_{12}; very long chain acyl-CoA dehydrogenase, VLCAD, active on C_{12} to C_{24}). SCAD, and MCAD are homo-tetrameric enzyme complexes in the matrix, while the homo-dimeric VLCAD is localized in the inner membrane. Reaction products of the first step are enoyl-CoAs of different chain length and reduced prosthetic FAD groups of the acyl-CoA dehydrogenases. Reoxidation of $FADH_2$ is mediated by the electron transfer flavoprotein (ETF) that directly passes the electrons on to ubiquinone via an ETF: ubiquinone oxidoreductase (Figure 3) (see also RC-OXPHOS).

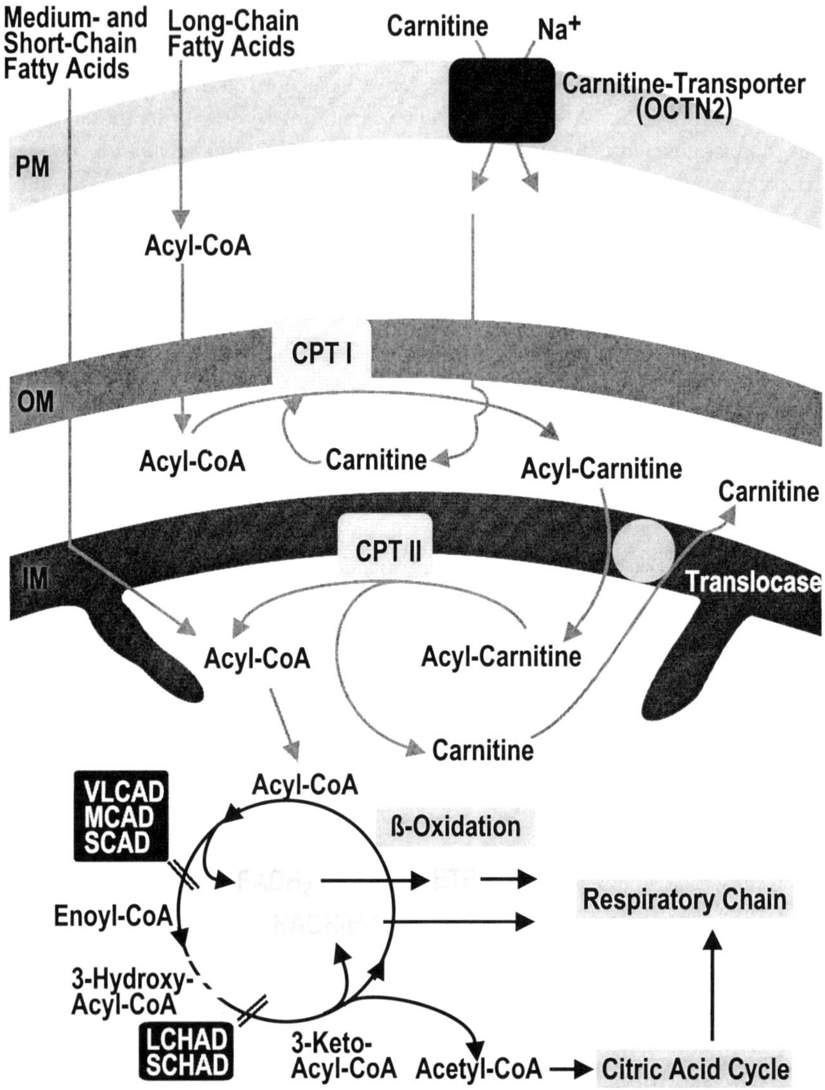

Figure 3. Schematic representation of the degradation pathways of fatty acids.

The second step of β-oxidation is a hydration reaction involving three 2-enoyl-CoA hydratases, which also exhibit substrate specificity. The third step, dehydrogenation of L-3-hydroxyacyl-CoA, is performed by two enzymes (short and long chain hydroxyacyl-CoA dehydrogenase, SCHAD and LCHAD) leading to NADH and 3-ketoacyl-CoAs which are subsequently cleaved by three different thiolases to give acetyl-CoA and a shortened acyl-CoA intermediate (Figure 3). Long chain thiolase together with long chain 2-enoyl-CoA hydratase and long chain 3-hydroxyacyl-CoA dehydrogenase form the so named "trifunctional protein" (Ijlst et al., 1994; Jackson et al., 1995) which is localized in the inner mitochondrial membrane in close proximity to CPT II, VLCAD and the RC-OXPHOS. The newly generated energy-rich reduction equivalents are completely conferred to the complex I and CoQ of the respiratory chain in order to utilize its energy potential to synthesize ATP.

The Mitochondrial Respiratory Chain

The RC-OXPHOS system is located within the inner mitochondrial membrane and consists of five multisubunit enzyme complexes (Figure 4) (for review see Nicholls and Ferguson, 1992). Within the electron-conducting cascade of the respiratory chain two complexes act as electron acceptors: first, Complex I (NADH: ubiquinone oxidoreductase or NADH dehydrogenase) serves as an electron acceptor of several NADH producing reactions, while Complex II (succinate: ubiquinone oxidoreductase or succinate dehydrogenase) takes over electrons from succinate via $FADH_2$. From both complexes electrons are transferred to ubiquinone (Coenzyme Q, CoQ). The transfer of electrons from Complex I to ubiquinone is associated with proton translocation across the inner membrane. In addition to Complex I and Complex II, two other redox pathways directly feed electrons to ubiquinone; flavoprotein (ETF), supplying electrons from the flavoprotein-linked step of fatty acid β-oxidation; and mitochondrial glycerophosphate dehydrogenase. The first is located on the matrix face of the inner mitochondrial membrane and the second is on the outer face.

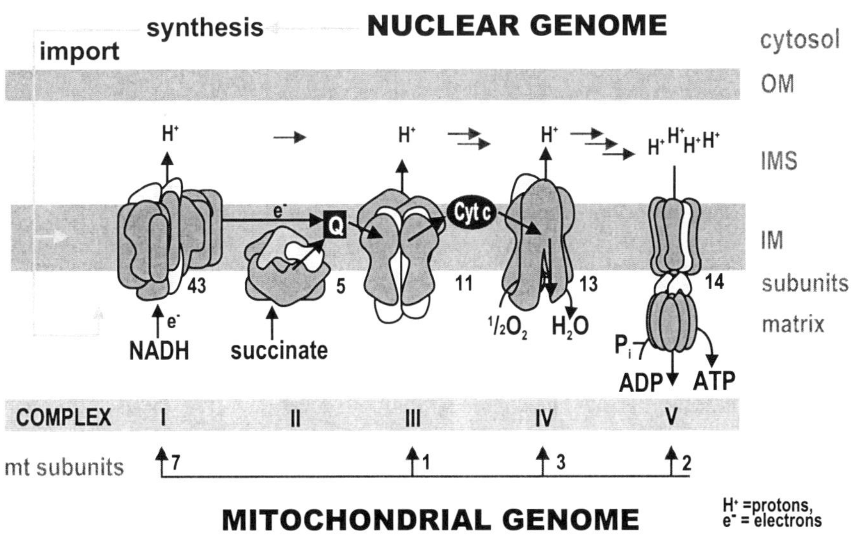

Figure 4. The mitochondrial respiratory chain in the mitochondrial inner membrane. The respiratory chain complexes I, III and IV and the ATP synthase (complex V) are assemblies of nuclear and mitochondrial encoded subunits. Abbreviations: Q, coenzyme Q (ubiquinone); Cyt c, cytochrome c.

The transfer of electrons from ubiquinol to cytochrome *c* is mediated by Complex III, which is also termed the cyt bc_1 complex or ubiquinol: ferricytochrome-c oxidoreductase. Cytochrome *c* has not been isolated as a component of a complex, although it can bind stoichiometrically to cytochrome *c* oxidase (COX; ferrocytochrome-c: oxygen oxidoreductase) which constitutes Complex IV. In both, Complex III and IV electron transfer drives the translocation of protons from the matrix to the intermembrane space with a stoichiometry of about $1H^+/1e^-$. As a hydrophilic protein, cytochrome *c* may be readily released from the outer face of the inner membrane being present as a soluble protein in the intermembrane space. Recent evidence has shown that cytochrome *c* released from the intermembrane space to the cytoplasm plays a crucial role in activation of the caspase protease family acting during the common death pathway in cellular apoptosis (Green and Reed, 1998).

The final step in the electron-transport chain of mitochondria is the sequential transfer of four electrons from the reduced cytochrome *c* pool to oxygen, forming $2H_2O$ in a four-electron reaction catalyzed by the cytochrome *c* oxidase (COX). In contrast to complex I, complexes II, III, IV and V are also found in lower eukaryotes and many bacteria.

The redox components in the respiratory chain are ordered in such a way that their operating redox potential (E_h) increase sequentially from NADH to O_2. The energy gained along this cascade is coupled to intrinsic proton pumps generating an electrochemical gradient across the inner membrane. This gradient provides a proton-motive force which drives ATP synthesis by the F_1F_0-ATP synthase, also named complex V.

Recently, the resolution of the atomic structures of the respiratory enzymes from mammalian mitochondria has contributed to the understanding of the molecular basis of human diseases involving their dysfunction (Walker, 1995). The greatest progress has been made with the ATP synthase; the structure of its catalytic domain, F_1-ATPase, has been solved by X-ray crystallography at 2.8 Å resolution (Abrahams et al., 1994). The intact ATP synthase contains 14 different polypeptides (Walker et al., 1995). The sequence of the proteins of ATP-synthase has provided information about amino acids that are important for its function. For example, amino acids contributing to nucleotide binding sites have been identified. Also, they provide the basis of models of secondary structure of membrane components that constitute the transmembrane proton channel. There is now compelling evidence in support of a rotary catalytic mechanism which is necessary to couple the proton translocation to the ATP synthesis (Leslie et al., 1999). Newly generated ATP is exported to the cytosol in exchange for the spent ADP by the adenine nucleotide translocator ANT. The ANT belongs to a class of hydrophobic mitochondrial inner membrane proteins which are imported via the TIM22•54 translocase (for review see Bauer et al., 2000). Recently, the TIM22•54 complex has been described to be involved in the pathogenesis of a neurodegenerative disorder, named Mohr-Traenjebaerg syndrome (Koehler et al., 1999), providing a new example for a mitochondrial disorder caused by a nuclear gene mutation.

Mitochondria possess their own genome (Figure 4 and 5). However, the mtDNA of mammals encodes only a few polypeptides, all of which are components of the oxidative phosphorylation cascade in the inner membrane translated by the mitochondrial translation apparatus. In addition, the sequences for RNA species necessary for mitochondrial protein biosynthesis are found in the mammalian mtDNA. Within the inner membrane, subunits encoded by the mtDNA assemble into hetero-oligomeric complexes and directly interact with those translated from the nuclear DNA (Figure 4).

Mitochondrial Genetics

Each mitochondrion contains up to ten copies of mtDNA. The mitochondrial genome, fully sequenced in 1981 (Anderson et al., 1981), is a circular and double stranded molecule 16569 nucleotide pairs in length (Figure 5). All 13 polypeptides encoded by the mtDNA assemble with nuclear encoded subunits to form the respiratory complexes I, III, IV and V (Figure 4). Due to its compact structure, its lack of histone protection, its insufficient repair mechanisms and its exposure to oxygen radicals generated along the respiratory chain, the mtDNA is highly vulnerable (Ames et al., 1993; Brown et al., 1992; Wallace et al., 1992). As a result its mutation rate is about 10 to 20 times higher than in the nuclear DNA. The mtDNA exhibits several unique features: it shows strict *maternal inheritance,* thus being exclusively transmitted from a mother to her offspring (Giles et al., 1980). All cells usually contain identical copies of mtDNA (*homoplasmy*).

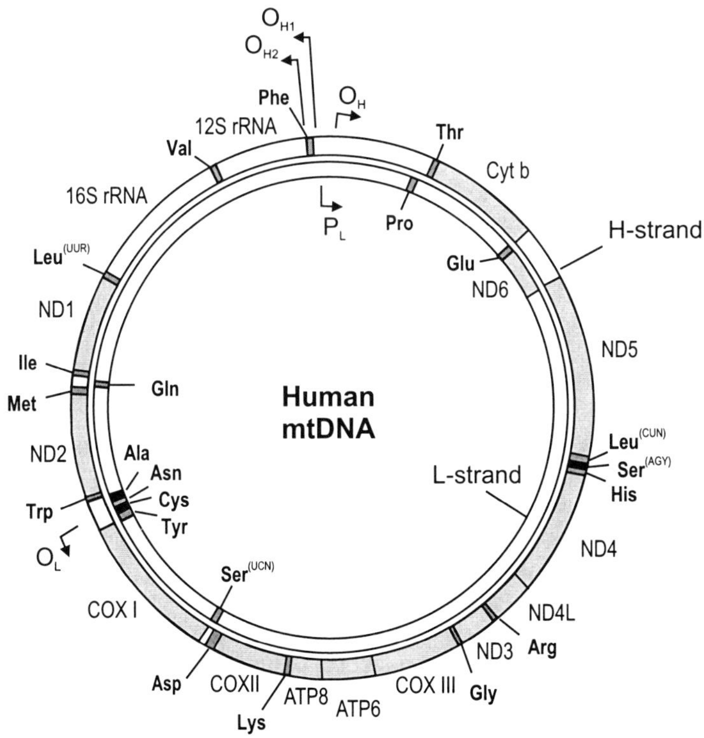

Figure 5. Map of the human mitochondrial DNA. O_H and O_L indicate the origins of replication, and P_H and P_L denote the transcriptional promoters of the heavy (H) and the light (L) strand, respectively.

The occurrence of two or more mtDNA genotypes in an individual, in a single cell, or in discrete organelles is referred to as *heteroplasmy* (Wallace, 1994b). Once a heteroplasmic mutation is inherited from the mother, or is occasionally acquired in one single mtDNA molecule during early embryogenesis, both normal and mutant mtDNA will segregate randomly during cytokinesis to the daughter cells (*mitotic segregation*) (Wallace, 1994b). So far, no recombination events between different mtDNA molecules have been discovered; the accumulation of mtDNA mutations and the increase of heteroplasmy is, therefore, assumed to occur along radiating maternal lineages (Wallace, 1994b). The time point at which the mutation occurs and the developmental fate of the affected progenitor cell will influence the degree of heteroplasmy and its distribution among different organs. Manifestation of a clinical phenotype depends, however, on the amount of mutated mtDNA and first occurs upon exceeding a critical *threshold* level. Threshold levels vary widely between different organs in the same organism. In many cases this fact renders the interpretation of mutational analyses from blood or muscle samples of patients more difficult (Wallace, 1992, 1994a; Shoffner, 1996).

To date, more than 50 point mutations in protein coding regions, tRNA or rRNA genes and over 100 length mutations including deletions, insertions or rearrangements of the mitochondrial genome have been associated with human mitochondrial diseases. Not all of these mutations are inherited along the maternal lineage. Non-inherited *somatic* mutations occur sporadically during life-time and affect only single individuals. Most of the *somatic mutations* of mtDNA are length mutations flanked by identical repeats. It seems likely that free radical damage is important for the occurrence of these mutations (Ames et al., 1993). There is an ongoing debate as to whether the *somatic* mutations accumulate during lifetime and contribute to various degenerative disorders and to aging itself (Lightowlers et al., 1999).

Maternally inherited mtDNA mutations range from neutral polymorphisms to severe or even lethal mutations. The latter ones usually do not persist in the maternal lineage and will not be transmitted to the next generation. Therefore, it is accepted that deleterious mutations are recent events. In contrast, mildly deleterious mutations may not be expressed until late in life and are transmitted to the offspring. Thus, they may have arisen a long time ago and became fixed in the population as low-frequency polymorphisms. Neutral polymorphisms are ancient, having accumulated along mtDNA lineages, and thus correlate with ethnic and geographic origin which have permitted investigation of ancient migration patterns. For example, the phylogenetic analysis of mtDNA variations in Caucasian populations has led to the identification of major mtDNA haplogroups each characterized by lineage-specific mtDNA mutations (Hofmann et al., 1997c; Torroni et al., 1996).

MtDNA haplotype backgrounds have been suggested to establish disease susceptibility in such a way that mtDNA variants act synergistically together with additional, nuclear or mitochondrial encoded mutations to increase the risk for onset of the disease. One of the best-studied examples confirming such a hypothesis is Leber's hereditary optic neuropathy (LHON) caused by a maternally transmitted mtDNA mutation. The LHON mutation has been found at increased frequencies in association with a distinct mtDNA haplogroup (Hofmann et al., 1997a, b; Torroni et al., 1997). The findings suggest that the combination of haplogroup specific variants increase both the penetrance of the pathogenic LHON mutations and the risk of the disease manifestation. Similarly, mtDNA haplogroup backgrounds might create a susceptibility basis for complex, polygenic disorders like diabetes mellitus. In particular, the maternal effect in NIDDM transmission could be accounted for by the inheritance of mtDNA variants together with nuclear gene mutations to increase susceptibility to diabetes (Cox, 1994). So far, no association between diabetes and a specific mtDNA haplogroup has been detected. However, a single mtDNA variant in the non-coding regions (nucleotide position 16189) has been reported to be associated with insulin resistance in the English populations (Poulton et al., 1998). It has to be further clarified if this holds true also for diabetic patients with different ethnic backgrounds.

Cytosolic and Mitochondrial Pathways Involved in the Regulation of Insulin Secretion in Pancreatic ß-Cells

Insulin secretion by pancreatic β-cells is regulated by a specific signal transduction cascade which involves mitochondrial functions. Glucose uptake and metabolism is required for stimulation of insulin release and biosynthesis. Special features of glucose metabolism located in the cytosol and in the mitochondria are central to an understanding of the physiological role of β-cells in glucose homeostasis as summarized in Figure 6. Insulin secretion in response to postprandial glucose elevation depends on: a) a high-capacity transport system in the plasma membrane (GLUT2) (1); b) the glucose phosphorylation by the high-K_m glucokinase (GK), which is the rate-limiting step along

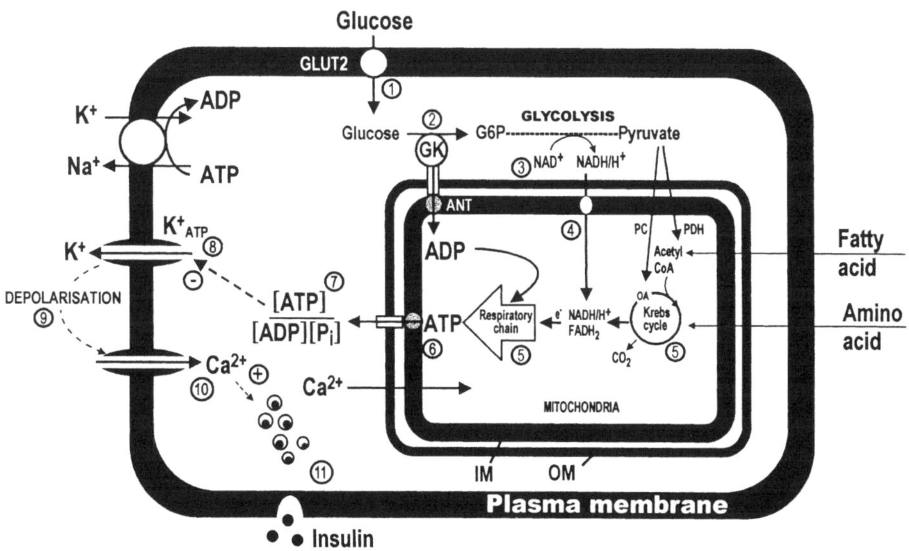

Figure 6. Crucial steps in glucose-induced insulin secretion in β-cells. The numbers indicated by () correspond to the explanations given in text.

the glycolytic pathway (2). Phosphorylation activity directly correlates with the glucose levels in the extracellular space and is therefore seen as a key factor in glucose sensing; c) production of $NADH/H^+$ utilized along the mitochondrial respiratory chain (3); d) the near absence of glycogen and fatty acid synthesis and of gluconeogenesis resulting in the primary catabolic intermediary metabolism; e) remarkably low activities in lactate dehydrogenase which avoid consumption of energy-rich reducing equivalents like $NADH/H^+$; f) high activity of hydrogen shuttle systems which allow efficient translocation of reducing equivalents to the mitochondrial matrix in order to be quantitatively oxidized (4); g) intra-mitochondrial energy generation along the citric acid cycle and the mitochondrial respiratory chain thereby indirectly coupling glycolytic flux rates to ATP synthesis (5); h) ANT mediated transport of ATP across the mitochondrial membranes (6) influencing the cytosolic [ATP/ADP] ratio (7); i) closure of the ATP-dependent K^+_{ATP}-conducting channel (8) leading to plasma membrane depolarization (9); j) triggering Ca^{2+}-influx through voltage-gated Ca^{2+} channels (10) as a signal for insulin release (11) (Figure 6).

Glucose Uptake and Metabolism in the β-Cell

The levels of glucose in the blood are maintained within narrow limits by regulation of its utilization and synthesis: Muscle, adipose tissue and brain are the primary sites of glucose utilization in the periphery; while glycogenolysis and gluconeogenesis is exerted by liver and kidney in fasted states. Both utilization and synthesis are under the control of the regulated secretion of insulin which is produced in β-cells of pancreatic islets. Insulin lowers the glucose levels by stimulating its uptake into insulin-sensitive tissues, and by inhibition of *de novo* glucose synthesis in the liver. In a series of elegant experiments (for review see Matschinsky, 1996), Matschinsky and colleagues have demonstrated that the rate of glucose utilization in β-cells and the subsequent changes in metabolite levels trigger the insulin secretory response. Accurate sensing of circulating blood glucose levels is therefore a prerequisite for the coordinated insulin secretion of β-cells.

Glucose is transported into the β-cell by low-affinity, high-capacity Na^+ independent carriers. GLUT2 is the predominant glucose translocator in human β-cells (Johnson et al., 1990b; Thorens et al., 1988). The K_m of the GLUT2 (17mM) (Johnson et al., 1990a) is significantly higher than of the ubiquitous GLUT1 and the fat- and muscle tissue-specific GLUT4. Since the uptake of glucose exceeds its metabolism, the GLUT2 carrier assures fast exchange between extra- and intracellular glucose in such a way that hyper- and hypoglycemia are rapidly mirrored by the intracellular glucose level. Glucose uptake into the β-cells is therefore not rate-limiting, suggesting the existence of an intracellular rather than a plasma membrane-associated glucose sensor. Phosphorylation of glucose by glucokinase (GK) was identified to play a key role in the sensing mechanism (Matschinsky, 1996) which directly couples glucose levels to the enzymatic activities of the metabolizing enzymes within the glycolysis. Glucokinase is a member of the hexokinase family; it differs in size, kinetic properties and tissue-specificity from the other family members. Glucokinase (or hexokinase IV isoform) exhibits a K_m (6-10 mM) close to the physiological range of blood glucose concentrations, while hexokinases I-III have their K_m within the micromolar range. In contrast to hexokinase I-III, GK of β-cells is apparently not controlled by direct or indirect product inhibition by glucose-6-phosphate. All other enzymes along the glycolytic cascade are present at high activities exceeding that of glucokinase by an order of magnitude. These features, together with its high capacity and its kinetics, make GK the pacemaking enzyme of the glycolysis. Taken together, GK appears to constitute the ideal glucose sensor element in all current concepts (Matschinsky, 1996; Trus et al., 1981). This is supported by observations made in patients with "mature" onset diabetes of the young (MODY2) harboring mutations in the glucokinase gene (Froguel et al., 1993). In these patients glucokinase activity is significantly reduced resulting in a subform of "non-insulin dependent diabetes mellitus" (NIDDM). The gene defect causes impaired glucose sensitivity as a consequence of decreased glucokinase activity in β-cells (Byrne et al., 1994; Velho et al., 1992).

Although liver and pancreatic islets share several similarities in respect to the expression of specific gene products such as the GLUT-2 transporter (Johnson et al., 1990a; Thorens et al., 1988) and glucokinase and glucose-6-phosphatase (Waddell and Burchell, 1988) the metabolism of glucose is quite different in both cell types. Glucokinase is found in liver and in pancreatic β-cells (Iynedjian et al., 1989; Magnuson and Shelton, 1989); but mRNA transcripts differ from each other at the 5'-end suggesting alternative splicing as a regulatory mechanism in these organs (Koranyi et al., 1992). In the liver, glucokinase is involved in utilization of excess blood glucose concentration after carbohydrate feeding. Only small parts of postprandial glucose (at a level between 6-8 mmol/l) enter the glycolytic pathway, whereas the vast amount is distributed into glycogen synthesis and the pentose monophosphate shunt (Newgard and McGarry, 1995). The enzymatic activity is increased after carbohydrate feeding or after insulin treatment, and decreased in diabetes. Thus, insulin appears to be the major regulator of glucokinase in liver. In contrast, glucokinase expression in β-cells seems to be constitutive and to be regulated mainly post-transcriptionally by glucose. As the glucose taken up by the β-cells feeds quantitatively into glycolysis thereby producing reducing equivalents, GK is able to "sense" glucose levels. The differential regulation of GK in liver and β-cells is mediated by tissue-specific promoters (Magnuson et al., 1994). The upstream promoter is responsible for expression of GK in β-cells, while the downstream promoter regulates the expression of liver GK.

In contrast to most other cells, glucose-6-phosphate is almost quantitatively metabolized along the glycolytic pathway in β-cells. Only small amounts of glucose are stored as glycogen (Malaisse et al., 1977). The pentose monophosphate shunt accounts for only a few percent of the total glucose uptake in β-cells (Ashcroft et al., 1972). Furthermore, gluconeogenesis is unlikely to occur in pancreatic β-cells, due to their lack of the two key enzymes, the phosphoenolpyruvate carboxykinase (PEP-CK) and the fructose-1,6-bisphosphatase (Meglasson and Matschinsky, 1986). Moreover, the known variety of positive and negative regulators seem to be of little consequence in the control of normal β-cell glycolysis, in particular because hexokinase I and glucose-6-phosphatase are nearly absent and GK is not inhibited by glucose-6-phosphate. Thus, all observations mentioned above support the role of glucokinase as the rate-limiting enzyme for the regulation of glycolysis and provide compelling evidence for its suitability as a sensor element in regulated insulin secretion.

Since all glycolytic intermediates are in an equilibrium, pyruvate is generated at a rate determined by the conversion of fructose-1,6-bisphosphate to phosphoenolpyruvate and the cytosolic ADP levels. Thus, its generation keeps pace with glucose phosphorylation. The generation of lactate from pyruvate is low because NADH/H^+ is efficiently removed from the cytosol by the hydrogen shuttle mechanism, α-glycero-phosphate shuttle and the malate/aspartate shuttle (see below).

Coordinated Regulation of Glucokinase and Mitochondrial Oxidative Metabolism

Hexokinases I-III are expressed in different mammalian tissues and have been shown to form complexes at the mitochondrial surface with the outer membrane protein porin thereby activating the enzyme (Kottke et al., 1988, 1994; Ohlendieck et al., 1986; Weiler et al., 1985). Targeting of GK to the mitochondrial outer membrane and its acitivation was also observed in β-cells. In contrast, is apparently absent from the mitochondrial outer membrane in liver cells (Malaisse-Lagae and Malaisse, 1988). This tissue-specific difference in glucokinase targeting may be the result of differential splicing leading to variations in the N-terminal portion of the enzyme. This is supported by the fact that the N-terminal part of hexokinase I was found to be responsible for binding to the mitochondrial surface (Polakis and Wilson, 1985). Targeting of glucokinase to the mitchondrial outer membrane may provide the molecular basis for the tissue-specific regulation of GK in β-cells at the protein level. Further investigation should be carried out to elucidate the binding mechanism of β-cell GK to the mitochondrial outer membrane.

Binding and Activation of Hexokinase Enzymes at the Mitochondrial Outer Surface

Hexokinase I, the ubiquitously expressed isoform, is predominantly found at sites where the mitochondrial inner and outer membrane are in close proximity (Kottke et al., 1988, 1994; Ohlendieck et al., 1986; Weiler et al., 1985). Mitochondrial porin as the specific binding site for hexokinases I-III is, however, randomly distributed in the outer membrane. It has been suggested that binding and activation of hexokinase are cooperative processes. Bound hexokinase was found in high molecular weight complexes with about 400 kDa consisting of hexokinase and porin in a tetrameric conformation (Brdiczka et al., 1994; Xie and Wilson, 1990). High affinity binding of hexokinase is presumably caused by the association of these porin-containing complexes to the adenine nucleotide translocator ANT in the mitochondrial inner membrane (Brdiczka, 1994; Bucheler et al., 1991; Knoll and Brdiczka, 1983). Bound hexokinase seems to have preferential access to intramitochondrially generated ATP subsequently utilized in the phosphorylation of cytosolic glucose (Gots et al., 1972). Recent investigations, however, indicate that mitochondrial binding of hexokinase is more important because ADP is efficiently supplied to the OXPHOS (Kabir and Nelson, 1991; Wicker et al., 1993) (Figure 6). Similarly, glucokinase bound to the mitochondria may function in a synergistic fashion by coupling cytosolic formation of intermediates along glycolysis directly to mitochondrial OXPHOS (Fiek et al., 1982; Gerbitz et al., 1996; Linden et al., 1982a, b).

Stimulation of Mitochondrial OXPHOS

As pointed out above, cytosolic glycolysis is coupled to mitochondrial energy metabolism at several points. (i) glucokinase is thought to be functionally coupled to porin/ANT complexes in the mitochondrial membranes thereby channeling ATP directly from the mitchondrial matrix to the kinase. As ATP is hydrolysed, ADP is efficiently redirected by the bound kinase to the mitochondrial matrix in order to stimulate its oxidative re-phosphorylation (Figure 6). (ii) reduction equivalents generated in the cytoplasm are translocated into the mitochondrial matrix by two shuttle systems thereby feeding the respiratory chain: the malate/aspartate shuttle (MacDonald, 1982) and the α-glycero-phosphate shuttle (MacDonald, 1990). The malate/aspartate shuttle couples $NADH/H^+$ to complex I (Figure 7) while the glycerophosphate cycle shuttles reducing equivalents to ubiqinone (Coenzyme Q) via an FAD-dependent mitochondrial glycerophosphate dehydrogenase (mGPDH) (Figure 7). mGPDH is ubiquitously expressed but seems to be most abundant in tissues with high glucose metabolism (MacDonald et al., 1996). (iii) Pancreatic islets are unique, as their content of pyruvate carboxylase (PC) is equivalent to that found in gluconeogenetic tissues, such as liver and kidney.

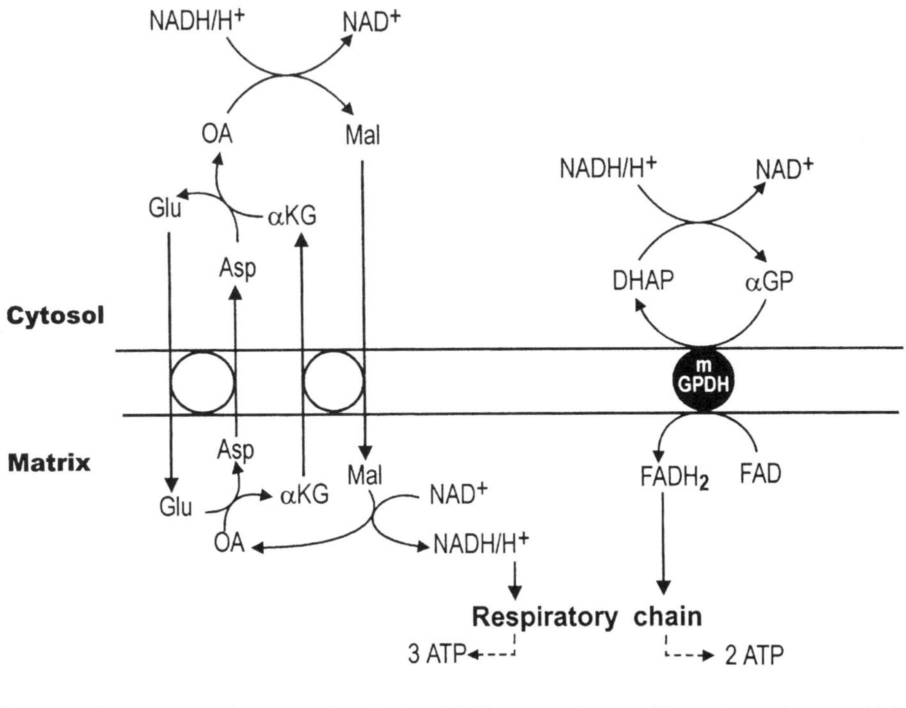

Figure 7. Hydrogen shuttles across the mitochondrial inner membranes. The α-glycerophosphate (right) and the aspartate/malate (left) shuttles allow efficient translocation of reduction equivalents from the cytsol into mitochondria. $NADH/H^+$ enters the respiratory chain at complex I, $FADH_2$ at ubiquinone. Abbreviations: OA, oxalacetate; Mal, malate; Glu, glutamate; Asp, aspartate; αKG, α-ketoglutarate; DHAP, dihydroxy acetonphosphate; αGP, α-glycerophosphate.

However, gluconeogenesis has only very little significance in β-cells. The latter observation prompted MacDonald (MacDonald, 1995a, b) to propose a pyruvate/malate shuttle (Figure 8) which determines the fate of cytosolic pyruvate. According to the model pruvate is metabolized in two different ways within the mitochondria: it is either decarboxylated by pyruvate dehydrogenase to form acetyl-CoA or it is carboxylated by pyruvate carboxylase to oxalacetate, which ends up as malate (Figure 8). Malate then leaves the mitochondria and is decarboxylated by a cytosolic malic enzyme to regenerate pyruvate. It is very likely that increased availability of pyruvate through the pyruvate-malate shuttle contributes to the supply of substrates for the citric acid cycle during glucose stimulation of β-cells. As the rate of pyruvate carboxylation is directly proportional to the extracellular glucose concentration, it seems likely that these *anaplerotic* reactions serve as an additional coupling site between glycolytic substrate flux and mitochondrial metabolism. The functional significance of the pyruvate/malate shuttle, however, is controversial (see below).

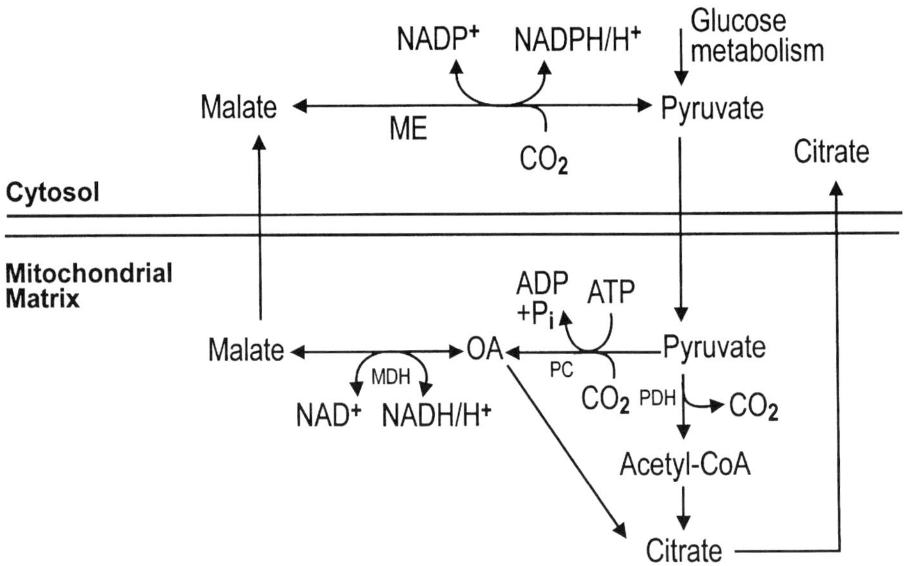

Figure 8. The pyruvate/malate shuttle.

The current model of the signal transfer in pancreatic islets outlined above is, in certain respects, hypothetical and several questions still have to be answered. However, the model proposes the coupling between glycolytic pathways and mitochondrial OXPHOS in nutrient-stimulated insulin secretion (Newgard and McGarry, 1995). One of the most intriguing open questions within this model is why are glucose, glyceraldehyde or mitochondrial fuels like 2-ketoisocaproate, potent insulin secretagogues; whereas pyruvate or lactate, which are also oxidized in the mitochondria is not effective in stimulating the insulin secretion (Hutton et al., 1980; MacDonald and

Fahien, 1988; Sener et al., 1978)? According to the model it seems likely that the energetically relevant coupling of glucose metabolism to the mitochondrial OXPHOS occurs *via* early steps along the glycolytic pathway rather than by further metabolism of pyruvate in mitochondria. While the oxidation of pyruvate plays an important role in non-β-cells, it has been proposed that in β-cells, a large proportion of pyruvate is eliminated from the cytosol or is transaminated, and does, therefore, not efficiently fuel the citric acid cycle (Matschinsky, 1996). The β-cell glycolytic reducing equivalents are nearly quantitatively shuttled into the mitochondria with the hydrogen shuttle systems being the main mechanism of coupling to the mitochondrial OXPHOS. This is supported by recent animal studies (Eto et al., 1999; Postic et al., 1999), in which the effect of functional defects in these shuttle systems on glucose-induced insulin secretion was investigated. Knock-out mice were generated by disruption of the gene encoding the mitochondrial GLPDH (mGPDH), the rate limiting enzyme of the glycerophosphate shuttle. Homo- and heterozygous knock-out mice ($mGPDH^{-/-}$, $mGPDH^{-/+}$) were normoglycemic and insulin secretion in response to an intraperitoneal glucose tolerance test was found to be at wild type levels. A drastic effect on the glucose-dependent insulin secretion and blood glucose levels could only be detected when the malate/aspartate shuttle was inhibited by aminooxyacetate (AOA) (MacDonald, 1981; Malaisse et al., 1982). When the inhibitor was added to isolated islets of wild-type and $mGPDH^{-/+}$ knock-out mice, only a slight suppression of high glucose stimulated insulin release could be observed. Administration of AOA to islets from $mGPDH^{-/-}$ mice, however, led to a dramatic reduction of glucose-induced insulin secretion. Direct measurements of ATP and ADP in AOA-treated islets in these mice revealed a decrease of about 40% of both the cellular ATP concentration and the ATP/ADP ratio. In contrast, the production of pyruvate from glucose in $mGPDH^{-/-}$ islets treated with AOA appears not to differ from that in wild-type mice. These observations suggest that both shuttle systems are able to at least in part functionally substitute for each other in providing reducing equivalents to the mitochondrial OXPHOS thereby triggering the glucose-stimulated insulin release from β-cells. This trigger mechanism is completely blocked when both shuttles are disabled.

So far, it has not been possible to estimate the relative importance of both shuttle systems in β-cells. However, it seems likely from the animal models that the malate/aspartate shuttle contributes more efficiently to the generation of ATP along the respiratory chain than the α-glycerophosphate shuttle which is more favorable thermodynamically (McCormack and Denton, 1994) (Figure 7). The assumption is based on two observations: $NADH/H^+$ is introduced into the respiratory chain via complex I, while $FADH_2$ enters via complex II (Figure 5). Observation of humans suffering from intoxication with the rodenticide Vacor®, i.e.1-(4-nitrophenyl)-3-(3-pyridyl-methyl) urea, demonstrated a similar phenotype as observed in the AOA treated $GLDH^{-/-}$ knock-out mice. Intoxication with the rodenticide was often lethal. Survivors, however, invariably presented with diabetes, mostly complicated with several severe neurological symptoms. Recently it could be shown that Vacor® specifically inhibits complex I of the respiratory chain while leaving the other respiratory chain complexes unaffected. Taken together this indicates that complex I is the main entrance point for reduction equivalents, and supports the hypothesis that the malate/aspartate shuttle constitutes the major coupling site between glycolysis and mitochondrial OXPHOS in glucose-stimulated insulin secretion.

The Cellular Energy Network

Direct measurements of oxygen consumption in various β-cell systems allowed the estimation of the rate of ATP generation as a function of glucose concentration and its relation to insulin secretion (Matschinsky, 1996). The ATP production along the respiratory chain has to reach a certain threshold before insulin release is triggered. It has been hypothesized that this threshold is mainly determined by the ratio of energy supply and demand. As non-hydrolyzed ATP begins to predominate, the energy potential, or ΔG of ATP hydrolysis, increases sharply thereby triggering the insulin secretion. It has, however, to be realized that cellular energy supply and demand are generally balanced and tightly regulated for economy and efficiency of energy use. The creatine kinase/phosphocreatine system plays a key role in the regulation and buffering of the energy metabolism of cells with intermittently high and fluctuating energy requirements, such as skeletal and cardiac muscle, brain or neuro-endocrine tissues. Creatine kinase (CK) is found in cytosolic and mitochondrial isoforms which are highly expressed in excitable tissues. The cytosolic isoforms occur as dimers. In cells of the neuroendocrinium, BB-CK is found, MM-CK in muscle and the hetero-dimer of both of these, MB-CK, is mainly present in cardiac muscle. BB-CK is also found in pancreatic islet cells (White et al., 1992). CK isoenzymes catalyze the reversible transfer of the energy-rich phosphoryl group from phosphocreatine (PCr) to ADP thereby regenerating ATP. Compared to ATP ($\Delta G^{o\prime}$ 30 kJ mol^{-1}), PCr exhibits a higher free energy of hydrolysis ($\Delta G^{o\prime}$ 44 kJ mol^{-1}). It is therefore thermodynamically more favorable to store the free energy as phospho-creatine. Increased biosynthetic and secretory activity, e.g. insulin secretion, and ion pumping are assumed to be sites of enhanced ATP consumption. Their demand is sufficiently supplied by the interplay between the cytosolic CK isozymes bound to the energy consuming ion pumps and the mitochondrial CK (Mi-CK). The CK/PCr system provides an energy supply mechanism (Figure 9), for which a variety of distinct properties have been proposed (Wallimann and Hemmer, 1994): (i) the CK/PCr system is thought to function as a temporal energy buffer which keeps the [ATP] and [ADP] constant. ATP is constantly regenerated via PCr thereby preventing a rapid ATP decrease and, as a consequence, also the accumulation of ADP; (ii) the CK/PCr system is assumed to improve the thermodynamic efficiency of ATP hydrolysis by keeping the intracellular [ADP] low. At subcellular sites where the CK is functionally coupled to ATP consuming processes like ion-channels (see below), the [ATP]/[ADP] ratio is locally maintained. Since ion pumps such as the sarcoplasmatic Ca^{2+}-pump are working close to their thermodynamic equilibrium, a locally high [ATP]/[ADP] ratio is a prerequisite for efficient ion pumping. In this respect, the Cr/PCr pool which reaches concentrations (25-30 mM) approximately five times higher than the cytosolic adenine nucleotide pool is suitable to overcome diffusion limitations and provides sufficient hydrolysable energy.

While the temporal energy buffering and metabolic regulatory roles of CK are widely accepted, the proposed spatial buffering or energy transport function of the CK/PCr shuttle system is still being debated. In this role, PCr is thought to function as an 'energy carrier' connecting sites of energy production, e.g. the mitochondrial oxidative phosphorylation, to sites of energy consumption, where the mitochondrial CK isoform plays a crucial role. CK functions as an energy buffer and transport system as indicated by the subcellular micro-compartmentalization of PCr/Cr, ATP/ADP and phosphate, and by the structural and functional properties of the mitochondrial CK isoform. The octomeric conformation of the mi-CK favors the metabolite channeling across the mitochondrial membranes. Moreover, this is also supported by the observation that cytosolic CK isozymes show specific subcellular compartmentalization in a variety of tissues.

In skeletal muscle, for example, cytosolic MM-CK is functionally coupled to ATP hydrolyzing enzymes such as the Ca^{2+}-ATPase (Korge and Campbell, 1994; Wallimann et al., 1992), Na^+/K^+-ATPase (Saks et al., 1977) and actomyosin ATPase. In the latter case, MM-CK is specifically located in the M-line (Saks et al., 1977). MM-CK is found to co-localize with glycolytic and glycogenolytic enzymes in the I-band of glycolytic fast twitch fibers. The close proximity of CK and pyruvate kinase allows fast exchange between the glycolysis and CrP pool which can be rapidly mobilized during muscle contraction. The kinetic advantage of this linkage is demonstrated *in vitro* (Dillon and Clark, 1990). [^{31}P] NMR experiments showed that the phosphate is exchanged between phosphoenol-pyruvate and phospho-creatine without change in intermediate ATP. The energy of phosphoenol pyruvate (PEP) can be transferred to creatine by direct interaction of pyruvate kinase and cytosolic CK (Dillon and Clark, 1990).

Similarly, the high free energy of ATP at the surface of the mitochondrial inner membrane is stored as PCr through a complex between mi-CK and adenylate translocator (ANT) (Adams et al., 1988; Brdiczka and Wallimann, 1994; Gellerich et al., 1993) (Figure 9). This mitochondrial coupling of kinases may accomplish an important function in permanently working muscles, like the heart or nerve cells, that depend more or less on blood-borne substrates. In these tissues, the coupling mechanism may regulate the uptake of energy providing substrates (glucose) according to the mitochondrial activity. On the other hand, the same organization of kinases may be important in β-cells to couple mitochondrial metabolism to extramitochondrial substrate (glucose) concentrations (Brdiczka and Wallimann, 1994; Wallimann et al., 1992).

Regulatory Function of the CK/PCr Shuttle System in β-Cells and Mechanisms of Insulin Exocytosis

The classical concept of glucose-stimulated insulin secretion proposes that ATP is the actual metabolic coupling factor for triggering insulin secretion because of its ability to inhibit K^+_{ATP}-channels in the plasma membrane. The K^+_{ATP}-channel is formed by two membrane proteins, the pore-forming Kir6.2 and the sulfonylurea receptor type 1 (*SUR1*). Upon binding, the K^+_{ATP}-channel is closed and K^+ is trapped in the cytosol. On the other hand, K^+ is actively transported across cell membrane by the Na^+/K^+-ATPase thereby leading to the intracellular K^+ accumulation and depolarization of the membrane (Cook and Hales, 1984). The free ATP is increased at the K^+_{ATP} channels locally by phosphocreatine through cell membrane bound creatine kinase. The average cellular

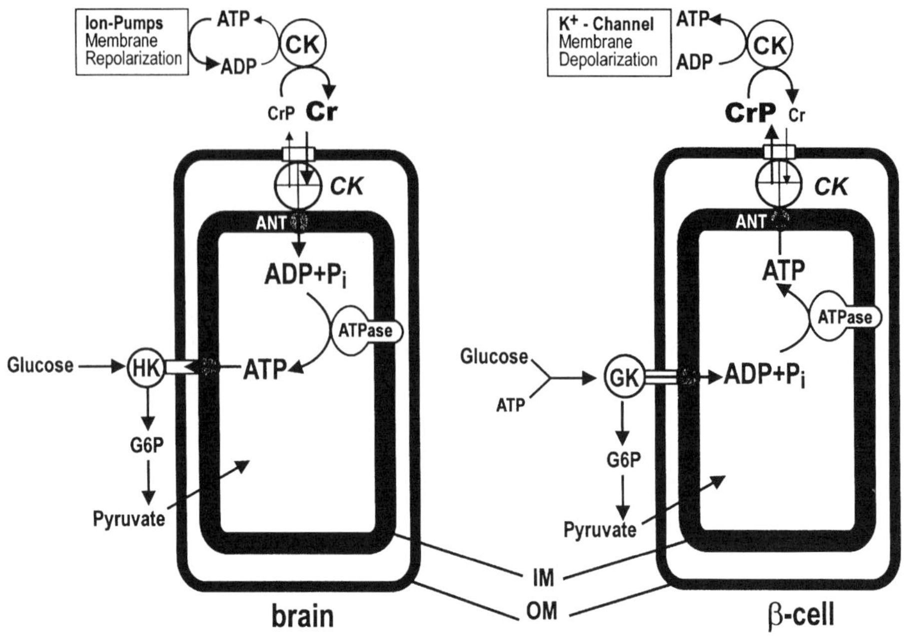

Figure 9. Schematic representation of the intracellular energy coupling by the CK/CrP shuttle system. The models outline the specific differences between brain and the insulin-secreting β-cell. In case of rapid ATP turnover upon stimulation of muscle or nerve cell activity, the level of Cr increases and stimulates OXPHOS by activating the MiCK (left). This leads to parallel increase of glucose phosphorylation and activation of hexokinase (HK). In β-cells, intramitochondrial ADP is increased via bound glucokinase (GK). This stimulates OXPHOS and leads to higher output of CrP. CrP increases the ATP levels in close proximity to the $K^+{}_{ATP}$ channel by interacting with the cell membrane associated CK (*adapted from Gerbitz et al. 1996*).

ATP concentration is between 5-8 mmo/l. Under these conditions the $K^+{}_{ATP}$ channel which has a K_i of 20-200 μmol/l would never open. However, as already stated by Ashcroft (Ashcroft, 1988), ATP may be compartmentalized in the cell in such a way that its local concentration close to the channel-binding site is kept very low leading to a low [ATP/ADP] ratio. Such compartmentalization could be performed by linking the $K^+{}_{ATP}$ channel to glycolysis and OXPHOS through the creatine phosphate/creatine kinase system which acts as the main regulator of channel activity via the [ATP/ADP] ratio (Figure 6). The proposed mechanism by which mitochondrial metabolism is involved in the regulation of the $K^+{}_{ATP}$-channel is schematically outlined in Figure 6.

At least four intracellular targets are thought to be influenced by the [ATP/ADP] coupling factor (Matschinsky, 1996): i) the $K^+{}_{ATP}$-channel, as described above (Hopkins et al., 1992; Cook, 1984; ii) the Ca^{2+}-ATPase of the endoplasmic reticulum; iii) ATP-sensitive and Ca^{2+}-independent phospholipase, which generates arachidonic acid as a second messenger (Turk et al., 1993), and iv) possibly the exocytotic process itself (Gembal et al., 1993). Each of these targets alone, or in combination, finally are able to initiate the intracellular and mitochondrial Ca^{2+} elevation that is thought to trigger insulin release (Figure 6). While the intracellular Ca^{2+}-reservoirs play a minor role, most of the Ca^{2+} increase is derived from the influx through voltage-gated Ca^{2+}-channels in the plasma membrane which are opened by the "glucose-induced" membrane depolarization (Ashcroft et al., 1994; Rorsman, 1997b). At low glucose concentrations (< 5mmol/l) the plasma membrane potential of the β-cell is negative, i.e. around -60mV. "Glucose-induced" closure of more than 90% of the K^{+}_{ATP}-channels leads to depolarization of the β-cell membrane (Rorsman, 1997b). The delay between the opening of the Ca^{2+}-channels and the onset of insulin exocytosis is less than 50 msec (Rorsman, 1997a, b). Thus, exocytosis of insulin is a rapid process which is assumed to be ATP-dependent at certain steps as withdrawal of cytoplasmic ATP nearly completely blocks insulin release (Regazzi et al., 1995). Functional and ultrastructural studies have shown that insulin is released in two steps from distinct granular subpopulations stored in the β-cells (Rorsman, 1997b). During a first phase insulin is immediately set free upon the release trigger; the second phase requires the mobilization from intracellular stores.

Animal Model for K^{+}_{ATP} Channel Function

The key role of the β-cell K^{+}_{ATP} channel was recently demonstrated in mice with a disruption of the Kir6.2 component of the channel (Miki et al., 1998). Homozygous Kir6.$2^{-/-}$ mice lacked K^{+}_{ATP} channel activity in all tissues tested. They had normal basal insulin levels, a higher resting plasma membrane potential and higher intracellular calcium concentration than control mice (Kir6.$2^{+/+}$). The mutants are unable to secrete significant amounts of insulin in response to high glucose or the sulfonylurea tolbutamide. Interestingly, the genetic defect does not lead to basal hyperglycemia. This is in contrast to another genetic animal model with a K^{+}_{ATP} channel defect only in β-cells (the Kir6.2G132S Tg mouse) but normal channel activity in other tissues like skeletal muscle (Miki et al., 1997). These latter mice developed marked hyperglycemia. Thus the ubiquitous K^{+}_{ATP} defect in the Kir6.$2^{-/-}$ mice seems to sensitize peripheral tissues for insulin action preventing hyperglycemia in this way.

Alternative Pathways

There is evidence that glucose can control insulin secretion independently of its action on ATP-sensitive K^{+}-channels. As proposed by Prentki and co-workers, malonyl-CoA and long-chain acyl-CoA esters might act as metabolic coupling factors in nutrient-stimulated insulin secretion. The cross-talk between glucose metabolism and fatty acids is well established, in that a short exposure of islets to fatty acids can considerably enhance glucose-stimulated insulin secretion, while long-term exposure to fatty acids can inhibit β-cell function *in vitro* (Zhou and Grill, 1995a; 1994) and *in vivo* (Mason et al., 1999; Sako and Grill, 1990). In a variety of tissues, malonyl-CoA is involved in the regulation of fatty acid synthesis and oxidation. As outlined in Figure 10, malonyl-CoA is the substrate for the fatty acid synthase (FAS) the key enzyme in fatty acid biosynthesis. Moreover, it is known to inhibit the carnitine palmitoyltransferase I (CPT I) thereby controlling mitochondrial oxidation of fatty acids (McGarry and

Foster, 1979). Consistently, elevated glucose was found to suppress the oxidation of long chain fatty acids (Berne, 1975; Tamarit-Rodriguez et al., 1984; Vara and Tamarit-Rodriguez, 1986). In pancreatic β-cells, Prentki and co-workers have observed that malonyl-CoA is rapidly accumulated in response to glucose stimulation; and have therefore proposed that malonly-CoA might act as metabolic coupling factor in nutrient-stimulated insulin-secretion (Prentki et al., 1992). According to this concept, the glucose-stimulated accumulation of malonyl-CoA would increase the cytoplasmic long chain fatty acid acyl-CoA (Acyl-CoA) concentration; Acyl-CoA, is thought to act as a trigger of insulin release (Brun et al., 1996). Supporting evidence for this model comes from the observations that (i) citrate, the precursor of malonly-CoA, is also elevated upon glucose stimulation (Corkey et al., 1989; Prentki et al., 1992) and (ii) the key enzymes which are involved in formation of malonyl-CoA, i.e. pyruvate carboxylase (PC) and acetyl-CoA carboxylase (ACC) are found to be abundant in pancreatic islets (Brun et al., 1996). Based on this concept (Figure 10) an enhanced glycolytic flux would stimulate the anaplerotic formation of mitochondrial citrate via PC. Citrate leaves the mitochondria to be converted to malonyl-CoA to increase Acyl-CoA in the cytosol by CPT I inhibition. Acyl-CoAs can be converted to diacylglycerol (DAG), an potent activator of protein kinase C (PKC) and inositol phosphates which can release $Ca^{2+}$from intracellular stores (Prentki and Matschinsky, 1987).

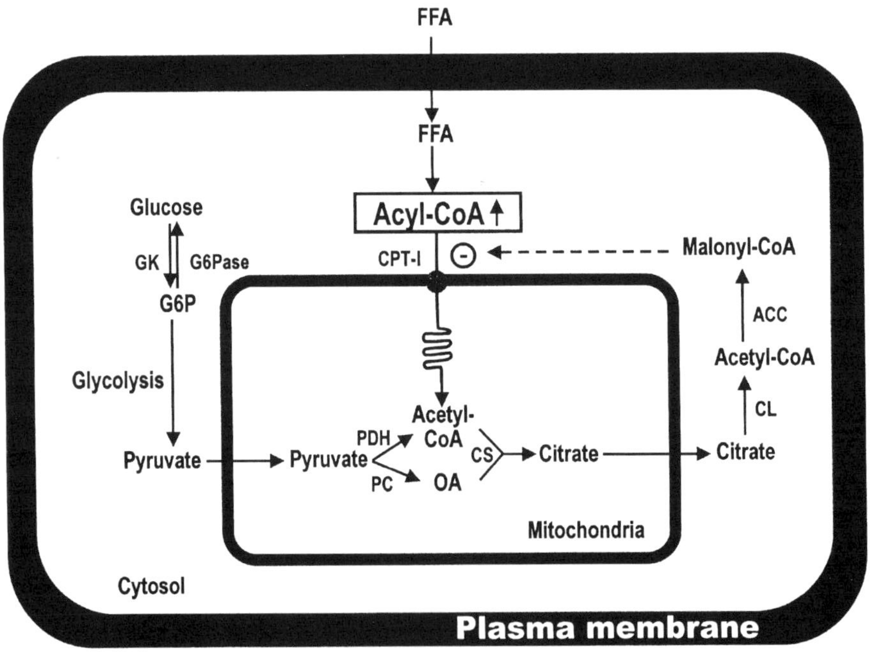

Figure 10. Regulation of cytosolic acyl-CoA levels. Glucose stimulation lead to a rapid accumulation of the cytosolic malonyl-CoA level via anaplerotic pathways. Inhibition of CPT-I by malonyl-CoA elevates cytosolic acyl-CoA concentrations, which are assumed to trigger insulin secretion. Abbreviations: CL, ATP-citrate lyase; FFA, free fatty acids; OA, oxalacetate *(Figure modified from Prentki and Corkey, 1996)*.

Recent observations which contradict this model show that long chain Acyl-CoA esters act as K^+_{ATP}-channel openers (Branstrom et al., 1997) at the pore-forming subunit (Kir6.2) thus effectively antagonizing the closing ATP effect (Branstrom et al., 1998). Thus, in contrast to the more firmly established role of adenine nucleotides, the intracellular signaling by Acyl-CoA and DAG in the course of glucose-stimulated insulin release remains speculative.

Most recently it was demonstrated that the β-cell is able to stimulate its insulin secretion by an autocrine mode that is activated through binding of insulin to the insulin receptors at the β-cell surface (Aspinwall et al., 1999). Insulin promotes the transcription of its own insulin gene via the insulin receptor (PI3-kinase/p70, S6 kinase and CaM kinase pathways) thereby increasing the intracellular insulin content (Leibiger et al., 1998). In mice a β-cell specific knockout of the insulin receptor (βIRKO) creates a defect of insulin secretion similar to that in type II diabetes (Kulkarni et al., 1999).

Mitochondrial Dysfunction in Diabetes

Although the exact mechanisms of insulin secretion are not entirely clear, it can be stated that both activation of glycolysis and stimulation of mitochondrial energy metabolism are two fundamental metabolic prerequisites. While the intramitochondrial energy generating and transducing pathways are more or less common to all cell types, specificity in the regulation of insulin secretion is conferred primarily at the "glucose sensing" steps upstream of the more unspecific intramitochondrial pathways. Mutations in the β-cell-specific "glucose-sensing" enzyme glucokinase are causative of non-syndromic diabetes. On the other hand, hyperglycemia is also a frequent condition in a large variety of monogenic syndromic disorders. The most prominent example is the maternally inherited mitochondrial diabetes, which is caused by mutations in the mtDNA (see below) and often presents as part of a multiorgan syndrome together with deafness and other neurological abnormalities.

The monogenic forms of diabetes account only for about 5% of all diabetes cases. The most frequent form of diabetes is the polygenic late-onset NIDDM, in which contributing gene defects have still to be elucidated. Furthermore the question whether mitochondrial dysfunction is a common risk factor for late-onset diabetes remains to be solved. In the following sections we will focus on the diabetes subforms in which distinct genetic defects were identified and an attempt is made to outline the possible mechanisms involved in the pathogenesis of diabetes associated with mitochondrial dysfunction.

Mitochondrial Diabetes (MDM)

Effect of mtDNA Depletion on β-Cell Function

Mitochondrial oxidative phosphorylation plays a major role in glucose-induced insulin secretion as extensively described above. It is therefore not surprising that β-cell function also depends on the structural and functional integrity of the mitochondrial genome. Cell culture experiments have now provided unambiguous evidence for the causative relationship between mtDNA abnormalities and insulin secretory defects. Using the DNA-intercalating agent ethidium bromide (EtBr), different mouse (Hayakawa et al., 1998; Soejima et al., 1996) and rat (Kennedy et al., 1998) insulinoma cell lines have been artificially depleted of their mitochondrial genome and studied for their ability to respond to glucose stimulation. Upon a glucose load, the mtDNA depleted β-cells completely failed to secrete insulin. The loss of a functional respiratory chain thereby specifically inhibits glucose-signaling steps, i.e. glucose-stimulated Ca^{2+} influx and glucose-stimulated membrane depolarization, whereas the effect of KCl on membrane depolarization and insulin exocytosis was completely preserved (Kennedy et al., 1998). The re-introduction of mtDNA into ρ^0 cells was able to fully restore glucose-stimulated insulin secretion in the hybrid cells (Soejima et al., 1996). Additional experiments have confirmed that early steps coupling glycolysis to mitochondrial signaling and not exocytosis is compromised when mtDNA is depleted (Kennedy et al., 1998). These data provide experimental evidence that mtDNA alterations may cause maternally inherited subforms of diabetes mellitus by "uncoupling" glycolysis from insulin secretion.

mtDNA Alterations in MDM

During the last decade, a number of publications have shown that length and point mutations of the mtDNA are also implicated in development of diabetes mellitus. Length mutations of the mtDNA in association with diabetes have been first described in 1988 in patients suffering from mitochondrial myopathies, such as "chronic progressive external ophthalmoplegia" (CPEO) or the complete Kearns-Sayre syndrome (KSS) (Gerbitz et al., 1990a, b; Holt et al., 1988; Lestienne and Ponsot, 1988; Moraes et al., 1989; Nelson et al., 1989). In 1993, the presence of more complex mtDNA rearrangements, rather than single deletions was suggested to underly certain forms of myopathy and diabetes (Dunbar et al., 1993; Poulton et al., 1993, 1994; Poulton and Holt, 1994). The proportion of each type of rearranged mtDNA molecule seems to vary between different tissues, and there is evidence that the balance of mtDNA rearrangements may be central to this form of mitochondrial diabetes (Dunbar et al., 1993; Poulton and Holt, 1994). In the family with diabetes and deafness described by Ballinger and co-workers (Ballinger et al., 1994, 1992) the relative amount of wild-type, duplicated and deleted mtDNA molecules varied considerably between and within family members. In EBV-transformed lymphoblastoid cell lines derived from different family members, the relative proportions of the three mtDNA types change drastically during culturing. While the deleted molecules were consecutively lost, the percentage of duplicated molecules increased up to 95% of total mtDNA (Ballinger et al., 1994). It was, therefore, suggested that duplicated rather than deleted mtDNA molecules may be transmitted in the germ line (Dunbar et al., 1993; Poulton and Holt, 1994).

The most commonly reported mtDNA point mutation associated with diabetes is an A/G exchange at nucleotide position (np) 3243 in the $tRNA^{Leu(UUR)}$. This mutation was first identified as cause of the MELAS (mitochondrial encephalopathy with lactic acidosis and stroke like episodes) syndrome (Goto et al., 1990). One year later, we described a family with the clinical features of MELAS together with deafness and diabetes mellitus carrying the mutation at np 3243 (Enter et al., 1991; Obermaier-Kusser et al., 1991). The specific association of the 3243 mutation with diabetes, deafness and other neurological symptoms, was confirmed by several other authors (Reardon et al., 1992; van den Ouweland et al., 1992, 1994). In nearly 70% of MDM, diabetes is found in combination with deafness (van den Ouweland et al., 1995, 1994), whereas the clustering with additional neurological symptoms such as MELAS is less common. This subform of MDM is, therefore, also described as maternally inherited diabetes and deafness (MIDD) (van den Ouweland et al., 1992).

Other less frequent mutations have also been described in association with diabetes. The $tRNA^{Leu(UUR)}$ gene thereby seems to be an etiological hot spot. Today, at least 11 disease-related mutations have been described in this gene, (Moraes et al., 1993); five of them (including the 3243 mutation) have been associated with syndromic diabetes (Figure 11): a mutation at np 3260 was found in a patient with maternally inherited myopathy, cardiomyopathy and diabetes (Zeviani et al., 1991); a C/T exchange at np 3256 in a patient with a MERRF-like disease (myoclonic-epilepsy and ragged red fibers) plus diabetes (Silvestri et al., 1992); a A/G transition at np 3252 in a patient suffering from mitochondrial encephalomyopathy, pigmentary retinopathy, dementia, diabetes and hypothyroidism (Morten et al., 1993) and a T/C exchange at np 3264 in a patient with diabetes, cerebellar ataxia, hearing loss and ophthalmoplegia (Suzuki et al., 1997). Diabetes is rarely found in association with the so-called MERRF mutations at np 8344 (Shoffner et al., 1990), np 8291 (Hirata et al., 1999) and at np 8296 (Kameoka et al., 1998) in the mitochondrial $tRNA^{Lys}$ gene (Figure 11). In Japanese populations, this latter mutation was found to account for 1% of diabetic cases (Kameoka et al., 1998). Recently, two new tRNA exchanges have been described,

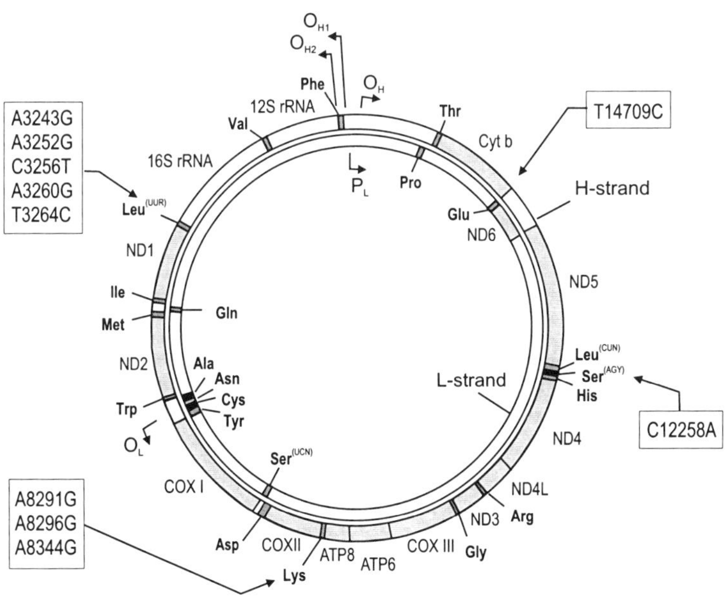

Figure 11. Map of pathogenic mtDNA point mutations which have been associated with diabetes. 12S and 16S denote the genes encoding the small and large rRNAs, ND1-6 are the genes encoding subunits of complex I, COI-III encode subunits of complex IV.

a T/C transition at np 14709 in the $tRNA^{Glu}$ gene in patients with diabetes and myopathy (Hao et al., 1995) and a C/A exchange in the $tRNA^{Ser}$ gene at np 12258 in association with diabetes, cerebellar ataxia, deafness and cataracts (Lynn et al., 1998).

How common is MDM? We have recently summarized reported prevalence rates from different countries (Gerbitz et al., 1996, 1995; van den Ouweland et al., 1995, 1994). The 3243 mutation accounts for 0.5 - 1.5% of unselected diabetic patients, independently of whether they are classified as type I or type II diabetes. In diabetic patients with familial history, the percentage increases up to 10%, in a population of diabetes and deafness even up to 60%. The prevalence seems not to be very different in various countries and populations (Gerbitz et al., 1995; Maassen and Kadowaki, 1996; Tsukuda et al., 1997). Taking into account a prevalence of diabetes of about 5% in the whole Western European or North American population, the frequency of the 3243 mutation is about 1: 1300.

Phenotypic Variability in MDM

The clinical phenotype of diabetic subjects harboring the 3243 mutation is extremely heterogeneous ranging from diabetes only, both diabetes and deafness, to more complex multisystemic manifestations (Gerbitz et al., 1995). Moreover, the penetrance of both diabetes and deafness is not complete. The penetrance of diabetes seems thereby not to correlate with the degree of heteroplasmy of the 3243 mutation in the peripheral lymphocytes (Velho et al., 1996). However, the degree of heteroplasmy might differ considerably between tissues and organs. Consistently, studies in autopsy pancreas specimens have demonstrated high proportions of mutated DNA in pancreatic islets while the mutations were undetectable in peripheral blood cells (Otabe et al., 1999). As the severity of a defect in mitochondrial OXPHOS in a specific tissue depends on the proportion of mutant mtDNA, differences in the degree of heteroplasmy between different cell types might explain the phenotypic diversity, even between members of the same family (Wallace, 1992). Other factors that could account for the phenotypic variability in MDM include differences in the mtDNA background or nuclear-encoded factors which might act synergistically with the mtDNA mutations (Jaksch et al., 1996).

Possible Pathomechanisms in MDM

Clinically, as well as pathogenetically, MDM is a distinct disease entity that can be clearly distinguished from the idiopathic forms of diabetes by several characteristic features (Gerbitz et al., 1995): MDM patients, compared to NIDDM patients, have an earlier onset of disease (<40 years), they are mostly lean, they are generally insulin-dependent, and are not insulin resistant. Compared to IDDM, MDM patients do not develop ketosis, and show no signs of anti-β-cell auto-antibodies. Measurements of the insulin secretory capacity of β-cells during glucose tolerance tests (Gerbitz et al., 1995), hyperglycemic clamp experiments and arginine infusion tests (Velho et al., 1996) revealed an abnormal, i.e. delayed and insufficient insulin and C-peptide response to a glucose-load. In contrast, a normal secretory response was obtained with arginine (Velho et al., 1996). The insulinotrophic action of arginine is thought to be mediated by the intracellular accumulation of positively charged arginine residues and subsequent depolarization of the plasma membrane. Thus, the insulin secretory defect in MDM might be the result from an impairment of glucose-sensing mechanisms rather than from an insulin production defect (Velho et al., 1996). A defect of glucose-regulated insulin secretion caused by the progressive reduction of oxidative phosphorylation has, therefore, been suggested as primary event in 3243 carriers.

Nevertheless, the pathomechanisms leading to MDM are controversally discussed. Several studies have suggested that the decreased insulin secretory capacity results from a selective and progressive β-cell loss rather than from a functional defect (Kobayashi et al., 1997; Otabe et al., 1999; Poulton et al., 1995). *In situ* characterization of islets from diabetic 3243 carriers yielded a marked decrease in the number of β-cells and α-cells, an extremely diminished COX activity (complex IV) and a compensatory elevated succinate dehydrogenase activity (complex II) (Kobayashi et al., 1997). The degree of heteroplasmy of the 3243 mutation was thereby highest in the β-cells compared to exocrine pancreatic tissue and peripheral polymorphonuclear cells. These findings have been recently confirmed by Otabe and co-workers (Otabe et al., 1999). In autopsy pancreas specimens from 3243 carriers diagnosed as having MELAS, they found severe reduction in total islet mass and in the number of both α- and β-cells. No evidences for insulitis or for apoptotic processes could be detected. The mechanism leading to reduction in the β-cell number remains to be determined. As proposed previously (Gerbitz, 1992), the increased production of reactive oxygen species (ROS) (see below) as a result of impaired oxidative phosphorylation might contribute to β-cell destruction in MDM.

Diabetes in Autosomally Inherited Mitochondrial Syndromes

Friedreich Ataxia

Friedreich ataxia (FRDA) is an autosomal recessive disorder with an estimated prevalence of 1:50000 and a deduced carrier frequency of 1:120 (Romeo et al., 1983) in European populations, making it the most common inherited ataxia. Clinically, FRDA is characterized by several neurological symptoms like unsteady gait, limp areflexia, muscle weakness and hypertrophic cardiomyopathy (Durr et al., 1996). Patients suffering from FRDA are highly susceptible to develop blindness, deafness and/or diabetes; glucose intolerance or diabetes mellitus is diagnosed in about 30% of cases. FRDA belongs to a class of neurodegenerative disorders caused by unstable trinucleotide-repeats. In FRDA, hyper-expansive GAA repeats in the first intron of the gene have been described to be causative for the disease (Campuzano et al., 1996). In healthy subjects, the number of GAA repeats is less than 39 repeats. If the GAA expansion exceeds a critical threshold length (>66), FRDA expression is inhibited at the level of gene transcription resulting in decreased frataxin protein levels (Bidichandani et al., 1997). The severity of the disease thereby correlates with the length of the expanded GAA repeats. In some FRDA cases, elongation of up to 1700 repeat units have been found. The vast majority of FRDA patients (about 95%) are homozygous for the GAA repeat expansion and the remaining 5% are compound heterozygous for the expansion and an additional point mutation in the FRDA gene.

The function of the frataxin protein is not entirely clear; the current knowledge is reviewed in Knight et al. (1999). By analogy to the yeast homologue Yfh1p, it has been shown that frataxin is a mitochondrial protein which is involved in mitochondrial iron metabolism. Yeast cells lacking the YFH1 gene exhibit a constitutive activity of the high affinity iron-uptake system, resulting in an about 10-fold elevated intramitochondrial iron levels and a breakdown of cellular iron homeostasis. Accumulation of iron is also observed in cardiac muscles from FRDA patients indicating similar functions of yeast yfh1p and human frataxin in the maintenance of mitochondrial iron homeostasis. Furthermore, yeast cells lacking the YFH gene exhibit hypersensitivity to oxidative stress. The observed loss of mtDNA integrity, i.e. depletion of mtDNA, probably results from increased formation of oxygen radicals due to an iron overload. Consistently, deficiencies of a number of iron-sulfur enzymes, such as aconitase, mitochondrial complex I, II and III have been found in heart muscles from FRDA patients (Rotig et al., 1997).

Mitochondria play a central role in cellular iron metabolism. As an essential cofactor of numerous enzymes, iron is required for the synthesis of heme and for the formation of cytosolic and mitochondrial iron-sulfur cluster containing enzymes like, for instance, the respiratory complexes I, II, and IV. It is therefore not surprising that blocking important steps in these pathways leads to disease. The discovery of further neurodegenerative diseases with abnormal mitochondrial iron balance seems likely in the future as mutations in two other yeast proteins, the heat shock protein Ssq1 and the cysteine desulphurase Nfs1p also result in massive intramitochondrial iron accumulation (personal communication, Kispal, G. 1999).

The high prevalence of diabetes in FRDA patients makes the frataxin gene an ideal candidate for a diabetes susceptibility gene. A recent report suggested that intermediate but significant GAA expansions in the FRDA gene are associated with NIDDM in Germany and the United States. The significance of this finding remains to be established since this association could not be confirmed by further studies in French, Danish (Dalgaard et al., 1999) and Dutch (t Hart et al., 1999) NIDDM populations.

Wolfram Syndrome

Wolfram syndrome is a rare autosomal-recessive inherited disorder characterized by early-onset diabetes mellitus and sensorineural deafness as essential features. Other less common neurological and endocrinological abnormalities include optic atrophy, diabetes insipidus, ataxia, hypogonadism, atonic bladder and a relatively high incidence of depression and psychotic behavior. As most of the clinical phenotypes are consistent with an ATP supply defect, Wolfram syndrome has proposed to be a mitochondrial disorder (Bezold et al., 1995; Bu and Rotter, 1993). Consistent with this hypothesis, Wolfram syndrome has been described to be sporadically caused by mtDNA deletions (Rotig et al., 1993). Similar to MDM, which is caused by mtDNA defects, a selective loss of islet β-cell has been found in postmortem specimens from patients with Wolfram syndrome (Karasik et al., 1989). Recently, we and others have identified the nuclear gene defect in Wolfram syndrome (Strom et al., 1998; Inoue et al. 1998). In the majority of cases diagnosed as having Wolfram syndrome we found loss-of-function mutations in a gene on chromosome 4p16, subsequently named *wolframin*. It encodes a putative transmembrane protein whose subcellular localization and function is not yet clear. A recent study described the formation of multiple mtDNA deletions in affected tissues as possible pathomechanism in Wolfram syndrome (Barrientos et al., 1996). As some of these patients were confirmed to carry also mutations in the nuclear-encoded wolframin gene, wolframin might be involved in the intergenomic signaling in such a way that a nuclear defect deleteriously interacts with the mitochondrial genome (Barrientos et al., 1996).

Diabetes has been described as a symptom in about 60 different genetic syndromes [see OMIM database]. Some of these diabetes syndromes show the characteristic clinical features typical for mitochondrial diseases and might, therefore, be caused by defects in mitochondrial proteins or in other factors influencing mitochondrial function. For example, *Rogers syndrome* is a rare autosomal-recessive inherited disorder characterized by diabetes, deafness and thiamine-responsive megaloblastic anemia, also named TRMA. The TRMA locus has been recently mapped to chromosome 1p23, however, the gene defect is not yet known[1]. Although the clinical phenotypes of TRMA resemble those observed in mitochondrial disorders, the TRMA gene might be involved in non-mitochondrial thiamine-metabolism (Stagg et al., 1999) thereby causing secondary abnormalities in thiamine-dependent mitochondrial enzyme complexes. Identification of underlying gene defects, and functional characterization of the respective gene products will give important insights into the pathomechanisms of neuro/endocrinological diseases.

Mouse Models of Mitochondrial Dysfunction Associated with Diabetes

Studies on patients have revealed some insight into the genetics of diabetes, but the pathophysiological mechanisms that underlie this complex disorder remain mysterious to a certain extent. New insights into these questions can now be obtained by the advanced technologies for gene manipulation in higher organisms, which allow creation and analysis of transgenic mouse models for diabetes associated diseases. In particular, the interrelationship of mitochondrial with other cellular functions involved in glucose regulation are now being studied in such models. The discussion of animal models is restricted to those that harbor mutations in genes known to contribute to mitochondrial function or those having obvious mitochondrial metabolic defects with so far unknown gene mutations.

Diabetic Animal Models with Affected Mitochondrial Metabolism

The targeted inactivation of the glycerophosphate shuttle by knock-out of the mitochondrial GPDH (mGPDH) was already discussed above.

A diabetic animal model which is also mainly defined by the observed alteration in the mtDNA content is the non-obese *Goto-Kakizaki (GK)* rat (Portha et al., 1991). This genetic animal model for NIDDM, is characterized by partial maternal inheritance. It could be demonstrated that in islets of the *GK* rats but not in liver, the content of mtDNA and mtRNA was greatly diminished compared to *Wistar* rats (Serradas et al., 1995). Whether this partial decrease is a result of an acquired mtDNA damage remains to be established. Besides the altered mtDNA content the *GK* rat exhibits only about 40% activity of mtGPDH and the glycerophosphate shuttle as well as low pyruvate carboxylase activity in β-cells. These findings seem to be a secondary phenomenon as they could be normalized by insulin treatment.

[1] Mutations in the thiamine transporter SLC1942 have been recently shown to underly TRMA (Labay et al., 1999; Fleming et al. 1999; Diaz et al. 1999).

Low mGPDH and pyruvate carboxylase activity was also demonstrated in islets isolated from *Zucker diabetic fatty rats* (MacDonald et al., 1996) and in the obese diabetic *db/db* mouse (Sener et al., 1993). However; these findings seem to be rather associated and not causative of diabetes in these animals (Brown et al., 1994; Sekine et al., 1994). In the *db/db* mice a fatty acid induced decrease in pyruvate dehydrogenase activity is reported to be an important feature of β-cell function (Zhou et al., 1996; Zhou and Grill, 1995b). Glucose-induced insulin release in *db/db* islets was markedly decreased compared with that from *db/+* controls. This effect in the *db/db* mice was however ameliorated by exposing the islets to etomoxir, an inhibitor of the carnitine palmityltransferase I (CPT I). When alpha-ketoisocaproate was used as a secretagogue, etomixir failed to impair insulin secretion in the *db/db* mice. While the metabolic defects in these animals have been known for a long time, the mutation in the autosomal recessive *diabetes* gene was first mapped to chromosome 4. The phenotype of *db/db* mice, which includes severe, early-onset obesity, extreme insulin resistance, and strain-specific susceptibility to diabetes, was shown to be identical to that of *ob/ob* mice in the same strain background (Coleman, 1978). The *fatty* gene (*fa*) in rats is thought to be a functional homologue of *db*, since *fa/fa* mutants exhibit an almost identical phenotype and both genes map to corresponding regions in rat and mouse. It could be shown that mouse *ob*, *db* and rat *fa* encode the same gene. It has been prosposed that *ob* and *db* encode the ligand and receptor, respectively, for a physiological pathway that regulates body weight. The cloning of *ob*, and the demonstration that it encodes a secretory protein, leptin, that binds specifically to a receptor (OB-R) in the brain, have validated critical aspects of this hypothesis (Chua et al., 1996).

The most prominent experimental animal model in diabetes research is the streptozotocin (STZ) rat. STZ is a potent diabetogenic agent with a high β-cell toxicity causing DNA strand breaks. Since mtDNA is more vulnerable to damage, gradual functional deterioration of the mitochondrial genome is assumed to increase with increasing age. In principle, STZ can lead to the formation of free radicals such as nitric oxide and hydroxyl radical ($OH^{\bullet}$) by alterations of the mtDNA thereby inhibiting mitochondrial OXPHOS. Free reactive oxygen species (ROS) are assumed to mainly contribute to STZ-specific β-cell destruction

Formation of Oxygen Radicals and their Effect on Pancreatic β-Cells

Mitochondrial OXPHOS is the major source of reactive oxygen radicals (ROS), which are toxic-by products of mitochondrial respiration. ROS are mainly produced along the mitochondrial respiratory chain upon electron transfer inhibition between the RC-complexes. Excess electrons accumulate within the early stage complexes (complex I and CoQ), from where they can be donated directly to molecular oxygen to generate the superoxide anion $O_2^{\bullet}$ (Wallace and Melov, 1998) (Figure 12). The excess of $O_2^{\bullet}$ is detoxified by reduction to hydrogen peroxide (H_2O_2) inside and outside the mitochondrium mediated by three different superoxide dismutases (SOD1, -2, -3). Two Cu^{2+}-, Zn^{2+}-dependent isoforms (SOD1 and -3) are located in the cytosol and the extracellular space, respectively, while the Mn^{2+}-dependent SOD2 is located in the mitochondrial matrix. The de-toxification product H_2O_2 is reduced to water by glutathione peroxidase (GPx)

(Figure 12), which is found in the cytosol and in mitochondria, and the peroxisomal enzyme catalase. In the presence of transition metals like Cu^{2+} and Fe^{2+}, H_2O_2 can be converted via the *Fenton* reaction to the highly reactive and toxic hydroxyl radical OH·. Free ROS are known to have acute and chronic effects on the integrity of mitochondrial and other cellular functions. While ROS chronically lead to damage of mtDNA and nuclear DNA, proteins and lipids, an acute increase of ROS may cause inactivation of the Krebs cycle and the respiratory chain complexes (see below) in all cells.

In β-cells, there is now compelling evidence that supports the role for ROS in cytokine-induced cell damage as one mechanism leading to autoimmune diabetes, and NIDDM. In 1985, Mandrup-Poulsen and Nerup could demonstrate for the first time that administration of interleukin-1 (IL-1) to rat islets causes inhibition of insulin secretion and leads to the islet destruction after prolonged exposure indicating that cytokines could directly influence the function of β-cells (Mandrup-Poulsen, 1996). The mechanism by which IL-1 causes β-cell destruction seems to result from expression of inducible nitric oxide synthase (iNOS) and the increased production of free radical nitric oxide (Welsh et al., 1995). Nitric oxide (*NO*) is generated from O_2, NADP and arginine by the action of *NO* synthases outside and inside the mitochondria (Bates et al., 1996). *NO* is also derived from several other donors. *In vivo* experiments indicated that IL-1 is sufficient to induce damage of islets in rats mediated by selective expression of iNOS. In NOD mice and human β-cells combined treatment of IL-1 and interferon- γ is required to yield the same effect.

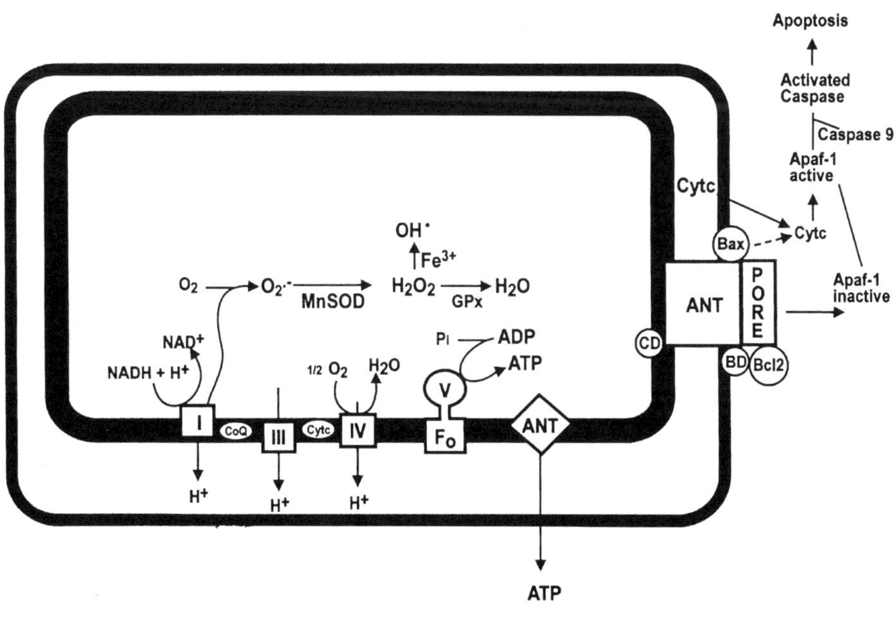

Figure 12. Generation of reactive oxygen species along the mitochondial respiratory chain and initial steps of apoptosis (Figure is adapted according to *Wallace et al.*, Science 1999).

Cytokine-induced β-cell destruction is mediated, at least in part, by *NO*-dependent inhibition of mitochondrial glucose oxidation and its coupling to the OXPHOS. In particular, *NO* leads to inhibition of the Krebs cycle enzyme *aconitase*, the impairment of electron transport in complex I and II of the respiratory chain and the formation of alkali-labile abasic sites mainly in the mtDNA. The alterations of mtDNA may be caused by both, *NO*-dependent and - independent mechanisms, while the IL-1 induced mitochondrial dysfunction seems strictly to be associated with nitric oxide induced inactivation of targeted enzymes. Observations in MnSOD-deficient mice indicate that the inactivation is probably due to the destruction of the *Fe-S* centers within the electron transferring proteins of complex I and II.

The involvement of free radical species in IL-1-dependent β-cell destruction has been controversial. *NO*, oxygen radicals $O_2^{\bullet}$ and/or peroxynitrite *ONOO*$^-$ a reaction product of *NO* and superoxide, have been implicated in both IL1- and IL-1 + IFN-γ-induced β-cell damage. However, sensitivity to one or the other nitrogen oxide species differs among mammalian species and is still under investigation. *NO* and *ONOO*$^-$ exert distinct effects by blocking COX activity (*NO*) (Cleeter et al., 1994) and inactivating complexes III (*ONNOO*$^-$). As the *NO* inhibition of COX is reversible and occurs at physiological concentrations (Cleeter et al., 1994) mitochondrial iNOS could act as an regulator of OXPHOS.

There is growing evidence that prolonged hyperglycemia can cause oxidative stress, especially in the β-cell where the expression of antioxidant enzymes is low (Lenzen et al., 1996; Tiedge et al., 1997). One of the mechanisms producing ROS might be the nonenzymatic glycation of proteins through the *Maillard* reaction and the resulting *Amadori* products (Brownlee et al., 1984; Njoroge and Monnier, 1989; Sakurai and Tsuchiya, 1988; Schleicher and Wieland, 1989, 1981). A number of publications have reported an increased level of 8-hydroxy-2'-deoxyguanosine (8-OHdG), a marker for oxidative stress (Kasai, 1997; Toyokuni et al., 1999) in urine and mononuclear blood cells from NIDDM patients (Dandona et al., 1996; Leinonen et al., 1997)

Most recently, the amounts of 8-OHdG and of 4-hydroxy-2-nonenal (HNE), an aldehyde formed through ROS-induced lipid peroxidation (Esterbauer et al., 1991; Uchida and Stadtman, 1993) was measured, for the first time, in β-cells of the *GK* rat (Ihara et al., 1999). Quantitative immunohistochemical analysis with specific antibodies revealed higher levels of both 8-OHdG and HNE-modified proteins in β-cell of *GK* rats compared to control Wistar rats. Levels of both ROS-markers could even be elevated by feeding the rats with a 30% sucrose solution, but were reduced by treatment with the alpha-glucosidase inhibitor *voglibose*. These findings suggest that glucose toxicity (Yki-Jarvinen, 1992) to the β-cell might result, at least in part, from hyperglycemia induced ROS attacks. Such a mechanism seems even more likely as islets from the *GK* rat utilize glucose at much higher rates than islets from control rats (Ling et al., 1998).

Oxidative Stress and Apoptosis as Possible Mechanisms in *β*-Cell Destruction

There seems to exist a mitochondrial balance between prooxidants and antioxidants to avoid damage of macromolecules within the mitochondria, especially alteration of the mtDNA. Consequently, it has been hypothesized that disturbance of this balance by increased substrate supply, for instance by prolonged hyperglycemia or inhibition of the respiratory chain would lead to an increased ROS production (Wallace, 1999). The imbalance would therefore, reduce the energy output from mitochondrial OXPHOS thereby presumably impairing multiple cellular processes and on the other hand lead to an increased oxidative stress as described above. As a consequence, increased oxidative stress and decreased energy levels might initiate endogenous mechanism leading to the activation of the mitochondrial permeability transition state as the initial step of apoptosis.

Observations in the recently established *Ant1*-deficient mouse seems to confirm the sequential interaction of mitochondrial oxidative phosphorylation, the generation of free ROS and the initiation of apoptosis as disease associated mechanism in the development of mitochondrial disorders in general and also in mitochondrial diabetes. Targeted inactivation of the nuclear-encoded *Antl* gene has provided a model for chronic ATP deficiency presenting with classical symptoms of a mitochondrial myopathy (Graham et al., 1997). While, ANT1 is expressed in skeletal muscle, heart and brain; a second translocator ANT2 is expressed in all tissues but the skeletal muscle. Thus, ANT1$^{-/-}$ mice are completely deficient in ANT in skeletal muscle, partially deficient in heart, and have normal ANT levels in liver. The ANT1 knock out mice demonstrated several interesting features: i) in the skeletal muscle an increased SDH and COX staining of the Type I oxidative muscle fibers was found in combination with a massive proliferation of giant mitochondria. ii) both nuclear and mitochondrial OXPHOS genes are upregulated in *Ant1*$^{-/-}$-mice. Moreover, the mRNA for Mcl-1, a muscle homolog of Bcl-2 was found to be upregulated (Murdock et al., 1999). This implies that ANT1 deficiency may affect not only mitochondrial OXPHOS, but also oxidative stress and possibly apoptosis. iii), as Ant1$^{-/-}$ mice show elevated serum levels of lactate, alanine, and succinate, it was concluded that consistent with inhibition of the respiratory chain and Krebs cycle as assumed to result from the inhibition of *aconitase*. Their skeletal muscle mitochondria are completely resistant and their heart mitochondria are partially resistant to the stimulation of oxygen consumption by exogenous ADP. This is consistent with the reductions in ANT levels and the associated inhibition of ADP/ATP exchange. Thus, inhibition of ADP supply to the mitochondrial proton-translocating ATP synthase (complex V) consequently seems to inhibit proton-pumping by the complexes I, III and IV and to block the electron flux along the respiratory chain. The electrons would then accumulate around complex I where they would be available to form toxic $O_2^{\bullet}$ (Esposito et al., 1999).

Indeed, increased H_2O_2 production is found mainly in the skeletal muscle, heart and brain. As a consequence the skeletal muscle tries to compensate for the increased ROS production by inducing SOD2 and glutathione peroxidase (Esposito et al., 1999). In muscles cells of the heart only an moderate increase in antioxidant enzymes was found. In line with these results, mtDNA damage and rearrangements did accumulate more rapidly in the heart than in skeletal muscle.

The role of increased mitochondrial ROS production in causing impaired cell function have been confirmed by studying mice carrying a knock-out mutation in the nuclear gene encoding the mitochondrial MnSOD (Lebovitz et al., 1996; Li et al., 1995). MnSOD-deficient mice die from dilated cardiomyopathy at about 8 days of age and they exhibit striking lipid deposits in the liver (Li et al., 1995; Melov et al., 1998). Complex II and complex III function were extremely reduced compared to wild-type animals in skeletal muscle and heart and of complex I activity in heart. Mitochondrial aconitase activity was deteriorated in heart and brain. Moreover, DNA from the heart and brain of the mice shows accumulation of oxidative DNA damage (Melov et al., 1999). These observations indicate that the high levels of mitochondrial $O_2^{\bullet}$ anion resulting from the MnSOD-deficiency can inhibit the respiratory chain complexes I and II and the Krebs cycle via aconitase. While all these effects were severely present in homozygous MnSOD$^{-/-}$ mice, the heterozygous mutant showed less severe phenotypes and could, therefore, provide a better model for the gradual decline of mitochondrial functionality involved in some neurodegenerative diseases and diabetes (Williams et al., 1998). The reduction of glutathione levels and activity in aconitase and complex I was found to be about 30%. However, the mice still exhibit features of oxidative damage to mitochondrial DNA and protein. ADP-stimulated respiration appears to be significantly reduced and mitochondria showed increased sensitivity to Ca^{2+} and oxidative stress (Williams et al., 1998). The observation confirmed that chronic exposure of β-cells to ROS impairs mitochondrial function and possibly potentiates apoptosis thereby leading to diabetes.

Acknowledgments

We like to thank K. Gempel for helpful discussion and P. Delly for excellent graphical artwork. This work was supported by grants from the Deutsche Forschungsgemeinschaft (Ba1438/2) (Ge213/7), the Thyssen-Stiftung (1990/2-45 and 1993/1-98 to K.-D.G.); the Wilhelm-Sander Stiftung (91.046.1 to K.-D.G. and 97.061.1 to M.F.B.), the Münchner Medizinische Wochenschrift, and the Friedrich-Baur Stiftung to M.F.B. and by the Bundesministerium für Bildung und Forschung (BMBF)(01KW9605/4) to S.H and K.-D.G.

References

Abrahams, J.P., Leslie, A.G., Lutter, R. & Walker, J.E. (1994). Structure at 2.8 A resolution of F1-ATPase from bovine heart mitochondria. Nature 370, 621-8.

Adams, V., Bosch, W., Hammerle, T. & Brdiczka, D. (1988). Activation of low Km hexokinases in purified hepatocytes by binding to mitochondria. Biochim. Biophys. Acta 932, 195-205.

Alcolado, J.C. & Alcolado, R. (1991). Importance of maternal history of non-insulin dependent diabetic patients. B.M.J. 302, 1178-1180.

Ames, B.N., Shigenaga, M.K. & Hagen, T.M. (1993). Oxidants, antioxidants, and the degenerative diseases of aging. Proc. Natl. Acad. Sci. USA 90, 7915-7922.

Anderson, S., Bankier, A.T., Barrell, B.G., de Bruijn, M.H., Coulson, A.R., Drouin, J., Eperon, I.C., Nierlich, D.P., Roe, B.A., Sanger, F., Schreier, P.H., Smith, A.J., Staden, R. & Young, I.G. (1981). Sequence and organization of the human mitochondrial genome. Nature 290, 457-465.

Ashcroft, F.M. (1988). Adenosine 5'-triphosphate-sensitive potassium channels. Annu Rev Neurosci 11, 97-118.

Ashcroft, F.M., Proks, P., Smith, P.A., Ammala, C., Bokvist, K. & Rorsman, P. (1994). Stimulus-secretion coupling in pancreatic beta cells. J. Cell Biochem. 55, 54-65.

Ashcroft, S.J., Weerasinghe, L.C., Bassett, J.M. & Randle, P.J. (1972). The pentose cycle and insulin release in mouse pancreatic islets. Biochem. J. 126, 525-532.

Aspinwall, C.A., Lakey, J.R. & Kennedy, R.T. (1999). Insulin-stimulated insulin secretion in single pancreatic beta cells. J. Biol. Chem. 274, 6360-6365.

Ballinger, S.W., Shoffner, J.M., Gebhart, S., Koontz, D.A. & Wallace, D.C. (1994). Mitochondrial diabetes revisited. Nat. Genet. 7, 458-459.

Ballinger, S.W., Shoffner, J.M., Hedaya, E.V., Trounce, I., Polak, M.A., Koontz, D.A. & Wallace, D.C. (1992). Maternally transmitted diabetes and deafness associated with a 10.4 kb mitochondrial DNA deletion. Nat, Genet, 1, 11-15.

Barrientos, A., Volpini, V., Casademont, J., Genis, D., Manzanares, J.M., Ferrer, I., Corral, J., Cardellach, F., Urbano-Marquez, A., Estivill, X. & Nunes, V. (1996). A nuclear defect in the 4p16 region predisposes to multiple mitochondrial DNA deletions in families with Wolfram syndrome. J. Clin. Invest. 97, 1570-1576.

Bates, T.E., Loesch, A., Burnstock, G. & Clark, J.B. (1996). Mitochondrial nitric oxide synthase: a ubiquitous regulator of oxidative phosphorylation? Biochem. Biophys. Res. Commun. 218, 40-44.

Bauer, M.F., Gempel, K., Reichert, A.S., Rappold, G.A., Lichtner, P., Gerbitz, K.D., Neupert, W., Brunner, M. & Hofmann, S. (1999). Genetic and Structural Characterization of the Human Mitochondrial Inner Membrane Translocase. J. Mol. Biol. 289, 69-82.

Bauer, M.F., Hofmann, S. & Brunner, M. (2000). The Tim10/DDP protein family. Trends Cell Biology 10.

Bauer, M.F., Sirrenberg, C., Neupert, W. & Brunner, M. (1996). Role of Tim23 as voltage sensor and presequence receptor in protein import into mitochondria. Cell 87, 33-41.

Berne, C. (1975). The metabolism of lipids in mouse pancreatic islets. The oxidation of fatty acids and ketone bodies. Biochem. J. 152, 661-666.

Bezold, R., Hofmann, S., Jaksch, M., Kaufhold, P. & Gerbitz, K.D. (1995). DIDMOAD or Wolfram syndrome. A mitochondrial-mediated disorder. Diabetes Care 18, 583-584.

Bidichandani, S.I., Ashizawa, T. & Patel, P.I. (1997). Atypical Friedreich ataxia caused by compound heterozygosity for a novel missense mutation and the GAA triplet-repeat expansion. Am. J. Hum. Genet. 60, 1251-1256.

Branstrom, R., Corkey, B.E., Berggren, P.O. & Larsson, O. (1997). Evidence for a unique long chain acyl-CoA ester binding site on the ATP- regulated potassium channel in mouse pancreatic beta cells. J. Biol. Chem. 272, 17390-17394.

Branstrom, R., Leibiger, I.B., Leibiger, B., Corkey, B.E., Berggren, P.O. & Larsson, O. (1998). Long chain coenzyme A esters activate the pore-forming subunit (Kir6. 2) of the ATP-regulated potassium channel. J. Biol. Chem. 273, 31395-31400.

Brdiczka, D. (1994). Function of the outer mitochondrial compartment in regulation of energy metabolism. Biochim. Biophys. Acta 1187, 264-269.

Brdiczka, D. & Wallimann, T. (1994). The importance of the outer mitochondrial compartment in regulation of energy metabolism. Mol. Cell Biochem. 133-134, 69-83.

Brdiczka, D., Wicker, U. & Gellerich, F. (1994). The function of the outer membrane pore in the regulation of peripheral kinases and energy metabolism, M. Forte and M. Colombini, eds. (Berlin, Heidelberg: Sprinegr Verlag).

Brown, L.J., MacDonald, M.J., Lehn, D.A. & Moran, S.M. (1994). Sequence of rat mitochondrial glycerol-3-phosphate dehydrogenase cDNA. Evidence for EF-hand calcium-binding domains. J. Biol. Chem.

269, 14363-14366.

Brown, M.D., Torroni, A., Shoffner, J.M. & Wallace, D.C. (1992). Mitochondrial tRNA(Thr) mutations and lethal infantile mitochondrial myopathy. Am. J. Hum. Genet. 51, 446-447.

Brownlee, M., Vlassara, H. & Cerami, A. (1984). Nonenzymatic glycosylation and the pathogenesis of diabetic complications. Ann. Intern. Med. 101, 527-537.

Brun, T., Roche, E., Assimacopoulos-Jeannet, F., Corkey, B.E., Kim, K.H. & Prentki, M. (1996). Evidence for an anaplerotic/malonyl-CoA pathway in pancreatic beta-cell nutrient signaling. Diabetes 45, 190-198.

Bu, X. & Rotter, J.I. (1993). Wolfram syndrome: a mitochondrial-mediated disorder? Lancet 342, 598-600.

Bucheler, K., Adams, V. & Brdiczka, D. (1991). Localization of the ATP/ADP translocator in the inner membrane and regulation of contact sites between mitochondrial envelope membranes by ADP. A study on freeze-fractured isolated liver mitochondria. Biochim. Biophys. Acta 1056, 233-242.

Byrne, M.M., Sturis, J., Clement, K., Vionnet, N., Pueyo, M.E., Stoffel, M., Takeda, J., Passa, P., Cohen, D., Bell, G.I. & et al. (1994). Insulin secretory abnormalities in subjects with hyperglycemia due to glucokinase mutations. J. Clin. Invest. 93, 1120-1130.

Campuzano, V., Montermini, L., Molto, M.D., Pianese, L., Cossee, M., Cavalcanti, F., Monros, E., Rodius, F., Duclos, F., Monticelli, A. & et al. (1996). Friedreich's ataxia: autosomal recessive disease caused by an intronic GAA triplet repeat expansion. Science 271, 1423-1427.

Chua, S.C., Jr., Chung, W.K., Wu-Peng, X.S., Zhang, Y., Liu, S.M., Tartaglia, L. & Leibel, R.L. (1996). Phenotypes of mouse diabetes and rat fatty due to mutations in the OB (leptin) receptor. Science 271, 994-996.

Cleeter, M.W., Cooper, J.M., Darley-Usmar, V.M., Moncada, S. & Schapira, A.H. (1994). Reversible inhibition of cytochrome c oxidase, the terminal enzyme of the mitochondrial respiratory chain, by nitric oxide. Implications for neurodegenerative diseases. FEBS Lett. 345, 50-54.

Coleman, D.L. (1978). Obese and diabetes: two mutant genes causing diabetes-obesity syndromes in mice. Diabetologia 14, 141-148.

Cook, D.L. & Hales, C.N. (1984). Intracellular ATP directly blocks K+ channels in pancreatic B-cells. Nature 311, 271-273.

Corkey, B.E., Glennon, M.C., Chen, K.S., Deeney, J.T., Matschinsky, F.M. & Prentki, M. (1989). A role for malonyl-CoA in glucose-stimulated insulin secretion from clonal pancreatic beta-cells. J. Biol. Chem. 264, 21608-21612.

Cox, N.J. (1994). Maternal component in NIDDM transmission. How large an effect? Diabetes 43, 166-168.

Dalgaard, L.T., Hansen, T., Urhammer, S.A., Clausen, J.O., Eiberg, H. & Pedersen, O. (1999). Intermediate expansions of a GAA repeat in the frataxin gene are not associated with type 2 diabetes or altered glucose-induced beta-cell function in Danish Caucasians. Diabetes 48, 914-917.

Dandona, P., Thusu, K., Cook, S., Snyder, B., Makowski, J., Armstrong, D. & Nicotera, T. (1996). Oxidative damage to DNA in diabetes mellitus. Lancet 347, 444-445.

Dillon, P.F. & Clark, J.F. (1990). The theory of diazymes and functional coupling of pyruvate kinase and creatine kinase. J. Theor. Biol. 143, 275-84.

Dunbar, D.R., Moonie, P.A., Swingler, R.J., Davidson, D., Roberts, R. & Holt, I.J. (1993). Maternally transmitted partial direct tandem duplication of mitochondrial DNA associated with diabetes mellitus. Hum. Mol. Genet. 2, 1619-1624.

Durr, A., Cossee, M., Agid, Y., Campuzano, V., Mignard, C., Penet, C., Mandel, J.L., Brice, A. & Koenig, M. (1996). Clinical and genetic abnormalities in patients with Friedreich's ataxia. N. Engl. J. Med. 335, 1169-1175.

Eaton, S., Bartlett, K. & Pourfarzam, M. (1996). Mammalian mitochondrial beta-oxidation. Biochem. J. 320, 345-357.

Enter, C., Muller-Hocker, J., Zierz, S., Kurlemann, G., Pongratz, D., Forster, C., Obermaier-Kusser, B. & Gerbitz, K.D. (1991). A specific point mutation in the mitochondrial genome of Caucasians with MELAS. Hum. Genet. 88, 233-236.

Esposito, L.A., Melov, S., Panov, A., Cottrell, B.A. & Wallace, D.C. (1999). Mitochondrial disease in mouse results in increased oxidative stress. Proc. Natl. Acad. Sci. USA 96, 4820-4825.

Esterbauer, H., Schaur, R.J. & Zollner, H. (1991). Chemistry and biochemistry of 4-hydroxynonenal, malonaldehyde and related aldehydes. Free Radic. Biol. Med. 11, 81-128.

Eto, K., Tsubamoto, Y., Terauchi, Y., Sugiyama, T., Kishimoto, T., Takahashi, N., Yamauchi, N., Kubota, N., Murayama, S., Aizawa, T., Akanuma, Y., Aizawa, S., Kasai, H., Yazaki, Y. & Kadowaki, T. (1999). Role of NADH shuttle system in glucose-induced activation of mitochondrial metabolism and insulin secretion. Science 283, 981-985.

Fiek, C., Benz, R., Roos, N. & Brdiczka, D. (1982). Evidence for identity between the hexokinase-binding protein and the mitochondrial porin in the outer membrane of rat liver mitochondria. Biochim. Biophys. Acta 688, 429-440.

Froguel, P., Zouali, H., Vionnet, N., Velho, G., Vaxillaire, M., Sun, F., Lesage, S., Stoffel, M., Takeda, J., Passa, P. & et al. (1993). Familial hyperglycemia due to mutations in glucokinase. Definition of a subtype of diabetes mellitus. N. Engl. J. Med. 328, 697-702.

Gellerich, F.N., Wagner, M., Kapischke, M., Wicker, U. & Brdiczka, D. (1993). Effect of macromolecules on the regulation of the mitochondrial outer membrane pore and the activity of adenylate kinase in the inter- membrane space. Biochim. Biophys. Acta 1142, 217-227.

Gembal, M., Detimary, P., Gilon, P., Gao, Z.Y. & Henquin, J.C. (1993). Mechanisms by which glucose can control insulin release independently from its action on adenosine triphosphate-sensitive K+ channels in mouse B cells. J. Clin. Invest. 91, 871-880.

Gerbitz, K.D. (1992). Does the mitochondrial DNA play a role in the pathogenesis of diabetes? Diabetologia 35, 1181-1186.

Gerbitz, K.D., Gempel, K. & Brdiczka, D. (1996). Mitochondria and diabetes. Genetic, biochemical, and clinical implications of the cellular energy circuit. Diabetes 45, 113-126.

Gerbitz, K.D., Obermaier-Kusser, B., Lestienne, P., Zierz, S., Muller-Hocker, J., Pongratz, D., Paetzke-Brunner, I. & Deufel, T. (1990). Mutations of the mitochondrial DNA: the contribution of DNA techniques to the diagnosis of mitochondrial encephalomyopathies. J. Clin. Chem. Clin. Biochem. 28, 241-250.

Gerbitz, K.D., Obermaier-Kusser, B., Zierz, S., Pongratz, D., Muller-Hocker, J. & Lestienne, P. (1990). Mitochondrial myopathies: divergences of genetic deletions, biochemical defects and the clinical syndromes. J. Neurol. 237, 5-10.

Gerbitz, K.D., van den Ouweland, J.M., Maassen, J.A. & Jaksch, M. (1995). Mitochondrial diabetes mellitus: a review. Biochim. Biophys. Acta 1271, 253-260.

Giles, R.E., Blanc, H., Cann, H.M. & Wallace, D.C. (1980). Maternal inheritance of human mitochondrial DNA. Proc. Natl. Acad. Sci. USA 77, 6715-6719.

Goto, Y., Nonaka, I. & Horai, S. (1990). A mutation in the tRNA(Leu)(UUR) gene associated with the MELAS subgroup of mitochondrial encephalomyopathies. Nature 348, 651-653.

Gots, R.E., Gorin, F.A. & Bessman, S.P. (1972). Kinetic enhancement of bound hexokinase activity by mitochondrial respiration. Biochem. Biophys. Res. Commun. 49, 1249-1255.

Graham, B.H., Waymire, K.G., Cottrell, B., Trounce, I.A., MacGregor, G.R. & Wallace, D.C. (1997). A mouse model for mitochondrial myopathy and cardiomyopathy resulting from a deficiency in the heart/muscle isoform of the adenine nucleotide translocator. Nat. Genet. 16, 226-234.

Green, D.R. & Reed, J.C. (1998). Mitochondria and apoptosis. Science 281, 1309-1312.

Hao, H., Bonilla, E., Manfredi, G., DiMauro, S. & Moraes, C.T. (1995). Segregation patterns of a novel mutation in the mitochondrial tRNA glutamic acid gene associated with myopathy and diabetes mellitus. Am. J. Hum. Genet. 56, 1017-1025.

Hayakawa, T., Noda, M., Yasuda, K., Yorifuji, H., Taniguchi, S., Miwa, I., Sakura, H., Terauchi, Y., Hayashi, J., Sharp, G.W., Kanazawa, Y., Akanuma, Y., Yazaki, Y. & Kadowaki, T. (1998). Ethidium bromide-induced inhibition of mitochondrial gene transcription suppresses glucose-stimulated insulin release in the mouse pancreatic beta-cell line betaHC9. J. Biol. Chem. 273, 20300-20307.

Hirata, K., Nakagawa, M., Higuchi, I., Hashimoto, K., Hanada, K., Takahashi, K., Niiyama, T., Izumi, K., Sakoda, S., Yamada, H. & Osame, M. (1999). Adult onset limb-girdle type mitochondrial myopathy with a mitochondrial DNA np8291 A-to-G substitution. J. Hum. Genet. 44, 210-214.

Hofmann, S., Bezold, R., Jaksch, M., Kaufhold, P., Obermaier-Kusser, B. & Gerbitz, K.D. (1997a). Disease relevance of the so-called secondary Leber hereditary optic neuropathy mutations. Am. J. Hum. Genet. 60, 1539-1542.

Hofmann, S., Bezold, R., Jaksch, M., Obermaier-Kusser, B., Mertens, S., Kaufhold, P., Rabl, W., Hecker, W. & Gerbitz, K.D. (1997b). Wolfram (DIDMOAD) syndrome and Leber hereditary optic neuropathy (LHON) are associated with distinct mitochondrial DNA haplotypes. Genomics 39, 8-18.

Hofmann, S., Jaksch, M., Bezold, R., Mertens, S., Aholt, S., Paprotta, A. & Gerbitz, K.D. (1997c). Population genetics and disease susceptibility: characterization of central European haplogroups by mtDNA gene mutations, correlation with D loop variants and association with disease. Hum. Mol. Genet. 6, 1835-1846.

Holt, I.J., Cooper, J.M., Morgan-Hughes, J.A. & Harding, A.E. (1988). Deletions of muscle mitochondrial DNA. Lancet 1, 1462.

Hopkins, W.F., Fatherazi, S., Peter-Riesch, B., Corkey, B.E. & Cook, D.L. (1992). Two sites for adenine-nucleotide regulation of ATP-sensitive potassium channels in mouse pancreatic beta-cells and HIT cells. J. Membr. Biol. 129, 287-295.

Huizing, M., Iacobazzi, V., Ijlst, L., Savelkoul, P., Ruitenbeek, W., van den Heuvel, L., Indiveri, C., Smeitink, J., Trijbels, F., Wanders, R. & Palmieri, F. (1997). Cloning of the human carnitine-acylcarnitine carrier cDNA and identification of the molecular defect in a patient. Am. J. Hum. Genet. 61, 1239-1245.

Hutton, J.C., Sener, A., Herchuelz, A., Atwater, I., Kawazu, S., Boschero, A.C., Somers, G., Devis, G. & Malaisse, W.J. (1980). Similarities in the stimulus-secretion coupling mechanisms of glucose- and 2-keto acid-induced insulin release. Endocrinology 106, 203-219.

Ihara, Y., Toyokuni, S., Uchida, K., Odaka, H., Tanaka, T., Ikeda, H., Hiai, H., Seino, Y. & Yamada, Y. (1999). Hyperglycemia causes oxidative stress in pancreatic beta-cells of GK rats, a model of type 2 diabetes. Diabetes 48, 927-932.

Ijlst, L., Wanders, R.J., Ushikubo, S., Kamijo, T. & Hashimoto, T. (1994). Molecular basis of long-chain 3-hydroxyacyl-CoA dehydrogenase deficiency: identification of the major disease-causing mutation in the alpha-subunit of the mitochondrial trifunctional protein. Biochim. Biophys. Acta 1215, 347-350.

Indiveri, C., Iacobazzi, V., Giangregorio, N. & Palmieri, F. (1997). The mitochondrial carnitine carrier protein: cDNA cloning, primary structure and comparison with other mitochondrial transport proteins. Biochem. J. 321, 713-719.

Indiveri, C., Tonazzi, A., Stipani, I. & Palmieri, F. (1997). The purified and reconstituted ornithine/citrulline carrier from rat liver mitochondria: electrical nature and coupling of the exchange reaction with H+ translocation. Biochem. J. 327, 349-355.

Inoue, H., Tanizawa, Y., Wasson, J., Behn, P., Kalidas, K., Bernal-Mizrachi, E., Mueckler, M., Marshall, H., Donis-Keller, H., Crock, P., Rogers, D., Mikuni, M., Kumashiro, H., Higashi, K., Sobue, G., Oka, Y., Permutt, M.A. (1998). A gene encoding a transmembrane protein is mutated in patients with diabetes mellitus and optic atrophy (Wolfram syndrome). Nat Genet 20, 143-148

Iynedjian, P.B., Pilot, P.R., Nouspikel, T., Milburn, J.L., Quaade, C., Hughes, S., Ucla, C. & Newgard, C.B. (1989). Differential expression and regulation of the glucokinase gene in liver and islets of Langerhans. Proc. Natl. Acad. Sci. USA 86, 7838-7842.

Jackson, S., Schaefer, J., Middleton, B. & Turnbull, D.M. (1995). Characterisation of a novel enzyme of human fatty acid beta-oxidation: a matrix-associated, mitochondrial 2-enoyl-CoA hydratase. Biochem. Biophys. Res. Commun. 214, 247-253.

Jaksch, M., Hofmann, S., Kaufhold, P., Obermaier-Kusser, B., Zierz, S. & Gerbitz, K.D. (1996). A novel combination of mitochondrial tRNA and ND1 gene mutations in a syndrome with MELAS, cardiomyopathy & diabetes mellitus. Hum. Mutat. 7, 358-360.

Johnson, J.H., Newgard, C.B., Milburn, J.L., Lodish, H.F. & Thorens, B. (1990a). The high Km glucose transporter of islets of Langerhans is functionally similar to the low affinity transporter of liver and has an identical primary sequence. J. Biol. Chem. 265, 6548-6551.

Johnson, J.H., Ogawa, A., Chen, L., Orci, L., Newgard, C.B., Alam, T. & Unger, R.H. (1990b). Underexpression of beta cell high Km glucose transporters in noninsulin- dependent diabetes [published erratum appears in Science 1990 Nov 30;250(4985):1195]. Science 250, 546-549.

Kabir, F. & Nelson, B.D. (1991). Hexokinase bound to rat brain mitochondria uses externally added ATP more efficiently than internally generated ATP. Biochim. Biophys. Acta 1057, 147-150.

Kameoka, K., Isotani, H., Tanaka, K., Azukari, K., Fujimura, Y., Shiota, Y., Sasaki, E., Majima, M., Furukawa, K., Haginomori, S., Kitaoka, H. & Ohsawa, N. (1998). Novel mitochondrial DNA mutation in tRNA(Lys) (8296A-->G) associated with diabetes. Biochem. Biophys. Res. Commun. 245, 523-527.

Karasik, A., O'Hara, C., Srikanta, S., Swift, M., Soeldner, J.S., Kahn, C.R. & Herskowitz, R.D. (1989). Genetically programmed selective islet beta-cell loss in diabetic subjects with Wolfram's syndrome. Diabetes Care 12, 135-138.

Kasai, H. (1997). Analysis of a form of oxidative DNA damage, 8-hydroxy-2'- deoxyguanosine, as a marker of cellular oxidative stress during carcinogenesis. Mutat. Res. 387, 147-163.

Kennedy, E.D., Maechler, P. & Wollheim, C.B. (1998). Effects of depletion of mitochondrial DNA in metabolism secretion coupling in INS-1 cells. Diabetes 47, 374-380.

Kerscher, O., Holder, J., Srinivasan, M., Leung, R.S. & Jensen, R.E. (1997). The Tim54p-Tim22p complex mediates insertion of proteins into the mitochondrial inner membrane. J Cell Biol. 139, 1663-1675.

Knight, S.A., Kim, R., Pain, D. & Dancis, A. (1999). The yeast connection to Friedreich ataxia. Am. J. Hum. Genet. 64, 365-371.

Knoll, G. & Brdiczka, D. (1983). Changes in freeze-fractured mitochondrial membranes correlated to their energetic state. Dynamic interactions of the boundary membranes. Biochim. Biophys. Acta 733, 102-110.

Kobayashi, T., Nakanishi, K., Nakase, H., Kajio, H., Okubo, M., Murase, T. & Kosaka, K. (1997). In situ characterization of islets in diabetes with a mitochondrial DNA mutation at nucleotide position 3243. Diabetes 46, 1567-1571.

Koehler, C.M., Leuenberger, D., Merchant, S., Renold, A., Junne, T. & Schatz, G. (1999). Human deafness dystonia syndrome is a mitochondrial disease. Proc. Natl. Acad. Sci. USA 96, 2141-2146.

Koehler, C.M., Merchant, S., Oppliger, W., Schmid, K., Jarosch, E., Dolfini, L., Junne, T., Schatz, G. & Tokatlidis, K. (1998). Tim9p, an essential partner subunit of Tim10p for the import of mitochondrial carrier proteins. Embo. J. 17, 6477-6486.

Koranyi, L., Bourey, R., Turk, J., Mueckler, M. & Permutt, M.A. (1992). Differential expression of rat pancreatic islet beta-cell glucose transporter (GLUT 2), proinsulin and islet amyloid polypeptide genes after prolonged fasting, insulin-induced hypoglycaemia and dexamethasone treatment. Diabetologia 35, 1125-1132.

Korge, P. & Campbell, K.B. (1994). Local ATP regeneration is important for sarcoplasmic reticulum Ca2+ pump function. Am. J. Physiol. 267, C357-366.

Kottke, M., Adam, V., Riesinger, I., Bremm, G., Bosch, W., Brdiczka, D., Sandri, G. & Panfili, E. (1988). Mitochondrial boundary membrane contact sites in brain: points of hexokinase and creatine kinase location, and control of Ca2+ transport. Biochim. Biophys. Acta 935, 87-102.

Kottke, M., Wallimann, T. & Brdiczka, D. (1994). Dual electron microscopic localization of mitochondrial creatine kinase in brain mitochondria. Biochem. Med. Metab. Biol. 51, 105-117.

Kulkarni, R.N., Bruning, J.C., Winnay, J.N., Postic, C., Magnuson, M.A. & Kahn, C.R. (1999). Tissue-specific knockout of the insulin receptor in pancreatic beta cells creates an insulin secretory defect similar to that in type 2 diabetes. Cell 96, 329-339.

Lebovitz, R.M., Zhang, H., Vogel, H., Cartwright, J., Jr., Dionne, L., Lu, N., Huang, S. & Matzuk, M.M. (1996). Neurodegeneration, myocardial injury, and perinatal death in mitochondrial superoxide dismutase-deficient mice. Proc. Natl. Acad. Sci. USA 93, 9782-9787.

Leibiger, I.B., Leibiger, B., Moede, T. & Berggren, P.O. (1998). Exocytosis of insulin promotes insulin gene transcription via the insulin receptor/PI-3 kinase/p70 s6 kinase and CaM kinase pathways. Mol. Cell 1, 933-938.

Leinonen, J., Lehtimaki, T., Toyokuni, S., Okada, K., Tanaka, T., Hiai, H., Ochi, H., Laippala, P., Rantalaiho, V., Wirta, O., Pasternack, A. & Alho, H. (1997). New biomarker evidence of oxidative DNA damage in patients with non- insulin-dependent diabetes mellitus. FEBS Lett 417, 150-152.

Lenzen, S., Drinkgern, J. & Tiedge, M. (1996). Low antioxidant enzyme gene expression in pancreatic islets compared with various other mouse tissues. Free Radic. Biol. Med. 20, 463-466.

Leslie, A.G., Abrahams, J.P., Braig, K., Lutter, R., Menz, R.I., Orriss, G.L., van Raaij, M.J. & Walker, J.E. (1999). The structure of bovine mitochondrial F1-ATPase: an example of rotary catalysis. Biochem. Soc. Trans. 27, 37-42.

Lestienne, P. & Ponsot, G. (1988). Kearns-Sayre syndrome with muscle mitochondrial DNA deletion. Lancet 1, 885.

Li, Y., Huang, T.T., Carlson, E.J., Melov, S., Ursell, P.C., Olson, J.L., Noble, L.J., Yoshimura, M.P., Berger, C., Chan, P.H. & et al. (1995). Dilated cardiomyopathy and neonatal lethality in mutant mice lacking manganese superoxide dismutase. Nat. Genet. 11, 376-381.

Lightowlers, R.N., Jacobs, H.T. & Kajander, O.A. (1999). Mitochondrial DNA--all things bad? Trends Genet. 15, 91-93.

Linden, M., Gellerfors, P. & Nelson, B.D. (1982). Pore protein and the hexokinase-binding protein from the outer membrane of rat liver mitochondria are identical. FEBS Lett 141, 189-192.

Linden, M., Gellerfors, P. & Nelson, B.D. (1982). Purification of a protein having pore forming activity from the rat liver mitochondrial outer membrane. Biochem. J. 208, 77-82.

Ling, Z.C., Efendic, S., Wibom, R., Abdel-Halim, S.M., Ostenson, C.G., Landau, B.R. & Khan, A. (1998). Glucose metabolism in Goto-Kakizaki rat islets. Endocrinology 139, 2670-2675.

Lynn, S., Wardell, T., Johnson, M.A., Chinnery, P.F., Daly, M.E., Walker, M. & Turnbull, D.M. (1998). Mitochondrial diabetes: investigation and identification of a novel mutation. Diabetes 47, 1800-1802.

Maassen, J.A. & Kadowaki, T. (1996). Maternally inherited diabetes and deafness: a new diabetes subtype. Diabetologia 39, 375-382.

MacDonald, M.J. (1990). Elusive proximal signals of beta-cells for insulin secretion. Diabetes 39, 1461-1466.

MacDonald, M.J. (1982). Evidence for the malate aspartate shuttle in pancreatic islets. Arch. Biochem. Biophys. 213, 643-649.

MacDonald, M.J. (1995). Feasibility of a mitochondrial pyruvate malate shuttle in pancreatic islets. Further implication of cytosolic NADPH in insulin secretion [published erratum appears in J. Biol. Chem. 1996 Mar 1;271(9):5287]. J. Biol. Chem. 270, 20051-20058.

MacDonald, M.J. (1981). High content of mitochondrial glycerol-3-phosphate dehydrogenase in pancreatic islets and its inhibition by diazoxide. J. Biol. Chem. 256, 8287-8290.

MacDonald, M.J. (1995). Influence of glucose on pyruvate carboxylase expression in pancreatic islets. Arch Biochem. Biophys. 319, 128-132.

MacDonald, M.J. & Fahien, L.A. (1988). Glyceraldehyde phosphate and methyl esters of succinic acid. Two "new" potent insulin secretagogues. Diabetes 37, 997-999.

MacDonald, M.J., Tang, J. & Polonsky, K.S. (1996). Low mitochondrial glycerol phosphate dehydrogenase and pyruvate carboxylase in pancreatic islets of Zucker diabetic fatty rats. Diabetes 45, 1626-1630.

Magnuson, M.A., K.D., N. & C.C., P. (1994). Glucokinase gene expression and regulation, Drasnin B and L.D., eds. (New York: Humana Press).

Magnuson, M.A. & Shelton, K.D. (1989). An alternate promoter in the glucokinase gene is active in the pancreatic beta cell. J. Biol. Chem. 264, 15936-15942.

Malaisse, W.J., Malaisse-Lagae, F. & Sener, A. (1982). The stimulus-secretion coupling of glucose-induced insulin release: effect of aminooxyacetate upon nutrient-stimulated insulin secretion. Endocrinology 111, 392-397.

Malaisse, W.J., Sener, A., Koser, M., Ravazzola, M. & Malaisse-Lagae, F. (1977). The stimulus-secretion coupling of glucose-induced insulin release. Insulin release due to glycogenolysis in glucose-deprived islets. Biochem. J. 164, 447-454.

Malaisse-Lagae, F. & Malaisse, W.J. (1988). Hexose metabolism in pancreatic islets: regulation of mitochondrial hexokinase binding. Biochem. Med. Metab. Biol. 39, 80-89.

Mandrup-Poulsen, T. (1996). The role of interleukin-1 in the pathogenesis of IDDM. Diabetologia 39, 1005-1029.

Mason, T. M., Goh, T., Tchipashvili, V., Sandhu, H., Gupta, N., Lewis, G.F. & Giacca, A. (1999). Prolonged elevation of plasma free fatty acids desensitizes the insulin secretory response to glucose in vivo in rats. Diabetes 48, 524-530.

Matschinsky, F.M. (1996). Banting Lecture 1995. A lesson in metabolic regulation inspired by the glucokinase glucose sensor paradigm. Diabetes 45, 223-241.

McCormack, J.G. & Denton, R.M. (1994). Mammalian mitochondrial metabolism and its regulation mitochondria, V. Darley Usmar and A.H.V. Shapira, eds. (Chapel Hill: Portland).

McGarry, J.D. & Foster, D.W. (1979). In support of the roles of malonyl-CoA and carnitine acyltransferase I in the regulation of hepatic fatty acid oxidation and ketogenesis. J. Biol. Chem. 254, 8163-8168.

Meglasson, M.D. & Matschinsky, F.M. (1986). Pancreatic islet glucose metabolism and regulation of insulin secretion. Diabetes Metab. Rev 2, 163-214.

Melov, S., Coskun, P., Patel, M., Tuinstra, R., Cottrell, B., Jun, A.S., Zastawny, T.H., Dizdaroglu, M., Goodman, S.I., Huang, T.T., Miziorko, H., Epstein, C.J. & Wallace, D.C. (1999). Mitochondrial disease in superoxide dismutase 2 mutant mice. Proc. Natl. Acad. Sci. USA 96, 846-851.

Melov, S., Schneider, J.A., Day, B.J., Hinerfeld, D., Coskun, P., Mirra, S.S., Crapo, J.D. & Wallace, D.C. (1998). A novel neurological phenotype in mice lacking mitochondrial manganese superoxide dismutase. Nat. Genet. 18, 159-163.

Miki, T., Nagashima, K., Tashiro, F., Kotake, K., Yoshitomi, H., Tamamoto, A., Gonoi, T., Iwanaga, T., Miyazaki, J. & Seino, S. (1998). Defective insulin secretion and enhanced insulin action in KATP channel- deficient mice. Proc. Natl. Acad. Sci. USA 95, 10402-10406.

Miki, T., Tashiro, F., Iwanaga, T., Nagashima, K., Yoshitomi, H., Aihara, H., Nitta, Y., Gonoi, T., Inagaki, N., Miyazaki, J. & Seino, S. (1997). Abnormalities of pancreatic islets by targeted expression of a dominant-negative KATP channel. Proc. Natl. Acad. Sci. USA 94, 11969-11973.

Moraes, C.T., Ciacci, F., Bonilla, E., Jansen, C., Hirano, M., Rao, N., Lovelace, R.E., Rowland, L.P., Schon, E.A. & DiMauro, S. (1993). Two novel pathogenic mitochondrial DNA mutations affecting organelle number and protein synthesis. Is the tRNA(Leu(UUR)) gene an etiologic hot spot? J. Clin. Invest. 92, 2906-2915.

Moraes, C.T., DiMauro, S., Zeviani, M., Lombes, A., Shanske, S., Miranda, A.F., Nakase, H., Bonilla, E., Werneck, L.C., Servidei, S. & et al. (1989). Mitochondrial DNA deletions in progressive external ophthalmoplegia and Kearns-Sayre syndrome. N. Engl. J. Med. 320, 1293-1299.

Morten, K.J., Cooper, J.M., Brown, G.K., Lake, B.D., Pike, D. & Poulton, J. (1993). A new point mutation associated with mitochondrial encephalomyopathy. Hum. Mol. Genet. 2, 2081-2087.

Murdock, D.G., Boone, B.E., Esposito, L.A. & Wallace, D.C. (1999). Up-regulation of nuclear and mitochondrial genes in the skeletal muscle of mice lacking the heart/muscle isoform of the adenine nucleotide translocator. J. Biol. Chem. 274, 14429-14433.

Nelson, D.R., Felix, C.M. & Swanson, J.M. (1998). Highly conserved charge-pair networks in the mitochondrial carrier family. J. Mol. Biol. 277, 285-308.

Nelson, I., Degoul, F., Obermaier-Kusser, B., Romero, N., Borrone, C., Marsac, C., Vayssiere, J.L., Gerbitz, K., Fardeau, M., Ponsot, G. & et al. (1989). Mapping of heteroplasmic mitochondrial DNA deletions in Kearns-Sayre syndrome. Nucleic Acids Res. 17, 8117-8124.

Neupert, W. (1997). Protein import into mitochondria. Annu Rev Biochem 66, 863-917.

Newgard, C.B. & McGarry, J.D. (1995). Metabolic coupling factors in pancreatic beta-cell signal transduction. Annu. Rev. Biochem. 64, 689-719.

Nicholls, D. & Ferguson, S. (1992). Transport of macromolecules across membranes. In Bioenergetics 2: Academic Press), pp. 233-234.

Njoroge, F.G. & Monnier, V.M. (1989). The chemistry of the Maillard reaction under physiological conditions: a review. Prog. Clin. Biol. Res. 304, 85-107.

Obermaier-Kusser, B., Paetzke-Brunner, I., Enter, C., Muller-Hocker, J., Zierz, S., Ruitenbeek, W. & Gerbitz, K.D. (1991). Respiratory chain activity in tissues from patients (MELAS) with a point mutation of the mitochondrial genome [tRNA(Leu(UUR)]. FEBS Lett 286, 67-70.

Ohlendieck, K., Riesinger, I., Adams, V., Krause, J. & Brdiczka, D. (1986). Enrichment and biochemical characterisation of boundary membrane contact sites from rat liver mitochondria. Biochim. Biophys. Acta 860, 672-689.

Otabe, S., Yasuda, K., Mori, Y., Shimokawa, K., Kadowaki, H., Jimi, A., Nonaka, K., Akanuma, Y., Yazaki, Y. & Kadowaki, T. (1999). Molecular and Histological Evaluation of Pancreata from Patients with a Mitochondrial Gene Mutation Associated with Impaired Insulin Secretion. Biochem. Biophys. Res. Commun. 259, 149-156.

Palmieri, F., Bisaccia, F., Capobianco, L., Dolce, V., Fiermonte, G., Iacobazzi, V., Indiveri, C. & Palmieri, L. (1996). Mitochondrial metabolite transporters. Biochim. Biophys. Acta 1275, 127-132.

Polakis, P.G. & Wilson, J.E. (1985). An intact hydrophobic N-terminal sequence is critical for binding of rat brain hexokinase to mitochondria. Arch. Biochem. Biophys. 236, 328-337.

Portha, B., Serradas, P., Bailbe, D., Suzuki, K., Goto, Y. & Giroix, M. H. (1991). Beta-cell insensitivity to glucose in the GK rat, a spontaneous nonobese model for type II diabetes. Diabetes 40, 486-491.

Postic, C., Shiota, M., Niswender, K.D., Jetton, T.L., Chen, Y., Moates, J.M., Shelton, K.D., Lindner, J., Cherrington, A.D. & Magnuson, M.A. (1999). Dual roles for glucokinase in glucose homeostasis as determined by liver and pancreatic beta cell-specific gene knock-outs using Cre recombinase. J. Biol. Chem. 274, 305-315.

Poulton, J., Brown, M.S., Cooper, A., Marchington, D.R. & Phillips, D.I. (1998). A common mitochondrial DNA variant is associated with insulin resistance in adult life. Diabetologia 41, 54-58.

Poulton, J., Deadman, M.E., Bindoff, L., Morten, K., Land, J. & Brown, G. (1993). Families of mtDNA re-arrangements can be detected in patients with mtDNA deletions: duplications may be a transient intermediate form. Hum. Mol. Genet. 2, 23-30.

Poulton, J. & Holt, I.J. (1994). Mitochondrial DNA: does more lead to less?. Nat. Genet. 8, 313-315.

Poulton, J., Morten, K.J., Marchington, D., Weber, K., Brown, G.K., Rotig, A. & Bindoff, L. (1995). Duplications of mitochondrial DNA in Kearns-Sayre syndrome. Muscle Nerve 3, S154-158.

Poulton, J., Morten, K.J., Weber, K., Brown, G.K. & Bindoff, L. (1994). Are duplications of mitochondrial DNA characteristic of Kearns-Sayre syndrome? Hum. Mol. Genet. 3, 947-951.

Prentki, M. & Matschinsky, F.M. (1987). Ca2+, cAMP, and phospholipid-derived messengers in coupling mechanisms of insulin secretion. Physiol Rev. 67, 1185-1248.

Prentki, M., Vischer, S., Glennon, M.C., Regazzi, R., Deeney, J.T. & Corkey, B.E. (1992). Malonyl-CoA and long chain acyl-CoA esters as metabolic coupling factors in nutrient-induced insulin secretion. J. Biol. Chem. 267, 5802-5810.

Rabinovitch, A., Suarez-Pinzon, W., Strynadka, K., Ju, Q., Edelstein, D., Brownlee, M., Korbutt, G.S. & Rajotte, R. V. (1999). Transfection of human pancreatic islets with an anti-apoptotic gene (bcl-2) protects beta-cells from cytokine-induced destruction. Diabetes 48, 1223-1229.

Rassow, J., Dekker, P.J.T., van Wilpe, S., Meijer, M. & Soll, J. (1999). The preprotein translocase of the mitochondrial inner membrane: function and evolution. J. Mol. Biol. 286, 105-120.

Reardon, W., Ross, R.J., Sweeney, M.G., Luxon, L.M., Pembrey, M.E., Harding, A.E. & Trembath, R.C. (1992). Diabetes mellitus associated with a pathogenic point mutation in mitochondrial DNA. Lancet 340, 1376-1379.

Regazzi, R., Wollheim, C.B., Lang, J., Theler, J.M., Rossetto, O., Montecucco, C., Sadoul, K., Weller, U., Palmer, M. & Thorens, B. (1995). VAMP-2 and cellubrevin are expressed in pancreatic beta-cells and are essential for Ca(2+)-but not for GTP gamma S-induced insulin secretion. Embo. J. 14, 2723-2730.

Romeo, G., Menozzi, P., Ferlini, A., Fadda, S., Di Donato, S., Uziel, G., Lucci, B., Capodaglio, L., Filla, A. & Campanella, G. (1983). Incidence of Friedreich ataxia in Italy estimated from consanguineous marriages. Am. J. Hum. Genet. 35, 523-529.

Rorsman, P. (1997a). Electrophysiological studies of insulin secretion, physiology and pathophysiology. Nord. Med. 112, 164-168.

Rorsman, P. (1997b). The pancreatic beta-cell as a fuel sensor: an electrophysiologist's viewpoint. Diabetologia 40, 487-495.

Rotig, A., Cormier, V., Chatelain, P., Francois, R., Saudubray, J.M., Rustin, P. & Munnich, A. (1993). Deletion of mitochondrial DNA in a case of early-onset diabetes mellitus, optic atrophy, and deafness (Wolfram syndrome, MIM 222300). J. Clin. Invest. 91, 1095-1098.

Rotig, A., de Lonlay, P., Chretien, D., Foury, F., Koenig, M., Sidi, D., Munnich, A. & Rustin, P. (1997). Aconitase and mitochondrial iron-sulphur protein deficiency in Friedreich ataxia. Nat. Genet. 17, 215-217.

Sako, Y. & Grill, V.E. (1990). A 48-hour lipid infusion in the rat time-dependently inhibits glucose-induced insulin secretion and B cell oxidation through a process likely coupled to fatty acid oxidation. Endocrinology 127, 1580-1589.

Saks, V.A., Lipina, N.V., Sharov, V.G., Smirnov, V.N., Chazov, E. & Grosse, R. (1977). The localization of the MM isozyme of creatine phosphokinase on the surface membrane of myocardial cells and its functional coupling to ouabain-inhibited (Na+, K+)-ATPase. Biochim. Biophys. Acta 465, 550-558.

Sakurai, T. & Tsuchiya, S. (1988). Superoxide production from nonenzymatically glycated protein. FEBS Lett. 236, 406-410.

Schleicher, E. & Wieland, O.H. (1989). Protein glycation: measurement and clinical relevance. J. Clin. Chem. Clin. Biochem. 27, 577-587.

Schleicher, E. & Wieland, O.H. (1981). Specific quantitation by HPLC of protein (lysine) bound glucose in human serum albumin and other glycosylated proteins. J. Clin. Chem. Clin. Biochem. 19, 81-87.

Sekine, N., Cirulli, V., Regazzi, R., Brown, L.J., Gine, E., Tamarit-Rodriguez, J., Girotti, M., Marie, S., MacDonald, M.J., Wollheim, C.B. & et al. (1994). Low lactate dehydrogenase and high mitochondrial glycerol phosphate dehydrogenase in pancreatic beta-cells. Potential role in nutrient sensing. J. Biol. Chem. 269, 4895-4902.

Sener, A., Herberg, L. & Malaisse, W.J. (1993). FAD-linked glycerophosphate dehydrogenase deficiency in pancreatic islets of mice with hereditary diabetes. FEBS Lett. 316, 224-227.

Sener, A., Kawazu, S., Hutton, J.C., Boschero, A.C., Devis, G., Somers, G., Herchuelz, A. & Malaisse, W.J. (1978). The stimulus-secretion coupling of glucose-induced insulin release. Effect of exogenous pyruvate on islet function. Biochem. J. 176, 217-232.

Serradas, P., Giroix, M.H., Saulnier, C., Gangnerau, M.N., Borg, L.A., Welsh, M., Portha, B. & Welsh, N. (1995). Mitochondrial deoxyribonucleic acid content is specifically decreased in adult, but not fetal, pancreatic islets of the Goto-Kakizaki rat, a genetic model of noninsulin-dependent diabetes. Endocrinology 136, 5623-5631.

Shoffner, J.M. (1996). Maternal inheritance and the evaluation of oxidative phosphorylation diseases. Lancet 348, 1283-1288.

Shoffner, J.M., Lott, M.T., Lezza, A.M., Seibel, P., Ballinger, S.W. & Wallace, D.C. (1990). Myoclonic epilepsy and ragged-red fiber disease (MERRF) is associated with a mitochondrial DNA tRNA(Lys) mutation. Cell 61, 931-937.

Silvestri, G., Moraes, C.T., Shanske, S., Oh, S.J. & DiMauro, S. (1992). A new mtDNA mutation in the tRNA(Lys) gene associated with myoclonic epilepsy and ragged-red fibers (MERRF). Am. J. Hum. Genet. 51, 1213-1217.

Sirrenberg, C., Bauer, M.F., Guiard, B., Neupert, W. & Brunner, M. (1996). Import of carrier proteins into the mitochondrial inner membrane mediated by Tim22. Nature 384, 582-585.

Sirrenberg, C., Endres, M., Becker, K., Bauer, M.F., Walther, E., Neupert, W. & Brunner, M. (1997). Functional cooperation and stoichiometry of protein translocases of the outer and inner membranes of mitochondria. J. Biol. Chem. 272, 29963-29966.

Sirrenberg, C., Endres, M., Folsch, H., Stuart, R.A., Neupert, W. & Brunner, M. (1998). Carrier protein import into mitochondria mediated by the intermembrane proteins Tim10/Mrs11 and Tim12/Mrs5. Nature 391, 912-915.

Soejima, A., Inoue, K., Takai, D., Kaneko, M., Ishihara, H., Oka, Y. & Hayashi, J.I. (1996). Mitochondrial DNA is required for regulation of glucose-stimulated insulin secretion in a mouse pancreatic beta cell line, MIN6. J. Biol. Chem. 271, 26194-26199.

Stagg, A.R., Fleming, J.C., Baker, M.A., Sakamoto, M., Cohen, N. & Neufeld, E.J. (1999). Defective high-affinity thiamine transporter leads to cell death in thiamine-responsive megaloblastic anemia syndrome fibroblasts. J. Clin. Invest. 103, 723-729.

Strom, T.M., Hortnagel, K., Hofmann, S., Gekeler, F., Scharfe, C., Rabl, W., Gerbitz, K.D. & Meitinger, T. (1998). Diabetes insipidus, diabetes mellitus, optic atrophy and deafness (DIDMOAD) caused by mutations in a novel gene (wolframin) coding for a predicted transmembrane protein. Hum Mol Genet 7, 2021-2028.

Suzuki, Y., Suzuki, S., Hinokio, Y., Chiba, M., Atsumi, Y., Hosokawa, K., Shimada, A., Asahina, T. & Matsuoka, K. (1997). Diabetes associated with a novel 3264 mitochondrial tRNA(Leu)(UUR) mutation. Diabetes Care 20, 1138-1140.

t Hart, L.M., Ruige, J.B., Dekker, J.M., Stehouwer, C.D., Maassen, J.A. & Heine, R.J. (1999). Altered beta-cell characteristics in impaired glucose tolerant carriers of a GAA trinucleotide repeat polymorphism in the frataxin gene. Diabetes 48, 924-926.

Tamarit-Rodriguez, J., Vara, E. & Tamarit, J. (1984). Starvation-induced changes of palmitate metabolism and insulin secretion in isolated rat islets stimulated by glucose. Biochem. J. 221, 317-324.

Taylor, K.D. & Piko, L. (1987). Patterns of mRNA prevalence and expression of B1 and B2 transcripts in early mouse embryos. Development 101, 877-892.

Thomas, F., Balkau, B., Vauzelle-Kervroedan, F. & Papoz, L. (1994). Maternal effect and familial aggregation in NIDDM. The CODIAB Study. CODIAB-INSERM-ZENECA Study Group. Diabetes 43, 63-67.

Thorens, B., Sarkar, H.K., Kaback, H.R. & Lodish, H.F. (1988). Cloning and functional expression in bacteria of a novel glucose transporter present in liver, intestine, kidney & beta-pancreatic islet cells. Cell 55, 281-290.

Tiedge, M., Lortz, S., Drinkgern, J. & Lenzen, S. (1997). Relation between antioxidant enzyme gene expression and antioxidative defense status of insulin-producing cells. Diabetes 46, 1733-1742.

Torroni, A., Huoponen, K., Francalacci, P., Petrozzi, M., Morelli, L., Scozzari, R., Obinu, D., Savontaus, M.L. & Wallace, D.C. (1996). Classification of European mtDNAs from an analysis of three European populations. Genetics 144, 1835-1850.

Torroni, A., Petrozzi, M., D'Urbano, L., Sellitto, D., Zeviani, M., Carrara, F., Carducci, C., Leuzzi, V., Carelli, V., Barboni, P., De Negri, A. & Scozzari, R. (1997). Haplotype and phylogenetic analyses suggest that one European-specific mtDNA background plays a role in the expression of Leber hereditary optic neuropathy by increasing the penetrance of the primary mutations 11778 and 14484. Am. J. Hum. Genet. 60, 1107-1121.

Toyokuni, S., Iwasa, Y., Kondo, S., Tanaka, T., Ochi, H. & Hiai, H. (1999). Intranuclear Distribution of 8-hydroxy-2'-deoxyguanosine. An immunocytochemical study. J. Histochem. Cytochem. 47, 833-836.

Trus, M.D., Zawalich, W. S., Burch, P.T., Berner, D.K., Weill, V.A. & Matschinsky, F. M. (1981). Regulation of glucose metabolism in pancreatic islets. Diabetes 30, 911-922.

Tsukuda, K., Suzuki, Y., Kameoka, K., Osawa, N., Goto, Y., Katagiri, H., Asano, T., Yazaki, Y. & Oka, Y. (1997). Screening of patients with maternally transmitted diabetes for mitochondrial gene mutations in the tRNA[Leu(UUR)] region. Diabet. Med. 14, 1032-1037.

Turk, J., Gross, R.W. & Ramanadham, S. (1993). Amplification of insulin secretion by lipid messengers. Diabetes 42, 367-374.

Uchida, K. & Stadtman, E.R. (1993). Covalent attachment of 4-hydroxynonenal to glyceraldehyde-3-phosphate dehydrogenase. A possible involvement of intra- and intermolecular cross-linking reaction. J. Biol. Chem. 268, 6388-6393.

van den Ouweland, J.M., Lemkes, H.H., Gerbitz, K.D. & Maassen, J.A. (1995). Maternally inherited diabetes and deafness (MIDD): a distinct subtype of diabetes associated with a mitochondrial tRNA(Leu)(UUR) gene point mutation. Muscle Nerve 3, S124-130.

van den Ouweland, J.M., Lemkes, H.H., Ruitenbeek, W., Sandkuijl, L.A., de Vijlder, M.F., Struyvenberg, P.A., van de Kamp, J.J. & Maassen, J.A. (1992). Mutation in mitochondrial tRNA(Leu)(UUR) gene in a large pedigree with maternally transmitted type II diabetes mellitus and deafness. Nat. Genet. 1, 368-371.

van den Ouweland, J.M., Lemkes, H.H., Trembath, R.C., Ross, R., Velho, G., Cohen, D., Froguel, P. & Maassen, J.A. (1994). Maternally inherited diabetes and deafness is a distinct subtype of diabetes and associates with a single point mutation in the mitochondrial tRNA(Leu(UUR)) gene. Diabetes 43, 746-751.

Vara, E. & Tamarit-Rodriguez, J. (1986). Glucose stimulation of insulin secretion in islets of fed and starved rats and its dependence on lipid metabolism. Metabolism 35, 266-271.

Velho, G., Byrne, M.M., Clement, K., Sturis, J., Pueyo, M.E., Blanche, H., Vionnet, N., Fiet, J., Passa, P., Robert, J.J., Polonsky, K.S. & Froguel, P. (1996). Clinical phenotypes, insulin secretion, and insulin sensitivity in kindreds with maternally inherited diabetes and deafness due to mitochondrial tRNALeu(UUR) gene mutation. Diabetes 45, 478-487.

Velho, G., Froguel, P., Clement, K., Pueyo, M.E., Rakotoambinina, B., Zouali, H., Passa, P., Cohen, D. & Robert, J.J. (1992). Primary pancreatic beta-cell secretory defect caused by mutations in glucokinase gene in kindreds of maturity onset diabetes of the young. Lancet 340, 444-448.

Waddell, I.D. & Burchell, A. (1988). The microsomal glucose-6-phosphatase enzyme of pancreatic islets. Biochem. J. 255, 471-476.

Walker, J.E. (1995). Determination of the structures of respiratory enzyme complexes from mammalian mitochondria. Biochim. Biophys. Acta 1271, 221-227.

Walker, J.E., Collinson, I.R., Van Raaij, M.J. & Runswick, M.J. (1995). Structural analysis of ATP synthase from bovine heart mitochondria. Methods Enzymol 260, 163-190.

Wallace, D.C. (1992). Diseases of the mitochondrial DNA. Annu. Rev. Biochem. 61, 1175-1212.

Wallace, D.C. (1999). Mitochondrial diseases in man and mouse. Science 283, 1482-1488.

Wallace, D.C. (1994a). Mitochondrial DNA mutations in diseases of energy metabolism. J. Bioenerg. Biomembr. 26, 241-250.

Wallace, D.C. (1994b). Mitochondrial DNA sequence variation in human evolution and disease. Proc. Natl. Acad. Sci. USA 91, 8739-8746.

Wallace, D.C., Lott, M.T., Shoffner, J.M. & Brown, M.D. (1992). Diseases resulting from mitochondrial DNA point mutations. J. Inherit. Metab. Dis. 15, 472-479.

Wallace, D.C. & Melov, S. (1998). Radicals raging. Nat. Genet. 19, 105-106.

Wallimann, T. & Hemmer, W. (1994). Creatine kinase in non-muscle tissues and cells. Mol. Cell. Biochem. 133-134, 193-220.

Wallimann, T., Wyss, M., Brdiczka, D., Nicolay, K. & Eppenberger, H.M. (1992). Intracellular compartmentation, structure and function of creatine kinase isoenzymes in tissues with high and fluctuating energy demands: the 'phosphocreatine circuit' for cellular energy homeostasis. Biochem. J. 281, 21-40.

Weiler, U., Riesinger, I., Knoll, G. & Brdiczka, D. (1985). The regulation of mitochondrial-bound hexokinases in the liver. Biochem. Med. 33, 223-235.

Welsh, M., Welsh, N., Bendtzen, K., Mares, J., Strandell, E., Oberg, C. & Sandler, S. (1995). Comparison of mRNA contents of interleukin-1 beta and nitric oxide synthase in pancreatic islets isolated from female and male nonobese diabetic mice. Diabetologia 38, 153-160.

White, K.C., Babbitt, P.C., Buechter, D.D. & Kenyon, G.L. (1992). The principal islet of the Coho salmon (Oncorhyncus kisutch) contains the BB isoenzyme of creatine kinase. J. Protein Chem 11, 489-494.

Wicker, U., Bucheler, K., Gellerich, F.N., Wagner, M., Kapischke, M. & Brdiczka, D. (1993). Effect of macromolecules on the structure of the mitochondrial inter-membrane space and the regulation of hexokinase. Biochim. Biophys. Acta 1142, 228-239.

Williams, M.D., Van Remmen, H., Conrad, C.C., Huang, T.T., Epstein, C.J. & Richardson, A. (1998). Increased oxidative damage is correlated to altered mitochondrial function in heterozygous manganese superoxide dismutase knockout mice. J. Biol. Chem. 273, 28510-28515.

Xie, G. & Wilson, J.E. (1990). Tetrameric structure of mitochondrially bound rat brain hexokinase: a crosslinking study. Arch. Biochem. Biophys. 276, 285-293.

Yki-Jarvinen, H. (1992). Glucose toxicity. Endocr. Rev. 13, 415-431.

Zeviani, M., Gellera, C., Antozzi, C., Rimoldi, M., Morandi, L., Villani, F., Tiranti, V. & DiDonato, S. (1991). Maternally inherited myopathy and cardiomyopathy: association with mutation in mitochondrial DNA tRNA(Leu)(UUR). Lancet 338, 143-147.

Zhou, Y.P., Berggren, P.O. & Grill, V. (1996). A fatty acid-induced decrease in pyruvate dehydrogenase activity is an important determinant of beta-cell dysfunction in the obese diabetic db/db mouse. Diabetes 45, 580-586.

Zhou, Y.P. & Grill, V. (1995a). Long term exposure to fatty acids and ketones inhibits B-cell functions in human pancreatic islets of Langerhans. J. Clin. Endocrinol. Metab. 80, 1584-1590.

Zhou, Y.P. & Grill, V.E. (1994). Long-term exposure of rat pancreatic islets to fatty acids inhibits glucose-induced insulin secretion and biosynthesis through a glucose fatty acid cycle. J. Clin. Invest. 93, 870-876.

Zhou, Y.P. & Grill, V.E. (1995b). Palmitate-induced beta-cell insensitivity to glucose is coupled to decreased pyruvate dehydrogenase activity and enhanced kinase activity in rat pancreatic islets. Diabetes 44, 394-399.

MITOCHONDRIAL SIGNALING AND CANCER

THOMAS J. PRESTON & GURMIT SINGH

Hamilton Regional Cancer Centre, 699 Concession Street, Hamilton, Ontario, CANADA, L8V 5C2

Table of Contents

Introduction

The central role of mitochondria in the regulation and optimization of the cellular environment is evident from their integral role in fundamental processes such as energy provision, redox status, as well as pH, calcium, and osmotic homeostasis. Due to their involvement in cellular metabolism, mitochondria facilitate both anabolic processes including synthesis of lipids, heme and nucleotides, and equally important degradative processes for the removal of toxic metabolites such as ammonium ions. The primary obligation of this organelle, however, is the provision of energy via oxidative phosphorylation mediated by the respiratory enzymes of the inner mitochondrial membrane. Through orchestration of these processes, mitochondria have the potential to influence signal transduction pathways and nuclear gene expression that ultimately affect the way a cell will differentiate, proliferate and die. Hence, a strong association between mitochondrial signaling and cancer is very likely.

In an attempt to address one aspect of cancer physiology, this chapter focuses on the potential role that mitochondrial signaling plays in the malignant cells. The primary aspects of mitochondrial involvement in tumor cell signaling discussed are: (i) mitochondria and cell transformation; (ii) nuclear/ mitochondrial regulation; (iii) regulation of mitochondrial enzymes in transformed cells; (iv) importance of mitochondrial ATP production in cell cycle control; (v) signals from mitochondria associated with tumor

Interorganellar Signaling in Age-Related Disease, Edited by Mark P. Mattson, 103—130

progression; (vi) redox balance associated with tumor progression; (vii) reactive oxygen species and NFκB signaling in cancer; and (viii) mitochondria and apoptosis.

Mitochondria and Cell Transformation

Changes in mitochondria including coupling of oxidative phosphorylation with enhanced glycolysis, altered calcium homeostasis, and the production of reactive oxygen species, have been associated with cell transformation (Baggetto, 1993; Pederson, 1987; Richter, 1993). Interestingly, all of these characteristics can be associated with the actions of the electron transport chain. The relationship between oxidative phosphorylation and glycolysis with malignancy has been investigated primarily in a rat hepatoma cell line. In this cell line, it has been observed that a high glycolytic rate is associated with a change in subcellular location of the hexokinase enzyme. Hexokinase is the initial enzyme in the glycolytic pathway where it converts glucose to glucose-6-phosphate, a process dependent on ATP hydrolysis. Normally this enzyme is located in the cytoplasm but in the rat hepatoma cell line, hexokinase is associated with the outer mitochondrial membrane. Altered localization of hexokinase is dependent on its N-terminal hydrophobic tail (Arora et al., 1990; Arora and Pederson, 1988). This association between hexokinase and mitochondria provides hexokinase with preferential access to mitochondrial produced ATP and thus enhances the glycolytic capacity of a cell (Nakashima et al., 1988).

Mitochondria are also capable of accumulating calcium and thus can regulate intracellular calcium concentrations. Calcium uptake is dependent on the mitochondrial membrane potential and the intramitochondrial concentration of inorganic phosphate. Since the electrochemical gradient is used to power the exchange of ATP for ADP and Pi, the loss of the membrane potential could limit intramitochondrial inorganic phosphate levels and thus the capacity to sequester calcium (Carafoli, 1987; McCormack et al., 1992). Regulation of cytosolic calcium by mitochondria, under normal conditions, is probably minor since the affinity for calcium and its uptake rate are significantly greater in other organelles such as the endoplasmic and sarcoplasmic reticulum (Carafoli, 1987). Under stressful conditions, where cytoplasmic calcium levels are increased due to an influx of calcium across the plasma membrane, mitochondria have the capacity to sequester large amounts of calcium, in bound and precipitated forms, with little effect on their function (Fiskum, 1985; Richter et al., 1991). Since intracellular calcium levels are important determinants of calcium-mediated signal transduction and possibly activation of endonucleases involved in programmed cell death, the magnitude of the mitochondrial membrane potential may influence these processes (Richter, 1993). In addition, three mitochondrial dehydrogenases, pyruvate dehydrogenase, NAD-dependent isocitric dehydrogenase and oxoglutarate dehydrogenases are regulated by matrix calcium content (McCormack et al., 1990, 1993). These three dehydrogenases are important in the citric acid (TCA) cycle and may affect the production of substrates (NADH or $FADH_2$) which are available to the electron transport chain. Activation of these dehydrogenases by calcium may augment rates of electron transport and possibly ATP production (Crompton et al., 1988; Gunter et al., 1990).

Oxidative phosphorylation consumes most of the body's oxygen and is the primary generator of reactive oxygen species (ROS) (Richter, 1992). In normal cells, mitochondria contain their own complement of superoxide dismutase and glutathione that are capable of scavenging most of the ROS produced (Grace, 1990; Sohal et al., 1992). In tumor cells, however, elevated activity of the electron transport chain may overwhelm these defense mechanisms, resulting in cellular damage such as lipid peroxidation, as well as damage to nuclear DNA and mitochondrial DNA (mtDNA) (Singh et al., 1985). Since genes for the electron transport chain are encoded by both the nuclear and mitochondrial genome, damage to either of these pools of genetic information may result in impaired electron transport, and ROS production can be augmented by such destabilization. This may permit protein complex dysfunction from oxygen or electrophiles causing further generation of reactive oxygen intermediates. ROS have been implicated in calcium release from mitochondria, leading to a variety of effects upon intracellular signaling (Richter and Kass, 1991).

A number of studies have shown that tumor cells accumulate and retain more lipophilic cations than normal cells, thus suggesting that there exists an elevated mitochondrial membrane potential ($\Delta\Psi m$) relative to normal cells. Since cells elevate their mitochondrial membrane potential during periods of activity, such as during cell migration and proliferation (Johnson et al., 1981; Darzynkiewicz et al., 1982), it is not surprising that tumor cells might maintain an elevated mitochondrial membrane potential since they are characterized by increased rates of proliferation and migration (Chen, 1988). A study of over 200 cells lines/types indicated that cancer cells, including these derived from tumors of the kidney, ovary, pancreas, lung, adrenal cortex, skin, breast, prostate, cervix, vulva, colon, liver, testis, esophagus, trachea, and tongue had elevated mitochondrial membrane potentials/rhodamine-123 accumulation compared to cells from normal tissue (Lampidis et al., 1983; Summerhayes et al., 1982). The magnitude of difference has been estimated to be at least 60 mV (Davis et al., 1985; Modica-Napolitano et al., 1987). This increase in mitochondrial membrane potential is not observed in cells from leukemias, lymphomas, neuroblastomas or osteosarcomas (Lampidis et al., 1982; Summerhayes et al., 1982). Therefore, this effect is not ubiquitous throughout the cancer models tested, however, the association of altered $\Delta\Psi m$ and cellular transformation is intriguing.

Nuclear/Mitochondrial Regulation

Since most of the mitochondrial proteins are encoded by nuclear genes and synthesized on cytoplasmic ribosomes, interactions between the nucleus and mitochondria must play an important role in mitochondrial regulation. Respiratory enzymes and mitochondrial membranes are normally synthesized and/or degraded in a coordinated fashion as the proportions of the various components are characteristic for a given cell (Grivell, 1995). The content of mitochondrial enzymes in mammalian cells is regulated by several environmental conditions including oxygen tension and hormones (Tzagoloff and Myers, 1986). It is undetermined whether most mitochondrial proteins, or even just those of the inner membrane, are regulated individually or as a set. Because various mitochondrial components have distinct turnover times and different degrees of nuclear and mitochondrial control, including specific genomic expressions, each protein appears to be individually regulated (Dieckmann and Staples, 1994) but, as discussed below, there are exceptions to this general rule. During hypoxic incubation of mouse lung macrophages and rat skeletal muscle L8 cells, the activities of various mitochondrial enzymes (both nuclear and mitochondrially encoded) decreased about 50-60% in tandem with no statistical difference between the various activities (Murphy et al., 1984). The kinetics of this decrease, and of the recovery of activity after re-exposure to normoxia, were also similar for the different enzyme activities. This suggests that, at least in certain cases, mitochondrial enzymes can be regulated as a unit.

It is known that nuclear gene expression is frequently regulated at the transcriptional level (Williams et al., 1986), however, the level(s) at which mtDNA gene expression is controlled is unknown. Some reports have addressed this question from the point of view of gene dose or mtDNA copy number (i.e., is the increase or decrease in mtDNA gene expression proportional to changes in the content of mtDNA?). In the experiment described above, although there were uniform decreases in mitochondrial enzyme activities after exposure to hypoxia, there was no change in the mtDNA copy number, a result that speaks against regulation of mitochondrial gene expression only by gene dosage.

Another experiment which sought to answer whether mtDNA gene dosage controls gene expression was conducted by Williams et al. (1986). They determined the concentrations of mtDNA, mitochondrial mRNA and cytochrome b mRNA (a mitochondrial gene product) in normal versus electrically stimulated rabbit striated muscles. This tissue was selected because the oxidative capacity of mammalian striated muscles can vary nearly 10-fold, a finding which presumably reflects major differences in the expression of genes (including those on mtDNA) that encode enzymes of oxidative metabolism. After electrical stimulation of the muscle, mtDNA, mtRNA, and cytochrome b levels were observed to vary in direct proportion to the oxidative capacity of the tissue. Because the expression of mitochondrial genes in mammalian striated muscle was proportional to their copy number, these results do support the hypothesis that amplification of the mitochondrial genome relative to chromosomal DNA (i.e., gene dosage) was responsible for the enhanced expression of mitochondrial genes in highly oxidative tissues. This interpretation, of course, differs from that of the hypoxia experiment described above. The different conclusions reached in the two experiments may relate to the different tissues and/or animals used. Also, the differences may reflect the different experimental designs: there were no changes in mtDNA copy number after 4 days of hypoxia, but neither were there changes in mtDNA content after the same period of muscle stimulation (the changes occurred after 7 days). In any event, the regulation of mitochondrial gene expression in differentiated cells, to date, is not well understood.

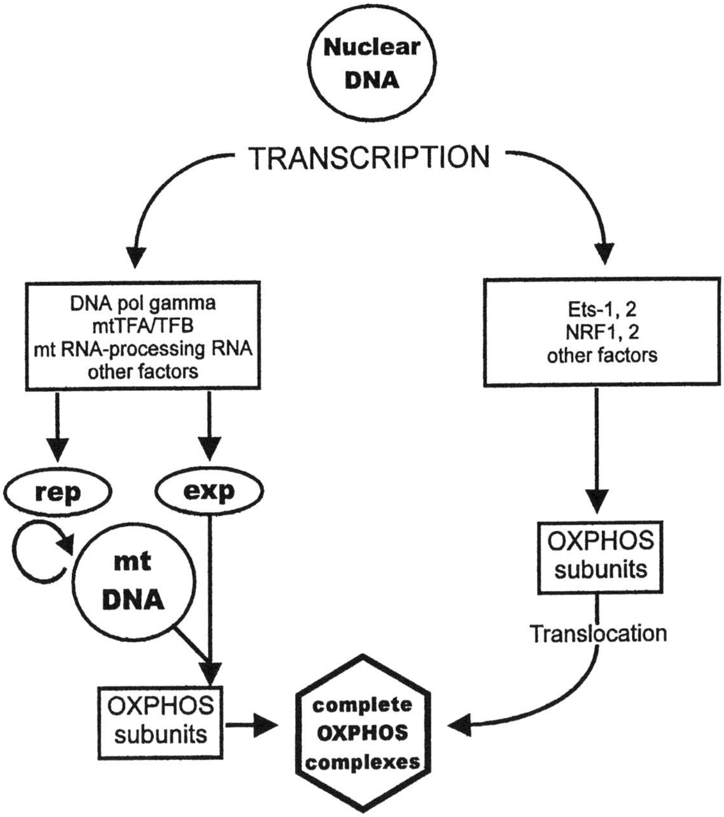

Figure 1. Interdependence of nuclear and mitochondrial regulation during oxidative phosphorylation (OXPHOS) complex synthesis. Different OXPHOS components are encoded by genes within the nuclear genome as well as mitochondrial DNA (mtDNA). Replication (rep) and expression (exp) of mtDNA is dependent upon nuclear-encoded protein products such as DNA polymerase (pol) γ (used for replication), mitochondrial transcription factors (mtTFA and mtTFB; used in replication and transcription) and the mitochondria RNA-processing RNA (used to prime replication). The nucleus contains the genes for components of OXPHOS (for example ATPsynβ and cytochrome C oxidase; COX) as well as the genes for transcription factors that control their expression, including *ets*-1, -2 and nuclear factors *nrf*-1, -2. mtDNA itself expresses essential subunits of OXPHOS, and the complete system cannot be formed until the nuclear components translocate into the mitochondrial inner membrane. Thus, both genetic systems are required in mitochondrial biogenesis. Functioning of the mitochondria, whether normal or aberrant as with malignancy, manipulates cell physiology in a number of ways which can then affect nuclear DNA.

Regulation of Mitochondrial Enzymes in Transformed Cells

The alterations in regulation of mitochondrial gene expression upon transformation of mammalian cells is also poorly characterized (Clayton, 1992). The pioneering manometric studies of Warburg in the 1920s (Warburg, 1923) first revealed that tumor cells are defective in respiration and have abnormally high rates of aerobic glycolysis. Warburg suggested that the cancer cell originates from heritable injury to the mitochondria in that the respiratory system of all cancer cells was thought to be damaged (Warburg, 1956 a, b). In the early 1960s, this hypothesis had lost much of its impact because elevated rates of glycolysis could not be associated in any consistent manner with changes in the structure or function of the mitochondrion (Weinhouse, 1956).

Nevertheless, numerous publications in the last two decades have catalogued a wide range of abnormalities in tumor cell mitochondria. For example, some investigators state that cancer cells apparently contain fewer mitochondria than normal cells and the organelles in the tumor cells may be structurally aberrant (Pederson, 1978; Singh and Maniccia-Bozzo, 1990; Singh and Veltri, 1991). Inner membrane alterations are also often observed in tumor mitochondria, such as changes in the amount and properties of ATPase (Mazurek et al., 1997). The levels of mitochondrial-associated enzymes such as hexokinase are also different in tumor cells (Mathupala et al., 1997). It is possible that these complex multiple changes in membrane characteristics result from rearrangements of mtDNA or from topological aberrations of the organellar genome (Clayton, 1991). Indeed, aberrations of mtDNA (mainly associated with an increase in the percentage of dimers and catenanes) have been found in tumor cells (Egilsson et al., 1979). Thus, mtDNA mutations may promote neoplasia in a manner resembling that proposed with nuclear gene mutation and may in fact cooperate to produce the phenotypic expression of such genomic irregularities (Singh et al., 1992).

Importance of Mitochondrial ATP Production in Cell Cycle Control

ATP is the primary utilizable source of high-energy phosphate within the cell and can be derived from both cytosolic glycolysis and mitochondrial oxidative phosphorylation. We are currently investigating the relationship between mitochondrially-derived ATP and progression through the cell cycle in an attempt to define the energetic dependence of proliferation, since most cell cycle research to date has focused on protein interactions that mediate cell division. One hypothesis is that cell cycle progression is sensitive to changes in the energy status of the cell, in particular, to the mitochondrial contribution to overall ATP availability (Sherratt and Turnbull, 1990). A sensory mechanism may exist that has the capacity to detect changes in the amount or availability of ATP and then signals the proteins of the cell cycle machinery to alter the rate of progression through the cell cycle. This would result in a redistribution of the population following energetic stress with an observable accumulation of cells at the proposed energetic checkpoints.

To first define the ATP profile of a cycling population, HL-60 human leukemic cells were separated by elutriation according to cell cycle phase and compared for total mitochondrial mass, mitochondrial membrane potential and ATP content. Our observations suggest that phase-specific fractions maintain different steady state levels of cellular ATP relative to mitochondrial mass, suggesting that different cell cycle phases have different energetic characteristics. The observed cycle-dependent difference in ATP content may highlight points during cycle transition which may be sensitive to ATP inhibition, particularly those points where steady state levels are relatively low, as they are in cells making the transition from G_1 to S (unpublished data). Further data from our laboratory suggest that an energy sensitive sensory mechanism exists and may be an additional property of the known cell cycle checkpoints which monitor changes in environmental factors and integrity of the genome. The HL-60 cell cycle profile was studied following disruption of mitochondrial energy production at the site of the ATP synthase in the inner mitochondrial membrane. Increasing doses of oligomycin b, an inhibitor of the complex V ATPase on the inner mitochondrial membrane, will reproducibly lower total cellular ATP and concurrently induce a significant redistribution of the cell population in a single generation. Using non-toxic doses of oligomycin to ensure that the observed redistribution is not a result of selective loss of cells from a given cycle phase, we were able to demonstrate an accumulation of cells in G_1 with the smallest measurable decrease in ATP (approx. 12%) and further decreases in ATP (20-30%) led to an increase in the G_2M population (Sweet and Singh, 1995). Therefore, we propose that energetic check points exist at G_1-S and G_2M boundaries which ensure that cellular ATP content is at or above the critical threshold required for successful passage through that cell cycle phase. In this way, a critical level of $\Delta\Psi m$ function would be necessary for efficient cell proliferation as observed with tumor cells.

The sensitivity of cells to arrest in G_1 with even small reductions in cellular ATP suggests that an active sensing mechanism is in operation to prevent cells from progressing through the cycle without a sufficient supply of ATP. We are currently examining the effect of ATP modulation on cell proliferation at the level of the cell cycle proteins, with the intention of further characterizing the molecular sensing mechanism for ATP depletion. An integral part of cell cycle control involves the formation of protein complexes between cyclins and their respective cyclin-dependent kinases (cdk's). Association of D-type cyclins with cdk4/6 is crucial for the initiation of a new cycle of proliferation while cyclin E/cdk2 complexes are essential for the transition from G_1 into S phase (Matsushime et al., 1991; Ohtsubo and Roberts, 1993). The best characterized substrate for phosphorylation by these complexes is the protein product of the Retinoblastoma tumor suppressor gene (Rbp) (Weinberg, 1995). When this protein is hypophosphorylated, it binds and inactivates the E2F transcription factor, thereby inhibiting its influence on the transcription of genes necessary for progression into S phase. Overexpression of these cyclins results in hyperphosphorylation of Rbp and rapid acceleration through G_1 while microinjection of antibodies directed against these cyclins causes cells to arrest prior to the onset of DNA replication (Resnitsky et al., 1994; MacLachlan et al., 1995). More recently, a new family of small regulatory proteins has been described which inhibit the kinase activity of cyclin/cdk complexes and so have potential roles as tumor suppressors (for review see Biggs and Kraft, 1995). These cyclin-dependent kinase inhibitors (cdki s: p16, p21, p27) which interact directly with the proteins controlling G_1 progression into S are of particular interest to us as they may represent the final mediators of G_1 accumulation in response to energetic stress.

Signals from Mitochondria Associated with Tumor Progression

Mutations within the cellular genome including those that have been connected to neoplasia can drastically affect the functioning of mitochondria. The nucleus supplies the polymerase enzymes and other factors responsible for the replication and expression of the mtDNA, and also encodes many of the proteins critical for such mitochondrial processes as glycolysis, oxidative phosphorylation and local redox balance (Grivell, 1995). However, genes specific to the mitochondria themselves have a profound influence upon cell physiology, through participation in metabolic pathways including lipid, heme, and pyridine nucleotide synthesis, transfer and ribosomal RNA expression, and of course energy provision (Sweet et al., 1998). Through these critical processes, as well as cellular redox regulation, the mitochondria have the ability to participate in signals associated with cancer and influence tumor cell phenotype by mechanisms which are both dependent and independent of the nucleus. As genomic mutation, chromosomal instability and aneuploidy have been linked with many cancers (Cohen and Ellwein, 1990; White, 1996), the mitochondrial dysfunction resulting from these nuclear events as well as mtDNA damage may play an important role during the progression of such diseases. It is important to note that as in normal tissues, the tumor is dependent upon mitochondria for the production of ATP (Sweet and Singh, 1995), however, dysfunction can exist which alters mitochondrial signaling without a debilitating loss of energy generation. For example, chaperone proteins of the heat shock protein (hsp) 60 and 70 families, as well as their cofactors which are located in the nucleus among other cellular compartments, may act as tumor suppressor proteins presumably owing to their control over the post-translational processing of other factors involved in cell cycle regulation. Loss of chaperone function is associated with tumor formation in *Drosophila* studies, and these findings may be repeated in mammalian studies (Kurzik-Dumke et al., 1995). mtDNA itself is highly sensitive to mutation, much more so than nuclear DNA, due to lack of protective histones, the high error rate during replication by DNA polymerase γ, and proximity to reactive oxygen species generated from O_2 respiration (Kunkel and Loeb, 1981; Sweet et al., 1998).

The alteration of mitochondrial genes can have substantial effects upon other compartments within the cell. For instance, after damage to mitochondrial structures by such agents as chemical carcinogens, mtDNA may actually enter the nucleus and act as transposable elements which modify the nuclear genome through integration (Cavalli and Liang, 1998). Signals may also be sent to the cytosol and nucleus from mitochondria in the form of peptides or ribosomal RNA, an observation described from studies using a variety of eukaryotic cells. Maternally transmitted factor (MTF) is a histocompatibility antigen found on the plasma membrane of mouse cells which has been identified as a peptide derived from a mitochondrial-encoded NADH dehydrogenase subunit (Loveland et al., 1990). Also, mitochondrial large ribosomal RNA (mtlrRNA) is a cytoplasmic factor that induces pole cell formation during *Drosophila melanogaster* development (Kobayashi et al., 1993). In response to impaired mitochondrial functioning from a deficiency in small mitochondrial ribosomal subunits, the expression of the *Neurospora crassa* nuclear *cyt*-21^{+} gene (producing a mitochondrial ribosomal protein) is upregulated 5-fold (Kuiper et al., 1988). One approach used to study the ability of mitochondrial signals to affect the behavior of transformed cells is to deplete mtDNA by serial cell passages in the presence of ethidium bromide or by inhibition of mRNA translation by mitochondrial ribosomes using chloramphenicol (Wang and Morais, 1997). After EtBr treatment, "ghost" mitochondria remain intact, with maintenance of a membrane potential ($\Delta\Psi m$) and ATP metabolism primarily via the glycolytic pathway. However there is no replication or expression of mtDNA. In avian cells transformed by the MC29 strain of avian leukosis virus (containing v-*myc*), loss of mtDNA blocked the anchorage-independent phenotype, which was rescued after reintroduction of wild type mitochondria by cybrid fusion (Zinkewich-Péotti et al., 1990). In the same model, lowered mtDNA expression resulted in the increased transcription of several genes including viral-encoded v-*myc* (Wang and Morais, 1997), and in a reciprocal manner the levels of mitogen-activated protein kinase kinase 2 (MEK2) were increased after treatment (Wang et al., 1997). So, by affecting the mitochondria, nuclear gene expression can be manipulated, and tumor physiology may be altered. But precisely how this occurs is not known. Mitochondrial rRNA and short peptides translated from mtDNA sequences may modify regulatory genomic sites, or the entire effect may be a response to the metabolic changes the cell undergoes along with mitochondrial dysfunction (Wang and Morais, 1997). Also, conflicting results from experimentation with human cells in these types of studies has stimulated questions regarding the effects of cell species, tissue type and state of differentiation on the penetrance of mitochondria-derived signals.

The contribution of mitochondria to the regulation of cellular redox conditions and response to oxidative stress in tumor cells has been another focus of research. A large proportion of the reactive oxygen species (ROS) produced in the cell such as hydrogen peroxide, superoxide and organic hydroperoxides are derived from oxidative respiration at the inner mitochondrial membrane (Boveris et al., 1972; Chance et al., 1979; Hunt et al., 1998). High levels of cellular ROS are toxic, therefore to maintain viability it is believed that the cell controls its redox environment by balancing antioxidant enzyme levels with oxidative species production, particularly at the mitochondria (Dorward et al., 1997). It is the deregulation of this balance that may be an important aspect of tumor biology: a decrease in electron transport chain efficiency may result in greater production of ROS, which could manipulate a host of cell signaling cascades, in the absence of cytotoxicity. This theory is supported by observations of an increased oxidative state of cancer cells in comparison to their untransformed counterparts (Remacle et al., 1995; Oberley and Oberley, 1997). If the mitochondria have any significant contribution to this difference, it could be realized by changes in the electron transport chain and/or local antioxidant proteins such as superoxide dismutase (SOD), glutathione/glutathione peroxidase (GSH/GPX), and catalase. In fact, lower levels of antioxidants such as catalase and GPX have been observed in human tumor lines established from the lung, prostate and kidneys (Oberley and Oberley, 1997) and in lung carcinoma biopsies (Coursin et al., 1996). It is these same enzymes which are upregulated in fibroblasts after chronic exposure to oxidative stress (Spitz et al., 1993; Hunt et al., 1998), therefore cancer cells which are resistant to the toxicity inherent with increased cellular oxidation have presumably acquired this characteristic by some other defense mechanism(s).

Because many cancer treatments, such as radiation and various chemotherapeutics, are targeted to destroy tumor cells ultimately via the intracellular generation of ROS, the ability of the cell to survive such an insult has important clinical consequences. Oberley et al. (1995) have studied the relationship between the level of expression of manganese-containing SOD (MnSOD), a nuclear-encoded superoxide radical scavenger which localizes to the mitochondrial matrix (Chance et al., 1979) and neoplastic transformation extensively. In immortalized, as well as malignant cell lines, overexpression of MnSOD resulted in growth inhibition (Li et al., 1998a,b; Oberley et al., 1995). This increase in MnSOD has also been demonstrated to suppress the malignant phenotype of human glioma, oral squamous carcinoma and breast cancer cells (Li et al., 1995; Liu et al., 1997; Zhong et al., 1997). A decrease in overall MnSOD level should translate into a higher concentration of its catalytic substrates, oxygen radicals, and lower product, hydrogen peroxide. However, the interplay between SODs and other antioxidant enzymes at the mitochondria and elsewhere among various cell types is an important consideration in this process. The investigators hypothesize that MnSOD may function as a tumor suppressor and possible marker of cell differentiation. If so, the endogenous regulation of the MnSOD gene may be affected by the changes to characterized tumor suppressors seen in a cancerous state. On the other hand, overexpression of MnSOD in pheochromocytoma cells increases their resistance to apoptosis induced by various insults (Keller et al., 1998).

Thus, one must consider effects of cellular oxidant status on cell proliferation and apoptosis. Here again we evoke a model of nuclear genome mutation in cancer exerting certain effects upon mitochondria, which then can lead to profound changes throughout the cell including reciprocal alterations to DNA metabolism and expression in the nucleus. The question of how reactive oxygen species may be involved in this type of signaling is intriguing. In theory, the response to ROS may be shifted from one of stress to growth in cells having constitutive activation of growth pathways and defects in the apoptotic program. Therefore, increased oxidation may confer a selective advantage to cancer cells *in vivo*, and the regulation of such an effect with therapies directed against cellular sources of ROS or aimed towards activation of scavenging enzymes might help control tumor growth.

Mitochondrial Contributions to Redox Balance and Cell Proliferation

Of the work directed at studying cellular redox state effects upon cancer initiation and progression, there is a growing focus on the ability of hydrogen peroxide (H_2O_2) to act as a specific mediator of both stress and proliferative signaling. The role of H_2O_2 in association with cellular toxicity during physiological processes such as phagocytic attack has long been characterized. Depending upon cell type, the concentrations of H_2O_2 sufficient to induce cell death in culture conditions range from 10^{-6} to 10^{-3} molar. However, low levels of endogenous cellular H_2O_2 production (10^{-10} to 10^{-7} molar) are generally non-toxic and, in fact, appear critical for efficient growth in culture (D'Souza et al., 1993; Lander, 1997; Szatrowski and Nathan, 1991; Tokoyuni et al., 1995). It is yet undetermined if this H_2O_2 dependency is intrinsic to cell functioning or an artifact of the tissue culture environment. However, reports implicating this compound as a promoter of several well-characterized proliferation pathways, as well as the previously discussed evidence of lowered H_2O_2 scavenger enzyme levels in tumors, suggest that the effects are real (Coursin et al., 1996; Rao, 1996; Sundaresan et al., 1995). Another important question concerning this research is whether the presence of elevated hydrogen peroxide levels in growth-stimulated cells is a critical component of this phenotype, or simply a downstream effect of increased metabolism. Indeed, cellular H_2O_2 levels have been observed to rise quickly in response to stimulation by growth factors (Bae et al., 1997; Chen et al., 1995), although such observations may be representative of a positive feedback-type mechanism for the maintenance of signal pathway activation. Data from our lab, which indicates that transformed cells in culture undergo growth arrest upon scavenging of extracellular H_2O_2 (unpublished results), is in support of such a model. This scavenging is extracellular, which raises some important questions: How can this affect intracellular signaling? Does the loss of extracellular H_2O_2 act as a "sink" into which cytosolic peroxides are lost? Or is the critical effect occurring at the plasma membrane? In some mammalian tissues, such as smooth muscle, H_2O_2 catabolizing-enzymes, such as catalase, appear to be able to cross the plasma membrane, and if so the site of function could be entirely within the cell (Sundaresan et al., 1995).

There exist a number of potential sources of hydrogen peroxide in the cell which function in normal metabolic pathways but may be deregulated in a neoplastic state. Two that have garnered attention as participators in proliferation and tumorigenesis are located at the plasma membrane and mitochondrial inner membrane. In certain mammalian nonphagocytic cells, such as hepatocytes and fibroblasts, a plasma membrane-localized multimolecular β-nicotinamide adenine dinucleotide (NADH)-oxidase complex has been identified. This system is thought to be distinct from the NADPH oxidase of neutrophils and is activated in response to a number of extracellular stimuli such as hormones, growth factors (Brightman et al., 1992), the cytokines interleukin-1 (IL-1), tumor necrosis factor α (TNF-α) and transforming growth factor β (TGF-β) (Thannickal and Fanburg, 1995). The role of this response is speculative, but with the growing acceptance of oxidative compounds as manipulators of extracellular signals, the authors suggest that increased H_2O_2 generation at this site may be required for known or unidentified post-receptor pathway transduction. It should be noted that this plasma membrane oxidase complex has not been clearly identified in cancer cells at this point. Interestingly, the small guanosine triphosphate (GTP)-binding protein Rac1 is a component of the plasma membrane NADPH oxidase in neutrophils and in rabbit fibroblasts Rac1 activation resulted in the generation of reactive oxygen species (Kheradmand et al., 1998), providing evidence for its possible involvement in the NADH-oxidase complex of nonphagocytic cells. Rac1 is critical for the growth factor-induced membrane ruffling and lamellipodia alterations observed in cancer cell motility and invasion (Anand-Apte et al., 1997; Keely et al., 1997; Michiels et al., 1995). It is therefore possible that certain levels of H_2O_2 may not only be required for the efficient transmission of growth signals, but may also participate in the anchorage-independent and migratory qualities indicative of the tumor cell at later stages of cancer progression.

At this time, such associations have not been characterized to any significant extent, and there is speculation as to the importance of the relative contribution of the reactive oxygen species produced at the plasma membrane to overall intracellular H_2O_2 levels.

The largest source of hydrogen peroxide in the cell, and thus the largest contributor to its putative effects upon cancer signaling, has long appeared to be the mitochondria. And as discussed in the previous section of this chapter, in a cancerous state the mitochondria can generate markedly higher amounts of H_2O_2. Superoxide anion radical ($O_2^{-\cdot}$) is generated during single electron transfer reactions, which occur in great abundance at the ubiquinone, Q cycle, complex III of the electron transport chain in the inner mitochondrial membrane (Boveris et al., 1972; Garcia-Ruiz et al., 1997). The subsequent catalytic conversion of $O_2^{-\cdot}$ in to H_2O_2 by MnSOD in the mitochondrion (Chance et al., 1979) is believed to account for much of the measured intracellular and secreted peroxide levels. For example, we have observed that extracellular generation of H_2O_2 by human cancer cell lines is significantly reduced by treatment with the electron transport chain inhibitor rotenone (unpublished data). These effects could be reflected in the growth kinetics of the tumor *in vivo*. Freyer (1998) has recently indicated that the quiescent cells within the inner region of tumor spheroids exhibit lower respiration rates by downregulating mitochondrial function. If H_2O_2 is needed for tumor growth, the mitochondrion is likely the most important contributor of this requirement.

The amount of evidence for the participation of hydrogen peroxide in a number of signal transduction pathways associated with human cancers has been increasing rapidly (Lander, 1997). Most of the information collected to date involves tissue culture models in which exogenous levels of H_2O_2 in the micromolar range are added to cells with subsequent investigation of cellular signaling, or the reciprocal study of H_2O_2 production responses to the introduction of activated oncogenes. All mammalian tissues in culture, whether transformed or not, generate variable levels of H_2O_2. However, lines exhibiting growth pathway upregulation such as those derived from human tumors have been shown to produce higher peroxide levels than non-transformed cells and also display increased resistance to cytotoxicity from oxidative stress (Szatrowski and Nathan, 1991; Yang and de Bono, 1997). One report postulates that in the human breast carcinoma line T47D, increases in reactive oxygen species generated by tumor necrosis factor α (TNF-α) in turn induce resistance to stress by modulating heat shock protein 27 (hsp27) activity (Mehlen et al., 1995). Thus, in the cancer cell where growth pathways can be inappropriately activated, and apoptosis avoided, the response to increased H_2O_2 may be shifted from one of stress into further proliferation and transformation.

A specific oncogenic signaling cascade of critical importance in many cancers is mediated by epidermal growth factor (EGF) stimulation (van der Geer and Hunter, 1994). EGF binds and activates members of the EGF receptor family at the plasma membrane, which then transmit signals to several cytoplasmic transducer molecules resulting in altered gene expression, usually directed towards cell growth promotion. Signal messenger proteins in this pathway which have been shown to be critical for its propagation are members of a family of small guanine nucleotide exchange factors including p21 Ras, Rho and Rac (Lowy, 1993; van der Geer and Hunter, 1994). A full description of Ras function as part of cancer signaling is beyond the focus of this paper. Briefly p21 Ras is located on the inner side of the plasma membrane and acts as a transducer of cell signals from tyrosine kinases such as receptors and the Src kinase family onto cytoplasmic serine/threonine kinases including Raf1 and the mitogen activated protein kinases (MAPKs) Erk1 and Erk2 (extracellular signal-regulated kinase 1 and 2) and

Jun-activating kinase (JNK) (Khosravi-Far and Der, 1994; Marshall, 1995). Upregulation of Ras activity transforms mammalian cells in culture through stimulation of growth responses, cell cycle manipulation and cytoskeletal changes (Mansour et al., 1994; Peeper and Bernards, 1997) and activating Ras mutations have been detected in approximately one third of all human cancers (Khosravi-Far and Der, 1994). Because of their importance in tumor cell biology, many investigators have focused on the responses of specific EGF and Ras pathway components to H_2O_2 treatment in order to elucidate the molecular mechanisms which underlie the cell proliferative effects observed. Addition of H_2O_2 (micromolar range) to a variety of cell types, including cultured rodent and human fibroblasts, human carcinoma cells and vascular smooth muscle has been demonstrated to cause an increase in tyrosine phosphorylation of receptor tyrosine kinases (RTKs) such as the epidermal growth factor receptor (EGFR), as well as an increase in Ras effector complex localization to these sites (Gamou and Shimizu, 1995; Rao, 1996). As would be expected, hydrogen peroxide and other peroxide compounds also promote MAPK family activation and translocation to the nucleus (Stevenson et al., 1994; Guyton et al., 1996). Specific members of the MAPK group which are involved in the cell response to stress (stress-activated protein kinases, SAPK) like JNK, that are associated with the processes of growth arrest and apoptosis, are also turned on after treatment with H_2O_2 (Wang et al., 1998).

How then can H_2O_2 cause cell proliferation when its presence also stimulates cell stress responses? The answer may relate to a balance of alternate signaling pathways: the directions to continue dividing and those pushing the cell into senescence or death work against one another, and the fate of the cell depends upon which signal is stronger. Here is where the interactions of H_2O_2 effects with other cancer-promoting stimuli or pathway alterations may be important in the outcome of cell response. Such interaction is demonstrated by reciprocal experiments to those using exogenous H_2O_2 addition, where the activation of specific growth pathway components by peptide growth factor addition or genetic mutation then measures H_2O_2 generation within the cell. For example, stimulation of starved carcinoma cells with EGF or vascular smooth muscle cells with platelet-derived growth factor (PDGF) induces a rapid increase of cellular H_2O_2 generation (Bae et al., 1997; Sundaresan et al., 1995), and the expression of activated v-Ras or Rac1 in non-transformed lines increases cytoplasmic levels of reactive oxygen species (Irani et al., 1997). Some progress has been made in analyzing these effects from the opposite direction, by blocking the H_2O_2 signal using the expression of inhibitory mutants of Src, Ras and Raf1 (Aikawa et al., 1997). Interestingly, this oxidative response appears to be required for the maintenance of these and other cell signals, and scavenging of H_2O_2 by antioxidants or antioxidant enzymes can in turn downregulate growth pathway activation. Pretreatment of a variety of cell types with the glutathione precursor *N*-acetyl-cysteine (NAC) has been shown to block the H_2O_2-dependent activation of MAPK (Guyton et al., 1998). As well, cytosolic catalase activity is inversely correlated with the stimulation of MAPK through ROS generation in the cell response to lysophosphatidic acid (LPA) (Chen et al., 1995). One recent report also indicates a role for the recently characterized peroxiredoxin (Prx) enzymes, peroxidases which use thioredoxin as a source of reducing equivalent, as regulators of EGF, PDGF and TNF-α signaling in mammalian cells (Kang et al., 1998).

Peroxiredoxin III (Prx III) is exclusively localized to the mitochondria, and may serve as a useful point of study in order to evaluate the contribution of mitochondrial-generated H_2O_2 in the propagation of the cell growth response.

From the information gathered so far, it is difficult to determine whether non-toxic ROS levels act to help initiate the activation of signals associated with cell proliferation and transformation or if they are required in an autocrine and/or paracrine manner for the maintenance of these signals after initiation. The latter model seems to be a more accurate description of these effects. For instance, *c-myc* expression in hepatoma cells starved from growth factors efficiently induces apoptosis, and this induction appears to be dependent upon oxidative stress by elevated intracellular H_2O_2 levels (Xu et al., 1997). Without prior growth and survival pathway activation, the H_2O_2 response becomes one of stress. Another point of controversy concerns how, specifically, reactive compounds like H_2O_2 function to assist in the transduction of signals such as the MAPK pathway in the cell. Cellular redox potential has been demonstrated to regulate cellular enzymatic activity (Dorward et al., 1997; Flohé et al., 1997). A number of investigators suggest that protein kinase activity may be maintained indirectly by H_2O_2 as a result of negative regulation of cytoplasmic protein phosphatase functioning. Candidate enzyme targets for this inhibition by H_2O_2 which are important in the regulation of growth signals include protein tyrosine phosphatases (PTPs) such as PTP1C (targeting RTKs and the Src kinase family) and protein serine/threonine phosphatases (PSPs) including PP1 (targets Raf and Src), PP2A (targets MAPKK, MAPK) and the MAPK-binding protein phosphatases (MBPs) (Sun et al., 1993; Hunter, 1995; Misra-Press et al., 1995; Muda et al., 1996). In mouse fibroblasts expressing human EGFR, addition of hydrogen peroxide decreases overall protein tyrosine phosphatase activity (Sullivan et al., 1994), and the role of H_2O_2 in the stimulation of nuclear factor kappa B (NFκB) is well documented (Siebenlist et al., 1994; Flohé et al., 1997). The importance of transcriptional regulation by NFκB in cancer signaling is a fast-growing topic of study which deserves further discussion.

Reactive Oxygen Species and NFκB Signaling in Cancer

ROS production by the mitochondria is also regulated by a number of stress stimuli and in this respect has been observed to be responsible for direct mitochondrial damage and cytotoxicity. Increases in ROS are observed in response to a number of mitochondrial "poisoning" agents, during hypoxia/reperfusion injury, and after stimulation of the inflammatory TNFα pathway. In these situations, the oxygen radicals and peroxides can directly damage the mitochondrial membranes and other sites within the cell, and also cause oxidation of mitochondrial-generated pyridine nucleotides which then stimulates Ca^{2+} release (Richter et al., 1995). This Ca^{2+} from mitochondrial stores contributes to overall Ca^{2+} cycling kinetics of the cell and is another mechanism by which damage occurs (Gunter et al., 1994). The TNFα pathway is of particular interest when studying cancer, as TNFα is an important cytokine utilized by the immune system for the targeted destruction of tumor cells through the induction of apoptosis. Upon binding to the subtypes of its receptor (TNFR), activation of the phosholipase A_2 and protein kinase C pathways occurs as well as generation of ceramide, a sphingolipid thought to

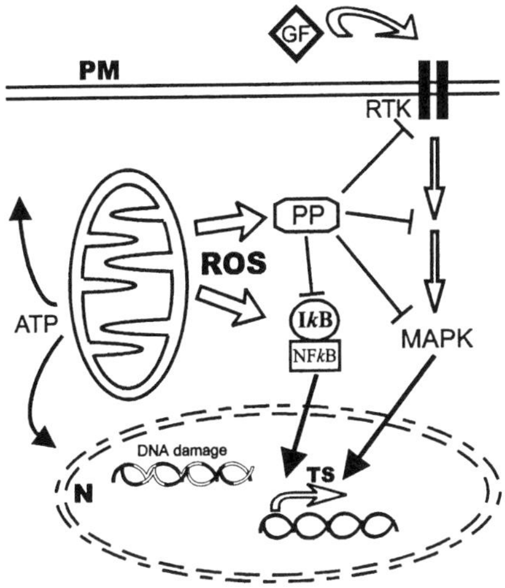

Figure 2. The mitochondria of a tumor cell are proposed to exhibit a loss of electron transport chain efficiency, resulting in an intracellular increase in reactive oxygen species (ROS) levels. ATP generation at the mitochondria is maintained at levels sufficient for cell proliferation presumably by greater reliance upon the glycolytic pathway. ROS can act to promote nuclear (N) and mitochondrial DNA damage and subsequent mutation directly, and may also function in a manner which positively regulates important cell growth and survival pathways, such as those originating from peptide growth factors (GF). One proposed mechanism for this latter effect is the inhibition of cytoplasmic protein phosphatase (PP) enzyme activity through redox regulation with subsequent kinase activation. In this manner, non-toxic ROS levels would participate in classical growth signals important in cancer progression such as those originating from receptor tyrosine kinases (RTK) at the plasma membrane (PM). Many of these signals are transmitted to the nucleus by the mitogen-activated protein kinases (MAPK) and Jun kinases, which act as transcription factors at specific DNA sites such as the AP-1 promoter and serum responsive elements. Protein phosphatase inhibition may also increase the phosphorylation and degradation of inhibitory factor kappa B (IκB), thereby releasing nuclear factor kappa B (NFκB) to the nucleus where it could act to manipulate gene transcription (TS) in favor of cell growth and transformation as well.

directly damage mitochondria by interference with complex III of the electron transport chain (Garcia-Ruiz et al., 1997). This process may be responsible for the large increases in ROS observed with TNF-mediated toxicity. In fact, the mitochondria seem to be intrinsic mediators of the TNF signal from the plasma membrane, as recent evidence indicates that the 55 kDa subtype of TNFR can induce mitochondrial clustering in fibrosarcoma lines via specific interaction with microtubules (De Vos et al., 1998). Another event observed after TNFR stimulation is the nuclear translocation of NFκB, which is involved with modulating the expression of a variety of inflammatory factors in lymphoid cells including TNFα itself (Siebenlist et al., 1994).

In order for cancer cells to avoid apoptosis by TNFα, they must block the propagation of its signal at some critical step, and the mitochondria are obvious candidates for such a resistance mechanism. One way that mitochondrial integrity may become resistant to

TNFα stimulation is by increased protection against the damaging effects of reactive oxygen species. Mitochondrial glutathione (GSH), sequestered for the cytoplasmic pool by a specific transporter, can protect from oxidative stress, as depletion of GSH results in cell sensitivity to the prooxidant effects of signals such as TNFα (Fernandez-Checa et al., 1997). As discussed previously, increased levels of this scavenger at the mitochondria may be an important factor promoting cell growth and transformation. Many tumor cells also present with an altered glucose metabolism typified by an increase in aerobic glycolysis (Brand, 1997). In theory, a reduction in mitochondrial oxygen consumption could lower the ROS produced as byproducts from the electron transport chain, thereby decreasing the effects of upstream signals which target this site to induce cell death. Products of the glycolytic pathway, such as pyruvate, may also contribute to this effect by acting as scavengers of ROS (Brand, 1997). How do these changes in the energy metabolism profile of the mitochondria occur? A likely cause is altered gene expression of components of the electron transport chain and glycolytic pathway. Reports on the study of cellular hypoxia and nutrient deprivation may provide evidence for such an effect. Hypoxia-inducible factor (HIF)-1 is a transcription factor involved in the response to hypoxia and hypoglycaemia, which activates glycolysis, erythropoiesis and angiogenesis in embryonic stem (ES) cell-derived tumors of nude mice (Carmeliet et al., 1998). Additional work indicates that the activation of certain oncogenes such as Src and Ras, which is observed in many cancers, may upregulate HIF-1 or HIF-1-like factors, leading to tumor survival and growth in stressful environments (Dang et al., 1997). How the cell tries to resist the TNFα signal may be very similar to this model, but not identical, as HIF-1 does not appear to affect apoptosis induced by cytokines (Carmeliet et al., 1998). However another transcription factor, NFκB, may play a critical role in this process.

The activation of NFκB after TNFα treatment of cells and along with cytoplasmic redox changes is well characterized and has been attributed to the function of NFκB as a downstream effector of the TNFα pathway resulting in toxicity (Siebenlist et al., 1994). However, it seems as though this view of NFκB may be too simplistic. Several studies have determined that the response of NFκB to these cell stresses may actually represent an antagonistic signal utilized by the cell as a protective mechanism against TNFα. As first demonstrated by Mattson and coworkers, activation of NF-kB in response to TNFα prevents apoptosis (Barger et al., 1995; Mattson et al., 1997). Subsequently, it was shown that fibroblasts and macrophages derived from knockout mice lacking both copies of the *rel A* gene, which encodes the p65 subunit of NFκB, display increased sensitivity to the TNFα death signal (Beg and Baltimore, 1996). This phenotype has also been shown to be manifested in primary and neoplastic human cell lines that express a dominant active form of IκB (inhibitory factor of κB), the regulator which binds NFκB in the cytosol to sequester it from the nucleus (Van Antwerp et al., 1996), and is also observed in response to other apoptotic activators such as ionizing radiation and cancer chemotherapeutics (Wang et al., 1996). Activating mutations and overexpression of NFκB are prevalent in breast and ovarian cancers, and inhibition of IκB is observed in ovarian carcinoma and ataxia telangiectasia (Luque and Gélinas, 1997). In an attempt to elucidate how NFκB might be priming the cell for survival, research has, as would be expected, focused on changes in genomic expression. Published reports so far indicate that some promoter regions targeted by NFκB are within anti-apoptotic genes including MnSOD (Mattson et al., 1997). Transcription of TRAF (TNFR-associated factor) and c-IAP (inhibitor of apoptosis) genes is induced by NFκB, and the protein products of this expression appear to be required for suppression of the TNF-induced apoptosis, by an unknown mechanism (Wang et al., 1998). An immediate early response gene, *IEX*-1L, has also been demonstrated to be activated by NFκB in a variety of transformed human cell lines after treatment with TNFα, and may be critical for increased resistance to cell death (Wu et al., 1998).

In addition to promoting survival, the NFκB pathway seems to be a contributing factor in tumor cell growth. The stimulation of NFκB activity during mitogenesis is well documented, and this activity can act synergistically with other growth-supporting transcription factors such as c-Jun and c-Fos (Luque and Gélinas, 1997; Siebenlist et al., 1994). Answers as to how these effects might be related to cancer could be provided by study of the Rel members of the NFκB family, which have similarity to the v-Rel oncogene product of the avian Reticuloendotheliosis Virus (REV-T) (Bose, 1992). v-Rel efficiently transforms avian lymphoid cells upon infection by the REV-T retrovirus, but does not produce the same effect when expressed in mammalian cell lines (Bose, 1992; Siebenlist, 1997). In spite of this apparent species specificity, alterations of the c-*rel* locus in human lymphomas (Siebenlist, 1997) suggest the existence of similar mechanisms of oncogenesis in mammalian cells. It was previously thought that these properties of v-Rel were a result of interference with normal NFκB signaling, however another model in which the activation of cellular genes by v-Rel is critical for its functioning has gained acceptance. The activation of the c-*jun* promoter by v-Rel through binding at a κB site and promoter region has been observed in HeLa cells, and in chicken fibroblasts the AP-1 complex of c-Jun and c-Fos are required for v-Rel-induced transformation (Fujii et al., 1996; Kravola et al., 1998). Another retroviral oncogene product, the Tax protein of the Human T cell Leukemia Virus-1 (HTLV-1), has recently been shown to transactivate the expression of cellular genes required for lymphocyte proliferation by acting on both AP-1 and NFκB-like sites on DNA (Mori et al., 1998). Thus, NFκB family members may participate in similar signaling pathways to those associated with AP-1 transcription factors, and this cooperation may be necessary for cell growth.

Mitochondria may act as critical regulators of NFκB signaling through their contribution to the local cytoplasmic redox environment. The increase in cellular H_2O_2 observed in response to stimulation by growth factors including TNFα is known to activate NFκB, but the molecular events leading to this activation have yet to be determined (Siebenlist et al., 1994; Lander, 1997). It is assumed that any NFκB signal promoter acts to promote the phosphorylation of IκB on amino-terminal serine residues 32 and 36, leading to its disassociation from the regulatory complex, ubiquitination, and degradation by the 26S proteosome (Flohé et al., 1997). A number of serine/threonine kinases linked to mitogenesis including protein kinase C (PKC), Ras, Raf, and MAPK kinases (MAPKKs or MEKs) have been proposed to directly phosphorylate Ser32,36 of IκB (Flohé et al., 1997; Remacle et al., 1995; Siebenlist, 1997), and as already mentioned the activity of these proteins may be controlled by reactive oxygen species. The apparent contribution of mitochondrial functioning to the regulation of cellular oxidation status, and the response to changes in this parameter, is certainly only part of an explanation as to how these organelles can influence tumor progression. An established model of the development of cancer involves loss of cell cycle control, coupled with a decreased apoptotic potential. A large body of evidence indicates that mitochondria are an essential component of all apoptosis signals, possibly by means of ROS generation, but also through other mechanisms intrinsic to cell physiology.

Mitochondrial Signaling and Apoptosis

The involvement of mitochondria as a critical component in the transduction of signals from apoptotic stimuli to molecular mediators of programmed cell death is an active area of research. Here we briefly summarize the current hypotheses regarding this role.

The inner mitochondrial membrane is impermeable and the trafficking of solutes is tightly regulated by specific transport system (Herman et al., 1990). A unique transport mechanism in mitochondria which permeabilizes the inner membrane to many unrelated small solutes and also to matrix proteins is referred to as mitochondrial permeability transition (MPT) (Zoratti and Szabo, 1995). Activation of the MPT is caused by the opening of a proteinaceous channel or pore which is observed with both normal tissue physiology and in pathological conditions (Gunter and Pfeiffer, 1990; Richter and Kass, 1991). It appears that this site functions as a sensor of matrix pH, divalent cations, adenine nucleotides, thiols, voltage and redox state of the pyridine nucleotide pool. Although the precise composition of the MPT pore complex is yet to be defined, it is currently thought to consist of the voltage-dependent anion channel (VDAC), the adenine nucleotide translocator (ANT), a matrix protein, such as cyclophilin D, and a receptor such as the peripheral benzodiazepine receptor on the outer membrane (Bernardi and Petronilli, 1996; Beutner et al., 1996; McEnery et al., 1992). The MPT pore complexes could also interact with other multiprotein complexes such as Tim (transporter of the inner membrane) complex, the Tom (transporter of the outer membrane) complex (Kinally et al., 1996) and the Bcl-2 complex (Kroemer, 1997).

Triggering of the MPT causes mitochondrial swelling due to the influx of small molecular weight solutes (up to 1500 daltons) which can be measured by decreased absorbance at 540 nm or increased forward and or side scatter by flow cytometry (Zamzami et al., 1996). The MPT is inhibited by the adenine nucleotide transporter ligand bongkrekic acid, which binds on the matrix side of the inner membrane and helps restore the mitochondrial potential, whereas atractyloside, a ligand which binds to adenine nucleotide transporter at the outer side of the inner membrane aggravates loss of potential (Kroemer et al., 1998). These effects are thought to occur by stabilizing the transporter in conformities associated with an open or closed state/conformation. One piece of evidence supporting this model is the activity of cyclosporin A, which binds to the matrix enzyme cyclophilin, prevents it maintaining the pore in an open conformation in the presence of Ca^{2+} and may act as an inhibitor of MPT (Halestrap and Davidson, 1990).

Signal transduction pathways triggered via intracellular or cell surface receptors are able to induce changes in mitochondrial permeability transition. Opening of the MPT is believed to result in a volume dysregulation of mitochondria which can disrupt the outer mitochondrial membrane and release caspase-activating proteins which are located within the intermembrane space in the cytosol. Thus, the release of cytochrome C, Apaf-l and procaspase-9 are thought to form the "apoptosome" which allows for the activation of caspase-3 and subsequent initiation of the execution caspase cascade (see Figure 3). Although apoptosis is independent of oxidative phosphorylation the central roles for mitochondria as the execution signal of apoptosis has now been confirmed in many cell systems. In general, there are at least three mechanisms which may be interrelated and are thought to be involved in apoptosis. They are (a) alteration in the energy metabolism via oxidative phosphorylation; (b) changes in redox potential dependent on mitochondria and (c) release of the "apoptosome" that triggers activation of caspase family proteases.

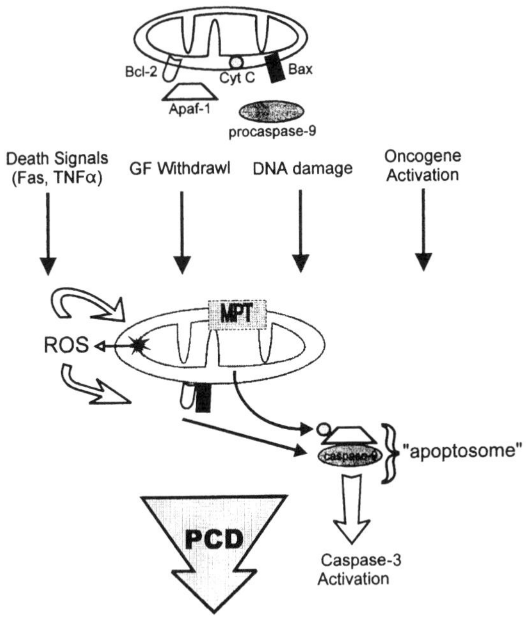

Figure 3. Mitochondria are critical regulators of the programmed cell death (PCD) signal of eukaryotic cells. Under normal physiological conditions, the mitochondrial inner transmembrane potential ($\Delta\Psi m$) is maintained, cytochrome c (Cyt C) is localized to the intermembrane space and the caspase pathway is inactive, possibly due to sequestering of apoptosis protease-activating factor 1 (Apaf-1) in an inactive conformation by members of the Bcl-2 subfamily of anti-apoptotic proteins. Upon initiation of the death signal from a variety of sources, the mitochondrial membrane permeability transition (MPT) pore opens, an event considered necessary for apoptosis to proceed. With pore opening $\Delta\Psi m$ collapses causing membrane swelling which could trigger the release of cytochrome c into the cytosol. Cytochrome (Cyt) C, Apaf-1 and procaspase-9 are thought to form the "apoptosome" in vertebrate cells, allowing for activation to caspase-9 and subsequent initiation of the execution caspase cascade, upon activation of the substrate procaspase-3. Membrane disruption also interferes with the electron transport chain, causing a sharp increase in reactive oxygen species (ROS) production. ROS may also contribute to activation of the apoptosome by the regulation of local protein function. Also, the mitochondrial response to apoptotic signaling may facilitate the interaction of certain members of the Bax subfamily of pro-apoptotic factors with Bcl-2 family members, thereby allowing for Apaf-1 release.

When these specific processes are blocked inappropriate cell survival and proliferation occur, leading to neoplasia. This can occur from nuclear-directed changes such as overexpression of antiapoptotic factors such as Bcl-2 or down regulation of caspase activity, or may also result from direct mitochondrial alterations that increase resistance to oxidative stress and MPT. Importantly, this block in apoptosis is thought to contribute to the clinical resistance of tumors to many types of anti-cancer therapies.

Summary

The role of nuclear signaling in cancer biology has been studied extensively, from the elucidation of growth factor pathway activation, induction of cell motility and invasive responses, to the inhibition of cell death programs. However, if one is to consider such processes in the context of the complete cellular environment, the influences of cytosolic components should be considered as well. It is becoming more apparent that mitochondria may be the primary players in this putative "cytoplasmic influence" upon tumor cell behavior. In accordance with the endosymbiotic origin proposed for mitochondria and eukaryotic precursors, there exists a large amount of cross talk between mitochondrial- and nuclear-derived signals within the cell. For quite some time the understanding that the coordinated expression of both genomes is required for proper mitochondrial function has been realized, and our realization that such communication has the potential to regulate other critical cellular activities is growing. The mitochondrial genome is more susceptible to mutagenic events than nuclear DNA, suggesting that mtDNA alterations may represent an initial promoting factor for the onset of neoplasia. This effect might be related to changes in cell Ca^{2+} homeostasis, local redox balances, metabolic pathways, energy production, apoptotic potential or other uncharacterized signals which are controlled at least in part by the mitochondria. Of course, this is a simplistic model. The physiological expression of these putative relationships between mitochondrial signaling and cancer would certainly depend highly upon other variables such as the local cellular environment, level of cellular differentiation and tissue type, all of which are regulated by nuclear signals. The approach to future cancer therapies may include the specific targeting of mitochondria. Drastic changes to this organelle would affect both normal and cancerous tissues, therefore, only more subtle manipulation of mitochondrial functioning would be effective. The alteration of mitochondrial ATP generation, reactive oxygen species production, or sensitization to programmed cell death as an adjunct therapy used along with conventional cancer treatment strategies might enhance chemical or radiation efficacy and reduce the acquired tumor resistance to these agents. As a better understanding of specific mitochondrial signaling is obtained, the usefulness of any such clinical applications can be more appropriately assessed.

Acknowledgements

This research was supported by the Medical Research Council of Canada 1998, studentship awarded to Tom Preston.

References

Aikawa, R., Komuro, I., Yamakazi, T., Zou, Y., Kudoh, S., Tanaka, M., Shiojima, I., Hiroi, Y. & Yazaki, Y. (1997). Oxidative stress activates extracellular signal-regulated kinases through Src and Ras in cultured cardiac myocytes of neonatal rats. J. Clin.Invest. 100, 1813-1821.

Anand-Apte, B., Zetter, B.R., Viswanathan, A., Qui, R.-G., Chen, J., Ruggier, R. & Symons, M. (1997). Platelet-derived growth factor and Fibronectin-stimulated migration are differentially regulated by the Rac and Extracellular Signal-regulated Kinase pathways. J. Biol.Chem. 272, 30688-30692.

Arora, K.K., Fanciulli, M. & Pederson, P.L. (1990). Glucose phosphorylation in tumor cells. J. Biol. Chem. 265, 6481-6488.

Arora, K.K. & Pederson, P.L. (1988). Functional significance of mitochondrial bound hexokinase in tumor cell metabolism. J. Biol. Chem. 263, 17422-17428.

Bae, Y.S., Kang, S.W., Seo, M.S., Baines, I.C., Tekle, E., Chock, P.B. & Rhee, S.G. (1997). Epidermal Growth Factor (EGF)-induced generation of hydrogen peroxide. J. Biol. Chem. 272, 217-221.

Baggetto, L.G. (1993). Role of mitochondria in carcinogenesis. Eur. J. Cancer 29A, 156-159.

Barger, S.W., Horster, D., Furukawa, K. Goodman, Y., Krieglstein J. & Mattson, M.P. (1995). Tumor necrosis factors α and β protect neurons against amyloid β-peptide toxicity: evidence for involvement of κB-binding factor and attenuation of peroxide and Ca2+ accumulation. Proc. Natl. Acad. Sci. USA 92, 9328-9332.

Beg, A.A. & Baltimore, D. (1996). An essential role for NF-κB in preventing TNF-α-induced cell death. Science 274, 782-784.

Bernardi, P. & Petronilli, V. (1996). The permeability transition pore as a mitochondrial calcium release channel: a critical appraisal. J. Bioenerg. Biomemb. 28, 129-136.

Beutner, G., Rück, A., Riede, B., Welte, W. & Brdiczkq, D. (1996). Complexes between kinases, mitochondrial porin, and adenylate translocator in rat brain resemble the permeability transition pore. FEBS Lett. 396, 189-195.

Biggs, J.R. & Kraft, A.S. (1995). Inhibitors of cyclin-dependent kinase and cancer. J. Mol. Med. 73, 509-514.

Bose, H.R. Jr. (1992). The Rel family: models for transcriptional regulation and oncogenic transformation. Biochimica et Biophysica Acta 1114, 1-17.

Boveris, A., Oshino, N. & Chance, B. (1972). The cellular production of hydrogen peroxide. Biochem. J. 128, 617-630.

Brand, K. (1997). Aerobic glycolysis by proliferating cells: protection against oxidative stress at the expense of energy yielD.J. Bioenerg. Biomemb. 29, 355-364.

Brightman, A.O., Wang, J., Kin-man Miu, R., Sun, I.L., Barr, R., Crane, F.L. & Morre, D.J. (1992). A growth factor- and hormone-stimulated NADH oxidase from rat liver plasma membrane. Biochimica et Biophysica Acta 1105, 109-117.

Carafoli, E. (1987). Intracellular calcium homeostasis. Ann. Rev. Biochem. 56, 395-433.

Carmeliet, P., Dor, Y., Hebert, J.-M., Fukumura, D., Brusselmans, K., Dewerchin, M., Neeman, M., Bono, F., Abramovitch, R., Maxwell, P., Koch, C.J., Ratcliffe, P., Moons, L., Jain, R.K., Collen, D. & Keshet, E.

(1998). Role of HIF-1α in hypoxia-mediated apoptosis, cell proliferation and tumor angiogenesis. Nature 394, 485-489.

Cavalli, L.R. & Liang, B.C. (1998). Mutagenesis, tumorigenicity, and apoptosis: are the mitochondria involved? Mutation Research 398, 19-26.

Chance, B., Sies, H. & Boveris, A. (1979). Hydroperoxide metabolism in mammalian organs. Physiol.Rev. 59, 527-605.

Chen, L.B. (1988). Mitochondrial membrane potential in living cells. Ann. Rev. Cell Biol. 4, 155-181.

Chen, Q., Olashaw, N. & Wu, J. (1995). Participation of reactive oxygen species in the lysophosphatidic acid-stimulated mitogen-activated protein kinase kinase activation pathway. J. Biol. Chem. 270, 28499-28502.

Clayton, D.A. (1991). Replication and transcription of vertebrate mitochondrial DNA. Annu. Rev. Cell Biol. 7, 453-78.

Clayton, D.A. (1992). Transcription and replication of animal mitochondrial DNAs. Int. Rev. Cytol. 141, 217-232.

Cohen, S.J. & Ellwein, L.B. (1990). Cell proliferation in carcinogenesis. Science 249, 1007-1011.

Coursin, D.B., Cihla, H.P., Sempf, J., Oberley, T.D. & Oberley, L.W. (1996). An immunohistochemical analysis of antioxidant and glutathione S-transferase enzyme levels in normal and neoplastic lung. Histol. Histopathol. 11, 851-860.

Crompton, M., Ellinger, H. & Costi A. (1988). Inhibition by cyclosporin A of a Ca2+-dependent pore in hear mitochondria activated by inorganic phosphate and oxidative stress. Biochem. J. 255, 357-360.

Darzynkiewicz, F., Traganos, L., Staiano-Coico, L., Kapuscinski, J. & Melamed, M.R. (1982). Interactions of rhodamine 123 with living cells studied by flow cytometry. Cancer Res. 42, 799-806.

Davis, S., Weiss, M.J., Wong, J.R., Lampidis, T.J. & Chen, L.B. (1985). Mitochondrial and plasma membrane potentials cause unusual accumulation and retention of rhodamine 123 by human breast adenocarcinoma-derived MCF-7 cells. J. Biol. Chem. 260, 13844-13850.

Dang, C.V., Lewis, B.C., Dolde, C., Dang, G. & Shim, H. (1997). Oncogenes in tumor metabolism, tumorigenesis, and apoptosis. J. Bioenerg. Biomembr. 29, 345-354.

De Vos, K., Goossens, V., Boone, E., Vercammen, D., Vancompernello, K., Vandenabeele, P., Haegeman, G., Fiers, W. & Grooten, J. (1998). The 55-kDa tumor necrosis factor receptor induces clustering of mitochondria through its membrane-proximal region. J. Biol. Chem. 273, 9673-9680.

Dieckmann, C.L. & Staples, R.R. (1994). Regulation of mitochondrial gene expression in Sacharomyces Cerevisial. Int. Rev. Cytol. 32, 145-181.

Dorward, A.M., Sweet, S., Moorehead, R. & Singh, G. (1997). Mitochondrial contributions to cancer cell physiology: redox balance, cell cycle, and drug resistance. J. Bioenerg.Biomembr. 29, 385-392.

D'Souza, R.J., Phillips, H.M., Jones, P.W., Strange, R.C. & Aber, G.M. (1993). Interactions of hydrogen peroxide with interleukin-6 and platelet-derived growth factor in determining mesangial cell growth: effect of repeated oxidant stress. Clin. Sci. 85, 747-751.

Egilsson, V., Evans, I.H., Wilkie, D. (1979). Toxic and mutagenic effects of carcinogens on the mitochondria of saccharomyces cerevisial. Mol. Gen. Genet. 174, 39-46.

Fernandez-Checa, J.C., Kaplowitz, N., Garcia-Ruiz, C., Colell, A., Miranda, M., Mari, M., Ardite, E. & Morales, A. (1997). GSH transport in mitochondria: defense against TNF-induced oxidative stress and alcohol-induced defect. Am. J. Physiol. 273, G7-G17

Fiskum, G. (1985). Mitochondrial respiration and calcium transport in tumor cells. In: Mitochondrial Physiology and Pathology. Fiskum, G. (editor). Van Nostrand Reinhold, New York, 180-201.

Flohé, L., Brigelius-Flohé, R., Saliou, C., Traber, M.T. & Packer, L. (1997). Redox regulation of NF-kappa B activation. Free Radic. Biol. Med. 22, 1115-1126.

Freyer, J.P. (1998). Decreased mitochondrial function in quiescent cells isolated from multicellular tumor spheroids. J. Cell. Physiol. 176, 138-149.

Fujii, M., Minamino, T., Nomura, M., Miyamoto, K.-I., Tanaka, J. & Seiki, M. (1996). Selective activation of the proto-oncogene *c-jun* promoter by the transforming protein v-Rel. Oncogene 12, 2193-2202.

Gamou, S. & Shimizu, N. (1995). Hydrogen peroxide preferentially enhances the tyrosine phosphorylation of epidermal growth factor receptor. FEBS Letters 357, 161-164.

Garcia-Ruiz, C., Colell, A., Mari, M., Morales, A. & Fernandez-Checa, J.C. (1997). Direct effect of ceramide on the mitochondrial electron transport chain leads to generation of reactive oxygen species. J. Biol. Chem. 272, 11369-11377.

Grace, S.C. (1990). Phylogenetic distribution of superoxide dismutase supports an endosymbiotic origin for chloroplasts and mitochondria. Life Sci. 47, 1875-1886.

Grivell, L.A. (1995). Nucleo-mitochondrial interactions in mitochondrial gene expression. Crit. Rev. Biochem. Mol. Biol 30(2), 121-164.

Gunter, T.E., Gunter, K.K., Sheu, S.-S. & Gavin, C.E. (1994). Mitochondrial calcium transport: physiological and pathological relevance. Am. J. Physiol. 267, C313-C339.

Gunter, T.E. & Pfeiffer, D.R. (1990). Mechanisms by which mitochondria transport calcium. Am. J. Physiol. 258, C755-C786.

Guyton, K.Z., Gorospe, M., Kensler, T.W. & Holbrook, N. (1996). Mitogen-activated protein kinase (MAPK) activation by butylated hydroxytoluene hydroperoxide: implications for cellular survival and tumor promotion. Cancer Res. 56, 3480-3485.

Guyton, K.Z., Liu, Y., Gorospe, M., Xu, Q. & Holbrook, N.J. (1998). Activation of mitogen-activated protein kinase by H_2O_2. J. Biol. Chem. 271, 4138-4142.

Halestrap, A.P. & Davidson, A.M. (1990). Inhibition of Ca^{2+}-induced large amplitude swelling of liver and heart mitochondria by cyclosporin is probably caused by the inhibitor binding to mitochondrial-matrix peptidyl-prolyl cis-trans insomerase and preventing it interacting with the adenine nucleotide translocase. Biochem. J. 268, 153-160.

Herman, B., Gores, G.J., Nieminen, A., Kawanishi, T., Harman, A. & Lemasters, J. (1990). Calcium and pH in anoxic and toxic injury. Crit. Rev. Tox. 21, 127-148.

Hunt, C.R., Sim, J.E., Sullivan, S.J., Featherstone, T., Golden, W., Von Kapp-Her, C., Hock, R.A., Gomez, R.A., Parisian, A.J. & Spitz, D.R. (1998). Genomic instability and catalase gene amplification induced by chronic exposure to oxidative stress. Cancer Res. 58, 3986-3992.

Hunter, T. (1995). Protein kinases and phosphatases: the yin and yang of protein phosphorylation and signaling. Cell 80, 225-236.

Irani, K., Xia, Y., Zweier, J.L., Sollott, S.J., Der, C.J., Fearon, E.R., Sundaresan, M., Finkel, T. & Goldschmidt-Clermont, P.J. (1997). Mitogenic signaling mediated by oxidants in Ras-transformed fibroblasts. Science 275, 1649-1652.

Johnson, L.V., Walsh, M.L., Bockus, B.J. & Chen, L.B. (1981). Monitoring of relative mitochondrial membrane potential in living cells by fluorescence microscopy. J. Cell Biol. 88, 526-535.

Kang, S. W., Chae, H.Z., Seo, M.S., Kim, K., Baines, I.C. & Rhee, S.G. (1998). Mammalian peroxiredoxin isoforms can reduce hydrogen peroxide generated in response to growth factors and tumor necrosis factor-α. J.Biol.l Chem. 273, 6297-6302.

Keely, P.J., Westwick, J.K., Whitehead, I.P., Der, C.J. & Parise, L.V. (1997). Cdc42 and Rac1 induce integrin-mediated cell motility and invasiveness through PI(3)K. Nature 390, 632-636.

Keller, J.N., Kindy, M.S., Holtsberg, F.W., St. Clair, D.K. Yen, H.-C., Germeyer, A., Steiner, S.M., Bruce-

Keller, A.J., Hutchins, J.B. & Mattson, M.P. (1998). Mitochondrial MnSOD prevents neural apoptosis and reduces ischemic brain injury: suppression of peroxynitrite production, lipid peroxidation and mitochondrial dysfunction. J. Neurosci. 18, 687-697.

Kheradmand, F., Werner, E., Tremble, P., Symons, M. & Werb, Z. (1998). Role of Rac1 and oxygen radicals in Collagenase-1 expression induced by cell shape change. Science 280, 898-902.

Khosravi-Far, R. & Der, C.J. (1994). The Ras signal transduction pathway. Cancer Metastasis Rev. 13, 67-89.

Kinnally, K.W., Lohret, J.A., Campo, M.L & Mannella, C.A. (1996). Perspectives on the mitochondrial multiple conductance channel. J. Bioenerg. Biomemb. 28, 115-123.

Kobayashi, S., Amikura, R. & Okada, M. (1993). Presence of mitochondrial large ribosomal RNA outside mitochondria in germ plasm of *Drosophila melanogaster*. Science 260, 1521-1524.

Kravola, J., Liss, A.S., Bargmann, W. & Bose, H.R. Jr. (1998). AP-1 factors play an important role in transformation induced by the v-*rel* oncogene. Molecular and Cellular Biology 18, 2997-3009.

Kroemer, G. (1997). Mode of action of the apoptosis-inhibitory proto-oncogene BCL-2. Nat. Med. 3, 614-620.

Kroemer, G., Dallaporta, B. & Resche-Rigon, M. (1998). The mitochondrial death/life regulator in apoptosis and necrosis: Ann. Rev. Physiol. 60, 619-642.

Kuiper, M. T. R., Akins, R.A., Holtrop, M., de Vries, H. & Lambowitz, A. M. (1988). Isolation and analysis of the *Neurospora crassa Cyt-21* gene. J. Biol. Chem. 263, 2840-2847.

Kunkel, T.A. & Loeb, L.A. (1981). Fidelity of mammalian DNA polymerases. Science 213, 765-767.

Kurzik-Dumke, U., Gundacker, D., Rentrop, M. & Gateff, E. (1995). Tumor suppresiion in *Drosophila* is causally related to the function of the *lethal(2)tumorous imaginal discs* gene, a *dnaJ* homolog. Dev. Genet. 16, 64-76.

Lampidis, T.J., Bernal, S.D., Summerhayes, I.C. & Chen, L.B. (1982). Rhoadamine-123 is selectively toxic and preferentially retained in carcinoma cells *in vitro*. Ann. New York Acad. Sci. 397, 299-302.

Lampidis, T.J., Bernal, S.D., Summerhayes, I.C. & Chen, L.B. (1983). Selective toxicity of rhodamine 123 in carcinoma cells *in vitro*. Cancer Res. 43, 716-720.

Lander, H.M. (1997). An essential role for free radicals and derived species in signal transduction. FASEB Journal 11, 118-124.

Li, J.-J., Oberley, L.W., St Clair, D.K., Ridnour, L.A. & Oberley, T.D. (1995). Phenotypic changes induced in human breast cancer cells by overexpression of manganese-containing superoxide dismutase. Oncogene 10, 1989-2000.

Li, N., Oberley, T.D., Oberley, L.W. & Zhong, W. (1998a). Inhibition of cell growth in NIH/3T3 fibroblasts by overexpression of manganese superoxide dismutase: mechanistic studies. J. Cell. Physiol.175, 359-369.

Li, N., Oberley, T.D., Oberley, L.W. & Zhong, W. (1998b). Overexpression of manganese superoxide dismutase in DU145 human prostate carcinoma cells has multiple effects on cell phenotype. The Prostate 35, 221-233.

Liu, R., Oberley, T. D. & Oberley, L.W. (1997). Transfection and expression of MnSOD cDNA decreases tumor malignancy of human oral squamous carcinoma SCC-25 cells. Hum. Gene Ther. 8, 585-595.

Loveland, B., Wong, C.-R., Yonekawa, H., Hermel, E. & Fischer-Lindahl, K. (1990). Maternally transmitted histocompatibility antigen of mice: a hydrophobic peptide of a mitochondrially encoded protein. Cell 60, 971-980.

Lowy, D.R. (1993). Function and regulation of ras. Ann. Rev. Biochem. 62, 851-891.

Luque, I. & Gélinas, C. (1997). Rel/NF–κB and IκB factors in oncogenesis. Seminars in Cancer Biology 8, 103-111.

MacLachlan, T.K., Sang, N. & Giordano (1995). Cyclins, cyclin-dependent kinases and Cdk inhibitors: Implications in cell cycle control and cancer. Crit. Rev, in Eukaryotic Gene Expression. 5, 127-156.

Mansour, S.J., Matten, W.T., Hermann, A.S., Candia, J.M., Rong, S., Fukasawa, K., Vande Woude, G.F. & Ahn, N.G. (1994). Transformation of mammalian cells by constitutively active MAP kinase kinase. Science 265, 966-970.

Marshall, C.J. (1995). Specificity of receptor tyrosine kinase signaling: transient versus sustained extracellular signal-regulated kinase activation. Cell 80, 179-185.

Mathupala, S.P., Rempel, A. & Pederson, P.L. (1997). Aberrant glycolytic metabolism of cancer cells: A remarkable coordination of genetic transcriptional, post-translational and mutational events that lead to a critical role for type II hexokinase. J. Bioenerg. Biomemb. 29(4), 339-343.

Matsushime, H., Roussel, M.F., Ashmun, R. & Sherr C.J. (1991). Colony-stimulating factor 1 regulates novel cyclins during the G1 phase of the cell cycle. Cell 65, 701-713.

Mattson, M.P., Goodman, Y., Luo, H., Fu, W. & Furukawa, K. (1997). Activation of NF-κB protects hippocampal neurons against oxidative stress-induced apoptosis: evidence for induction of Mn-SOD and suppression of peroxynitrite production and protein tyrosine nitration. J. Neurosci. Res. 49, 681-697.

Mazurek, S., Boschek, C.B. & Eigenbrodt, E. (1997). The role of phosphometabolites in cell proliferation, energy metabolism and tumor therapy. J. Bioenerg. Biomemb. 29(4), 315-330.

McCormack, J.G., Daniel, R.L., Osbaldeston, N.J., Rutter, G.A. & Denton, R.M. (1992). Mitochondrial Ca2+ transport and the role of matrix Ca2+ in mammalian tissues. Biochem. Soc. Trans. 20, 153-159.

McCormack, J.G. & Denton, R.M. (1990). The role of mitochondrial Ca2+ transport and matrix Ca2+ in signal transduction in mammalian tissues. Biochim. Biophys. Acta. 1018, 287-291.

McCormack, J.G. & Denton, R.M. (1993). The role of intramitochondrial Ca2+ in the regulation of oxidative phosphorylation in mammalian tissues. Biochem.Soc. Trans. 21, 793-799.

McEnery, M.W., Snowman, A.M., Trifiletti, R.R. & Snyder, S.H. (1992). Isolation of the mitochondrial benzodiazepine receptor: association with the voltage-dependent anion channel and the adenine nucleotide carrier. Proc. Nat. Acad. Sci., USA 89, 3170-3174.

Mehlen, P., Kretz-Remy, C., Briolay, J., Fostan, P., Mirault, M.-E. & Arrigo, A.-P. (1995). Intracellular reactive oxygen species as apparent modulators of heat-shock protein 27 (hsp27) structural organization and phosphorylation in basal and tumor necrosis factor α-treated T47D human carcinoma cells. Biochemistry Journal 312, 367-375.

Michiels, F., Habets, G.G.M., Stam, J.C., van der Kammen, R.A. & Collard, J.G. (1995). A role for Rac in Tiam-1-induced membrane ruffling and invasion. Nature 375, 338-340.

Misra-Press, A., Rim, C.S., Yao, H., Roberson, M.S. & Stork, P.J.S. (1995). A novel Mitogen-activated Protein Kinase phosphatase: structure, expression, and regulation. J. Biol. Chem. 270, 14587-14596.

Modica-Napolitano, J.S. & Aprille, J.R. (1987). Basis for the selective cytotoxicity of rhodamine 123. Cancer Res. 47, 4361-4365.

Mori, N., Mukaida, N., Ballard, D.W., Matsushima, K. & Yamamoto, N. (1998). Human T-cell Leukemia Virus type I Tax transactivates human *interleukin 8* gene through acting concurrently on AP-1 and nuclear factor-κB-like sites. Cancer Research 58, 3993-4000.

Muda, M., Boschert, U., Dickinson, R., Martinou, J.-C., Martinou, I., Camps, M., Schlegel, W. & Arkinstall, S. (1996). MKP-3, a novel cytosolic protein-tyrosine phosphatase that exemplafies a new class of Mitogen-activated Protein Kinase phosphatase. J. Biol. Chem. 271, 4319-4326.

Murphy, B.J., Robbin, E.D., Tapper, D.P., Wong, R.J. & Clayton, D.A. (1984). Hypoxic coordinate regulation of mitochondrial enzymes in mammalian cells. Science 223, 707-709.

Nakashima, R.A., Paggi L., Scott, J. & Pederson, P.L. (1990). Glucose cells. Science 223, 707-710. phosphorylation in tumor cells. J. Biol. Chem. 265, 6481-6488.

Oberley, T. D. & Oberley, L.W. (1997). Antioxidant enzyme levels in cancer. Histology and Histopathology

12, 525-535.

Oberley, T.D., Schultz, J.L., Li, N. & Oberley, L.W. (1995). Antioxidant enzyme levels as a function of growth state in cell culture. Free Rad. Biol. Med. 19, 53-65.

Ohtsubo, M. & Roberts, J.M. (1993). Cyclin-dependent regulation of G1 in mammalian fibroblasts. Science 259, 701-713.

Pederson, P.L. (1978). Tumor mitochondria and the bioenergetics of cancer cells. Prog. Exp. Tumor Res. 22, 190-274.

Peeper, D.S. & Bernards, R. (1997). Communication between the extracellular environment, cytoplasmic signaling cascades and the nuclear cell-cycle machinery. FEBS Letters 410, 11-16.

Rao, G.N. (1996). Hydrogen peroxide induces complex formation of SHC-Grb2-SOS with receptor tyrosine kinase and activates Ras and extracellular signal-regulated protein kinases group of mitogen-activated protein kinases. Oncogene 13, 713-719.

Remacle, J., Raes, M., Toussaint, O., Renard, P. & Rao, G. (1995). Low levels of reactive oxygen species as modulators of cell function. Mutat. Res. 316, 103-122.

Resnitsky, D., Gossen, M., Bujard, H. & Reed, S.I.. (1994). Acceleration of the G1/S phase transition by expression of cyclins D1 and E with an inducible system. Mol. Cell. Biol. 14, 1669-1679.

Richter, C. (1992). Reactive oxygen and DNA damage in mitochondria. Mutat. Res. 275, 249-255.

Richter, C. (1993) Pro-oxidant and mitochondrial CA^{2+}: their relationship to apoptosis and oncogenesis. FEBS Lett. 325, 104-107.

Richter, C., Gogvadze, V., Laffranchi, R., Schlapbach Sweitzer, M., Suter, M., Walter, P. & Yaffee, M. (1995). Oxidants in mitochondria: from physiology to diseases. Biochimica et Biophysica Acta 1271, 67-74.

Richter, C & Kass, G.E.N. (1991). Oxidative stress in mitochondria: its relationship to cellular Ca2+ homeostasis, cell death, proliferation, and differentiation. Chem. Biol. Interactions 77, 1-23.

Sherratt, H.S.A. & Turnbull, D.M. (1990). Mitochondrial oxidations and ATP synthesis in muscle. Bailliere's Clin. Endocrin. Metab. 4, 523-560.

Siebenlist, U. (1997). Meeting Report: NFκB/IκB proteins. Their role in cell growth, differentiation and development. Madrid, Spain, July 7-10, 1996. Biochimica et Biophysica Acta 1332, R7-R13

Siebenlist, U., Franzoso, G. & Brown, K. (1994). Structure, regulation and function of NF-κB. Ann. Rev. Cell Biol. 10, 405-455.

Singh, G., Hauswirth, W.W., Ross, W.E. & Neims, A.H. (1985). A method for assessing damage to mitochondrial DNA caused byradiation and epichlorohydrin. Mol. Pharmacol. 27,167-170.

Singh, G. & Maniccia-Bozzo, E. (1990). Evidence for lack of mitochondrial DNA repair follwoing cis-dichlorodiammine platinum treatment. Cancer Chemother. Pharamcol. 26, 97-100.

Singh, G., Sharkey, S.M. & Moorehead, R. (1992) Mitochondrial DNA damage by anticancer agents. Pharmac. Ther. 4, 217-230.

Singh, G. & Veltri, K. (1991). Effect of differentiation of embryonal carcinoma cells (P19) on mitochondrial DNA content. In Vitro Cell. Dev. Biol. 27A, 557-561.

Sohal, R.S. & Brunk, U.T. (1992). Mitochondrial production of pro-oxidants and cellular senescence. Mutat. Res. 275, 295-304.

Spitz, D.R., Phillips, J.W., Adams, D.T., Michael Sherman, C., Deen, D.F. & Li, G. C. (1993). Cellular resistance to oxidative stress is accompanied by resistance to cisplatin: the significance of increased catalase activity and total glutathione in hydrogen peroxide-resistant fibroblasts. J. Cell. Physiol. 156, 72-79.

Stevenson, M.A., Pollock, S.S., Coleman, C. N. & Culderwood, S. K. (1994). X-irradiation, phorbol esters, and H_2O_2 stimulate mitogen-activated protein kinase activity in NIH-3T3 cells through the formation of reactive oxygen intermediates. Cancer Res. 54, 12-15.

Sullivan, S.G., Chiu, D.T. Y., Errasfa, M., Wang, J.M., Qi, J.-S. & Stern, A. (1994). Effects of H_2O_2 on protein phosphatase activity in HER14 cells. Free Rad. Biol. Med. 16, 399-403.

Summerhayes, I.C., Lampidis, T.J., Bernal, S.D., Nadakavukaren K., Shepherd, E.L., Chen, L.B. (1982). Unusual retentionof rhodamine 123 by mitochondria in muscle and carcinoma cells. Proc. Natl. Acad. Sci. USA 79, 5292-5296.

Sun, H., Charles, C.H., Lau, L.F. & Tonks, N. K. (1993). MKP-1 (3CH134), an immediate early gene product, is a dual specificity phosphatase that dephosphorylates MAP Kinase in vivo. Cell 75, 487-493.

Sundaresan, M., Yu, Z.-X., Ferrans, V. J., Irani, K. & Finkel, T. (1995). Requirement for generation of H_2O_2 for platelet-derived growth factor signal transduction. Science 270, 296-299.

Sweet, S., Duivenvoorden, W. & Singh, G. (1998). Mitochondria as a critical target for toxicity. In Vitro Mol. Toxicol. 11, 73-81.

Sweet, S. & Singh, G. (1995). Accumulation of human promyelocytic leukemic (HL-60) cells at two energetic cell cycle checkpoints. Cancer Res. 55, 5164-5167.

Szatrowski, T. P. & Nathan, C. F. (1991). Production of large amounts of hydrogen peroxide by human tumor cells. Cancer Res. 51, 794-798.

Thannickal, V. J. & Fanburg, B. L. (1995). Activation of an H_2O_2-generating NADH oxidase system in human lung fibroblasts by transforming growth factor β1. J. Biol. Chem. 270, 30334-30338.

Tokoyuni, S., Okamoto, K., Yodoi, J. & Hiai, H. (1995). Persistent oxidative stress in cancer. FEBS Letters 358, 1-3.

Tzagoloff, A. & Myers, A.M. (1986). Genetics of mitochondrial biogenesis. Ann. Rev. Biochem. 55,249-285.

Van Antwerp, D.J., Martin, S.J., Kafri, T., Green, D.R. & Verma, I. M. (1996). Suppression of TNF-α-induced apoptosis by NF-κB. Science 274, 787-789.

van der Geer, P. & Hunter, T. (1994). Receptor protein-tyrosine kinases and their signal transduction pathways. Ann. Rev. Cell Biol. 10, 251-337.

Wang, C.-Y., Mayo, M. W. & Baldwin, A. S. Jr. (1996). TNF- and cancer therapy-induced apoptosis: potentiation by inhibition of NF-κB. Science 274, 784-787.

Wang, C.-Y., Mayo, M. W., Korneluk, R.G., Goeddel, D.V. & Baldwin, A. S. Jr. (1998). NF-κB antiapoptosis: induction of TFAF1 and TRAF2 and c-IAP1 and c-IAP2 to supress caspase-8 activation. Science 281, 1680-1683.

Wang, H., Meury, L. & Morais, R. (1997). Cloning and characterization of cDNAs encoding chicken mitogen-activated protein kinase kinase type 2, MEK2: downregulation of MEK2 in response to inhibition of mitochondrial DNA expression. Biochemistry 36, 15371-15380.

Wang, H. & Morais, R. (1997). Up-regulation of nuclear genes in response to inhibition of mitochondrial DNA expression in chicken cells. Biochimica et Biophysica Acta 1352, 325-334.

Wang, X., Martindale, J., Liu, Y & Holbrook, N. (1998).Opposing roles for ERK and JNK MAP kinases in the determination of cell fate following hydrogen peroxide (H_2O_2) treatment of HeLa cells. Proc. Am. Assoc. Cancer Res. 39, 142. Ref Type: Abstract

Warburg, O. (1923). Versuche an überleben dem carcinomgewebe. Biochem. Z. 142, 317-351.

Warburg, O. (1956a). On the origin of cancer cells. Science 123:309-314.

Warburg, O. (1956b). On respiratory impairment in cancer cells. Science 124, 269-270.

Weinberg, R.A. (1995). The retinoblastoma protein and cell cycle control. Cell 81, 323-330.

Weinhouse, S. (1956). On respiratory impairment in cancer cells. Science 124, 267-268.

White, E. (1996). Life, death, and the pursuit of apoptosis. Genes Dev. 10, 1-15.

Williams, R.S. (1986). Mitochondrial gene expression in mammalian striated muscle. J. Biol. Chem. 261,

12390-12394.

Williams, R.S., Salmons, S., Newsholme E.A., Kaufman, R.E. & Mellar, J. (1986). Regulation of nuclear and mitochondrial gene expression by contractile activity in skeletal muscle. J. Biol. Chem. 261, 376-380.

Wu, M.X., Ao, Z., Prasad, K.V.S., Wu, R. & Schlossman, S.F. (1998). *IEX*-1L, an apoptosis inhibitor involved in NF-κB-mediated cell survival. Science 281, 998-1001.

Xu, Q., Nguyen, Q., Lo, D.C. & Czaja, M.J. (1997). *c-myc*-dependent hepatoma cell apoptosis results from oxidative stress and not a deficiency in growth factors. J. Cell. Physiol. 170, 192-199.

Yang, W. & de Bono, D.P. (1997). A new role for vascular endothelial growth factor and fibroblast growth factors: increasing endothelial resistance to oxidative stress. FEBS Letters 403, 139-142.

Zamzami, N., Susin, S.A., Marchetti, P., Hirsch, T., Gomez-Monterray, I., Castedo, M. & Kroemer., G. (1996). Mitochondrial control of nuclear apoptosis. J. Exp. Med. 183, 1533-1544.

Zhong, W., Oberley, L.W., Oberley, T.D. & St Clair, D.K. (1997). Suppression of the malignant phenotype of human glioma cells by overexpression of manganese superoxide dismutase. Oncogene 14, 481-490.

Zinkewich-Péotti, K., Parent, M. & Morais, R. (1990). Mitochondrial DNA modulation of the anchorage-independent phenotype of transformed avian cells. Cancer Res. 50, 6675-6682.

Zoratti, M. & Szabo, I. (1995). The mitochondrial permeability transition. Biochim. Biophys. Acta 1241, 139-176.

MOLECULAR CHAPERONES AND AGE-RELATED DEGENERATIVE DISORDERS

ALBERTO J.L. MACARIO & EVERLY CONWAY DE MACARIO

Wadsworth Center, Room B-749, Division of Molecular Medicine, New York State Department of Health; and Department of Biomedical Sciences, School of Public Health, The University at Albany (SUNY); Empire State Plaza, P.O. Box 509, Albany, New York 12201-050, USA

Table of Contents

Introduction

Degenerative Diseases and Aging

Degenerative diseases of the CNS, including those associated with the prion agent, tend to appear in the 4th and 5th decades of life and progress with increasing age (Geller and

Interorganellar Signaling in Age-Related Disease, Edited by Mark P. Mattson, 131—162

Potter, 1999; Harris, 1999; Ross et al., 1998; Schweisguth, 1999; Tran and Miller, 1999; Wang et al., 1999). It is unclear whether these pathologic entities are primarily a cause or primarily a consequence of the aging process. Perhaps aging and the appearance of degenerative manifestations go hand in hand, helping each other as it were to drive the cell and the organism to the end of their lives.

The role of stress in the aging process and in the generation of degenerative diseases of the CNS is also unclear. However, there seems to be a correlation between cell stress, oxidation, cellular damage, aging, and death (Ballester and Harchelroad, 1999; Favatier and Polla, 1996; Geller and Potter, 1999; Halliwell and Gutteridge, 1990; Lin et al., 1998; McEwen, 1998; 1999; Rudolph et al., 1999; Rustin et al., 1999; Vendemiale et al., 1999).

In this Chapter, we will consider degenerative diseases of the CNS in aging individuals from the standpoint of protein quality and balance. A central tenet of our approach is that protein quality and balance play a key role in the cell's life, survival, and longevity. Protein quality and balance is represented by the set of different proteins, each with a functional configuration, that make up the cell, and that must be maintained within physiological ranges of concentrations and structural shapes.

Stress

Cell stress, caused by a variety of stressors (Table 1), perturbs protein balance, because it: (*i*) causes protein denaturation; (*ii*) down-regulates or shuts off genes that produce house-keeping proteins involved in the many vital, physiological functions that characterize a healthy cell, like ingestion and processing of nutrients, growth, differentiation, multiplication, and secretion of molecules useful to other cells; and (*iii*) induces accumulation of a set of proteins whose functions are to counteract the harmful effects of stress. The latter proteins are the stress proteins, also called heat-shock proteins or Hsp (Table 2). It follows that a stressed cell is profoundly disturbed with its protein balance broken and is in peril of dying. Whether or not death ensues depends on several factors, among which the action of stress proteins is paramount. In this regard, molecular chaperones (see Table 2, Group 1) play a central role in the folding of new polypeptides, preventing protein denaturation, and rescuing proteins partially denatured (Glover and Linquist, 1998; Hartl and Martin, 1995; Parsell et al. 1994). Therefore, it might be said that molecular chaperones are a major force in preventing the consequences of stress and in enhancing recovery after stress.

The Diseases

Mutated Genes — Defective Proteins — Aggregates

The diseases considered in this Chapter are characterized by the presence of one or more abnormal proteins with a tendency to misfold, and thereby to aggregate and form deposits inside and outside the cell. The molecular situation in these diseases resembles that observed in a stressed cell in as much as a protein molecule tends to aggregate and

Table 1. Cell Stressors

Type	Name Description[a]
Physical	Heat (including fever); several types of irradiation, including ultraviolet light.
Oxygen	Oxygen-derived free radicals (ROS); anaerobiosis to aerobiosis shift (e.g., reperfusion); hypoxia-anoxia (ischemia).
pH	Alkalosis; acidosis; pH shift.
Biological	Infection; inflammation; fever.
Psychological	Emotion; hormonal imbalance due to HPA axis dysfunction
Osmotic	Changes in the concentration of salt, sugars, other osmolytes (hyper- or hypo-osmotic shock).
Nutritional	Starvation: multiple; specific (carbon, glucose, nitrogen, phosphate, nitrate).
Antibiotics	Puromycin; tetracycline; nalidix acid.
Alcohols	Ethanol; methanol; butanol; propanol; octanol.
Metals	Cadmium; copper; chromium; zinc; tin; aluminum; mercury; lead; nickel.
Insecticides, pesticides	Lindane; diazinon; paraquat; thiram; tributyltin.
Mechanical	Compression; shearing.
Other	Benzene and derivatives; phenol and derivatives; teratogens; carcinogens; mutagens; arsenite; arsenate; amino acid analogs; nicotine; anesthetics; desiccation.

[a] Abbreviations are: ROS, reactive oxygen species; HPA axis, hypothalamic-pituitary-adrenal axis. These agents cause stress in cells from the three phylogenetic domains, Bacteria, Archaea (both prokaryotes), and Eucarya (eukaryotes). Modified from Macario and Conway de Macario, 2000, with permission from the copyright owner.

Table 2 Stress (Heat-Shock) Proteins

Group	Examples[a]
1	Molecular chaperones and chaperonins in organisms of the three phylogenetic domains: Bacteria, Archaea, and Eucarya. Eukaryotic representatives are listed in Table 7.
2	Regulators of stress (heat-shock) genes: positive regulators or activators (e.g., heat-shock factor or HSF, in eukaryotes); negative regulators (e.g., HrcA; HspR) sigma factors (e.g., σ^{32}, σ^{E}, σ^{B}) in bacteria.
3	Proteasome components and ubiquitin in archaea and eukaryotes. Other proteases: Clp family; Lon/La; HtrA in bacteria.
4	Proteins in the intercellular connective material of multicellular structures formed as a response to stress (e.g., in the archaeon *Methanosarcina mazeii*).
5	Transport proteins (e.g., TrkA in the archaeon *Methanosarcina mazeii*).
6	Enzymes and co-factors involved in the synthesis of other stress proteins (listed in 1-5 above), and other molecules such as thermo-protectants or "chemical" chaperones (e.g., trehalose in the yeast *Saccharomyces cerevisiae*; di-*myo*-inositol phosphate and cyclic diphosphoglycerate in some archaea).
7	All proteins, enzymes, co-factors, and gene-regulatory factors involved in sporulation (e.g., in the bacterium *Bacillus subtilis*).

[a] Modified from Macario and Conway de Macario, 2000, with permission from the copyright owner.

must be attended to by molecular chaperones. This is the reason why these diseases are sometimes grouped as proteinopathies.

Diseases pertinent (because of their pathology) to this Chapter are listed in Tables 3a-d, and 4. We have followed a classification that is found in textbooks and that is well known, although perhaps not universally accepted. We wished to provide a familiar picture as the basis for discussion, rather than embarking on a proposal for a novel classification based on the recent discoveries of genes, mutations, protein folding mechanisms, and forms of inheritance. An example of this latter type of classification is provided in Table 5, in which the diseases that have abnormal proteins with a polyglutamine expansion are listed. A similar type of classification will no doubt emerge in the near future also for other diseases now listed in Tables 3a-d and 4, as more knowledge is accumulated on their molecular and genetic mechanisms. In fact, it has been clear for a long time that classifications based on the anatomical distribution of the main lesions are not satisfactory (see for example Feany et al., 1996; Rosenberg 1995; Schneider et al., 1997; Wilhelmsen, 1999). Often, under the same name, there are different albeit related disorders, which happen to affect the same regions of the CNS (and thus appear quite similar in their signs and symptoms).

A common denominator of all these disease is that they show protein deposits in the affected tissues (Table 6). These deposits are a manifestation of the presence of an abnormal protein that tends to aggregate and precipitate. How much of this aggregation and precipitation depends on the abnormal protein itself and on other factors such as stress, and Hsp, remains to be determined for each disease. It may very well be that in some cases the abnormality of the protein leads readily to misfolding, even in the absence of other factors. In other diseases, however, the abnormal protein might have only a weak tendency to misfold, and will only do so and aggregate when other factors such as stress co-exist. Perhaps in the last analysis it will become clear that while sooner or later every cell and organism succumbs to persistent and/or repeated stress (life in fact), the presence of certain protein abnormalities will accelerate the appearance of the pathological manifestations typical of CNS degenerative diseases. These manifestations will be more pronounced in those regions of the CNS in which the normal version of the pathologic protein is more abundant and/or more critical to the cell in normal individuals, and in which it forms the earliest and most prominent deposits. Thus, a certain anatomical preference for the lesions characteristic of each disease will always be discernable.

It follows from the above that the stress response and the stress proteins, molecular chaperones in particular, are key players in the maintenance of health, and possibly in preventing degenerative disorders due to proteinopathies.

However, what happens if there are problems with the stress response, and the stress proteins are not of the quality needed for optimal function (i.e., when a molecular chaperone is defective because of a mutation)? Likewise, what happens when the stress proteins are not produced in the quantities necessary for the cell to cope with stress (e.g., when there is a malfunction of the gene-regulation mechanism)? Thus, we may envisage diseases caused by a combination of abnormal proteins (abnormal in quality and/or quantity), one characteristic of a given disease (see Tables 3a-d, 4, and 5) and the other a molecular chaperone.

Table 3a CNS Degenerative Diseases: Gray Matter

Disease	Histopathology	Proteins(s)[a]
Alzheimer's (AD)	Diffuse plaques; neuritic plaques; neurofibrillary tangles; amyloid angiopathy; Hirano bodies (with actin); vacuoles with argyrophilic granule	APP; Aβ; tau; presenilin 1; presenilin 2; apolipoprotein E; α_2-macroglobulin; caveolin; complement regulators[b]
Pick's (PD)	Pick cells with Pick bodies	Aβ; tau[c]

[a] Proteins (and genes) normal or abnormal, known or suspected to play a role in pathogenesis. Abbreviations are: APP, amyloid precursor protein; Aβ, amyloid β

[b] Crawford et al., 1999; Cruts et al., 1996; Jaikaran et al., 1999; Kurt et al., 1999; Licastro et al., 1999; Love et al., 1988; Nishiyama, et al., 1999; Poduslo et al., 1999; Schenk et al., 1999; Schweisguth, 1999; Small and McLean, 1999; Will et al., 1996; Xia et al., 1997; Yang et al., 1999; Yasojima et al., 1999.

[c] Feany et al., 1996; Love et al., 1988; Lowe et al., 1988; Schneider et al., 1997; Wilhelmsen,1999.

Table 3b CNS Degenerative Diseases: Basal Ganglia, Nigrostriatal System

Disease	Histopathology	Protein(s)[a]
Parkinson's (PD)	Substantia nigra, pallor; Lewy bodies	α-synuclein; parkin; UCH-L1[b]
Progressive supranuclear palsy (PSP)	Straight filaments (seen by EM)	Tau[c]
Corticobasal degeneration (CBD)	"Ballooned neurons"	Tau[d]
Multiple system atrophy (MSA):	Cytoplasmic inclusions	Tau; αB-crystallin[e]
striatonigral degeneration		
olivopontocerebellar atrophy		
Shy-Drager syndrome	Lewy bodies (occasionally)	
Huntington's (HD)	Neuronal intranuclear inclusions	Huntingtin[f]

[a] Proteins (and genes) normal or abnormal, known or suspected to play a role in pathogenesis. Abbreviations are: UCH-L1, ubiquitin-carboxy-terminal-hydrolase-L1; EM, electron microscopy.

[b] Harhangi et al., 1999; Tran and Miller, 1999; Wood et al., 1999.

[c] Hardman and Halliday, 1999.

[d] and [e] Feany et al., 1996; Schneider et al., 1997.

[f] Davies et al., 1997; DiFiglia et al., 1997; Kaemmer and Low, 1999; Li et al., 1999; Ross et al., 1998; Scherzinger et al., 1997; The Huntington's Disease Collaborative Research Group, 1993; Tran and Miller, 1999. See also Table 5.

Table 3c CNS Degenerative Diseases: Spinocerebellar System

Disease	Histopathology	Proteins(s)[a]
Spinocerebellar ataxias		
Friedrich's	Mitochondrial pathology; cardiomyopathy	Ataxin (frataxin)
Ataxia telangiectasia	Telangiectasias in CNS, skin, conjunctiva; thymus hypoplasia and other lymphoid system deficiencies; large nuclei in cells, "amphicytes", e.g., pituitary, Schwann, endothelial cells	ATM

[a] Proteins (and genes) normal or abnormal, known or suspected to play a role in pathogenesis (references in Lavin 1999; Ranum et al., 1995; Rosenberg, 1995; Rustin et al., 1999). See also Tables 5 and 6 for details and other references.

Table 3d CNS Degenerative Diseases: Motor Neurons

Disease	Histopathology	Protein(s)[a]
Amyotrophic lateral sclerosis (ALS) and familial ALS (FALS)	Loss of: anterior-horn cells in spinal cord, anterior-root myelinated fibers, and same for affected cranial nerve nuclei and fibers; cytoplasmic aggregates	α-synuclein; superoxide dismutase (SOD1)[b]
Bulbospinal atrophy (Kennedy's)	Testicular atrophy; oligospermia	Androgen receptor (AR) with poly-Q expansion[c]
Spinal muscular atrophy	Loss of motor neurons in children	?

[a] Proteins (and genes) normal or abnormal, known or suspected to play a role in pathogenesis.
[b] Penco et al., 1999; Tran and Miller, 1999.
[c] Hsiao et al., 1999; Li et al., 1998; Ross et al., 1998; Stenoien et al., 1999; Tran and Miller, 1999 (see also Tables 5 and 6).

Table 4 Transmissible Spongiform Encephalopathies (TSE): Human

Disease	Histopathology	Proteins(s)[a]
Creutzfeld-Jakob (CJD)	Spongiform appearance of CNS tissue; intracellular vacuoles and cysts; "spiked ball" plaques	PrP^{sc} (PrP^{res})[b]
Variant CJD (vCJD)	Same; "florid" plaques	Same
Gerstmann-Straussler-Schenker syndrome (GSS)	Similar to CJD	Same
Fatal familial insomnia (FFI)	Same (but no intracellular vacuoles)	Same
Kuru	Same, and kuru plaques	Same

[a] Proteins (and genes) normal or abnormal, known or suspected to play a role in pathogenesis. Abbreviation: PrP^{sc}, prion protein scrapie (or res, i.e., resistant to digestion with proteinase K).
[b] Bamborough et al., 1996; Borchelt et al., 1992; Harris, 1999; Harrison et al., 1999; Hourichi and Caughey, 1999; Huang et al., 1996; Jackson et al., 1999; Prusiner, 1982; Will et al., 1996.

There are several families of stress proteins, most of which are molecular chaperones or cochaperones that are important in eukaryotic cells. Those pertinent to humans and to organisms used as models to study problems pertinent to human pathology are listed in Table 7. It is not yet clear which of those proteins listed in the table are important *in vivo*, under "normal" conditions (which implies bombardment of the cell by a variety of stressors), and for what, and when they are needed, in regard to development of degenerative diseases.

Table 5 Proteins with Poly-Q Tract Expansion, Diseases in Which They are Found, and CNS Areas with the Earliest and/or the Most Prominent Lesions

Protein (Gene)[a]	Number of CAG Repeats in Abnormal Gene (Range)	Disease	Predominantly Affected Area
Huntingtin	36-121 (6-35)[b]	Huntington's	Nigrostriatal system
Ataxin 1	40-83 (6-39)	Spinocerebellar ataxia 1	Cerebellum and basal ganglia
Ataxin 2	33-77 (14-32)	Spinocerebellar ataxia 2	Cerebellum and brainstem
Ataxin 3	55-86 (12-40)	Spinocerebellar ataxia 3 (Machado-Joseph's)	Cerebellum and basal ganglia
P- or Q-type of Ca^{2+} channel	21-30 (4-18)	Spinocerebellar ataxia 6	Purkinje cells in cerebellum
Ataxin 7	38-130 (7-17)	Spinocerebellar ataxia 7	Cerebellum and retina
Androgen receptor	38-62 (9-36)	Bulbospinal atrophy (Kennedy's)	Spinal motor neurons
Atrophin 1	49-88 (3-36)	Dentatorubral pallidolusyan atrophy	Basal ganglia and cerebellum

[a] See also Tables 3b,c,d, and 6, for other details and references.
[b] Normal range within parentheses. Modified from Ross et al., 1998, and Tran and Miller, 1999, with permission from the copyright owners.

Table 6 Protein Deposits in Degenerative Diseases ("Proteinopathies")

Deposit Type	Component(s)[a]	Disease
Amyloid plaque; neuritic plaque (extracellular)[b]	Aβ; NACP (α-synuclein fragment); ubiquitin; activated C' components; α_1-chymotrypsin; apolipoprotein	Alzheimer's
Neurofibrillary tangles (intracellular)[b]	Tau; ubiquitin; MAP_2	
Pick body (intracellular)[c]	Neurofilaments; Aβ	Pick's
Lewy body (intracellular)[d]	α-synuclein; ubiquitin; UCH-L1; neurofilaments	Parkinson's
		Occasionally in Shy-Drager syndrome, Alzheimer's, and dementia with Lewy bodies (DLB)

Table 6 Continued.

Deposit Type[e]	Component(s)a	Disease
Cytoplasmic inclusions or aggregates (intracellular)[e]	Tau; ubiquitin; αB-crystallin; α-synuclein	Multiple systems atrophy
	Superoxide dismutase; ubiquitin; α-synuclein	Amyotrophic lateral sclerosis (AML) and familial AML (FAML)
	Mitochondria; SRC-1; Hsp70; Hsp90; HDJ-2; proteasome components	Bulbospinal atrophy (Kennedy's)
Straight filaments (intracellular)[f]	Tau	Progressive supranuclear palsy (PSP)
Nuclear inclusion (intranuclear)[g]	Protein with poly-Q tract expansion (e.g., huntingtin, ataxins); ubiquitin	Huntington's and other poly-Q tract expansion disorders (see Table 5)
Amyloid plaque (extracellular)[h]	PrP^{sc}; ubiquitin	Prion diseases (transmissible spongiform encephalopathies), particularly kuru
Corneal opacity (extracellular)[i]	α-crystallin	Cataract
Filaments (intracellular)[j]	Intermediate filaments; αB-crystallin; desmin	Desmin-related myopathy (DRM)
Rosenthal fiber (intracellular)[k]	Intermediate filaments; Hsp27; αB-crystallin	Alexander's
Mallory body (intracellular)[l]	Intermediate filaments; sHsp	Alcoholic hepatitis

[a] Abbreviations are: Aβ, amyloid β; C', complement; MAP_2, microtubule-associated protein 2; UCH-L1, ubiquitin carboxy-terminal-hydrolase-L1; SRC-1, steroid-receptor coactivator 1; Hsp, heat-shock protein; sHsp, small Hsp; HDJ-2, human Hsp40(DnaJ) homolog; PrP^{sc}, prion protein scrapie (or res, i.e., resistant to digestion with proteinase K).

[b] Kurt et al., 1999; Licastro et al., 1999; Lowe et al., 1988; Lowe et al., 1993; Mori et al., 1987; Perry et al., 1987; Schenk et al., 1999; Small and McLean, 1999; Yasojima et al., 1999.

[c] Feany et al., 1996; Love et al., 1988; Lowe et al., 1988; Schneider et al., 1997.

[d] Hardman and Halliday, 1999; Harhangi et al., 1999; Love et al., 1988; Lowe et al., 1988; 1993; Schneider et al., 1997; Tran and Miller, 1999.

[e] Hsiao et al., 1999; Li et al., 1998; Penco et al., 1999; Ross et al., 1998; Stenoien et al., 1999; Tran and Miller, 1999; Wood et al., 1999.

[f] Feany et al., 1996; Schneider et al., 1997.

[g] Cummings et al., 1998; Davies et al., 1997; DiFiglia et al., 1997; Li et al., 1999; Paulson et al., 1997; Ranum et al., 1995; Ross et al., 1998; Scherzinger et al. 1997; Skinner et al., 1997; Tran and Miller, 1999.

[h] Bamborough et al., 1996; Borchelt et al., 1992; Harris, 1999; Harrison et al., 1999; Hourichi and Caughey, 1999; Huang et al., 1996; Kato et al., 1992; Lowe et al., 1990; 1992.

[i] Linnik and Herscovitz, 1998; Litt et al., 1998; Muchowski et al., 1999; Perng et al., 1999.

[j] Vicart et al., 1998.

[k] Iwaki et al., 1989; 1993; Lowe et al., 1988.

[l] Lowe et al., 1988.

Table 7 Stress (Heat-Shock) Proteins and Chaperones in Eukaryotes

Family		Examples[a]		
Name(s)	Mass (kDa)	Cytosol; Nucleus	Mitochondria	Endoplasmic Reticulum (ER)
Heavy; High molecular weight; Hsp100	100 or higher	Hsp101(Hsp102); Hsp104; Hsp105α & β; Hsp110		Hsp170(Grp170)
Hsp90	81-99	Hsp90(Hsp82; Hsp83; Hsp90α & Hsp90β) Grp94		Hsp90(GRP94; rp94; endoplasmin; gp96)[b]
Hsp70 Chaperones	65-80	Hsp70(Hsp72) Hsc70(Hsp73)[c] SSA1-4; SSB1-2	mhsp70(SSC1)[d] mt-Hsp70(Grp75)	BiP(Grp78)[e] SSi1p; Kar2p; Lhs1p
Hsp60 Chaperonins	55-64	CCT(TRiC; TCP-1 complex) subunits (up to 9)[f]	Hsp60(Cpn60)	RBP
Hsp40 DnaJ	35-54	Hsp40; HDJ-1(MAS5) HDJ-2(HSDJ) Sis1p; Ydj1p[g]	Hsp40; Mdj1p Mdj2p	Hsp40; Hsp47 Sec63p; Scj1p; JEM1p
Small Hsp sHsp	34 or lower	Hsp10; PPIase; PDIase Prefoldin(Gim) subunits (up to 6) α-crystallin family[h]	Hsp10(Cpn10) PPIase; mt-GrpE (Yge1p;Mye1p)	Cpn20 PPIase PDIase

[a] Synonyms are within parentheses. Stress proteins and molecular chaperones of the chloroplast, typical of plants, are not included. General references pertinent to the various Hsp families (members, function, mechanism of action, partners, evolution) are: Bustard and Gupta, 1997; Gribaldo et al., 1999; Gupta 1995; 1995a; 1998; Gupta et al., 1994; Gupta and Singh, 1994; Hartl and Martin, 1995; Hunt et al., 1999; Macario, 1995; Oh et al., 1999; Ranson et al., 1998; Rassow et al., 1997; Ryan et al., 1997. Specific references for each Hsp family and cell locale are listed in the other footnotes.

[b] Grp94 (Glucose-Regulated Protein 94) interacts with one or more of these other chaperones: BiP(Grp78), ERp72 (a PDIase), calnexin, calreticulin, Hsp47, and hsp170. Interactions and participants vary with different Grp94 functions (Csermely et al., 1998; Ferrari and Soling, 1999; Mayer and Bukau, 1999; Trombetta and Helenius, 1998).

[c] Hsc70 interacts with cofactors Hip (for Hsc70-interacting protein; named also p48) and Hop (for hsp70/hsp90-organizing protein; named also IFF-SSP-3521, extendin, or p60 in mammals, or STI1 in yeast), and with CHIP (for carboxyl terminus of Hsc70-interacting protein), Hsp40, and BAG1 (Ballinger et al., 1999; Chen and Smith, 1998; Frydman and Hohfeld, 1997; Heine et al., 1999; Hohfeld et al., 1995; Netzer and Hartl, 1998).

[d] Mitochondrial Hsp70 interacts with mGrpE(Mge1p; Yge1p), Mdj1p, and members of the translocase system of the inner membrane (Tim proteins) for protein folding and transport. It also interacts with Mdj1p and proteases (e.g., PIM1) for degradation of misfolded proteins (Naylor et al., 1996;1998; Pfanner and Meijer, 1997; Ryan et al., 1997a; Wagner et al., 1994; Westermann and Neupert, 1997).

[e] BiP (Binding Protein, or Immunoglobulin Heavy Chain-Binding Protein) interacts with Sec63p, and other components of the ER for protein translocation; also other interactions occur between the ER chaperones (Craven et al., 1997; Hosokawa et al., 1998; Linnik and Herscovitz, 1998; Misselwitz et al., 1999; Nishikawa and Endo, 1997). See also Munro and Pelham, 1986; Saris et al., 1997; Zapun et al., 1999).

[f] CCT, for Chaperonin Containing TCP-1; also named TRiC for TCP-1 Ring Complex; TCP-1 stands for Tailess Complex Protein 1 (Rommelaere et al., 1999; Willison, 1999; Won et al., 1998). CCT interacts with Hop, and possibly with Prefoldin/Gim, to assist in protein folding (Gebauer et al., 1998; Geissler et al., 1998; Vainberg et al., 1998).

[g] Human and eukaryotic (mostly studied in yeast) homologs of the Hsp40 family are numerous, and they interact with other chaperones in the different cell compartments (Cheetham et al., 1992; Chellaiah et al., 1993; Cyr et al., 1994; Nishikawa and Endo, 1997; Oh et al., 1993).

[h] PPIase, for Peptidyl-Prolyl *cis-trans* Isomerase (Fisher et al., 1998); PDIase, for Protein-Disulfide Isomerase (Ferrari and Soling, 1999); prefoldin(Gim) (Geissler et al., 1998; Vainberg et al., 1998); α-crystallin family, and other sHsp (De Jong et al., 1998; Downs et al. 1999; Kim et al., 1998; 1998a; Lindner et al., 1998; Munchbach et al., 1998; Perng et al., 1999; Rogalla et al., 1999).

Protein Folding and Protein Degradation — Cell Compartments

In this Chapter, we will examine a few examples in which the products of molecular chaperone genes, mRNA and/or protein, have been found quantitatively altered in, or associated with, degenerative disorders of the CNS in tissues from humans or from experimental animals. It will be seen that in most if not all cases, the role of Hsps against the disease, or in promoting its development as a pathogenetic factor, was not elucidated. Mere detection of a stress response, manifested by increased levels of one or more Hsps (the proteins themselves, and/or the mRNA from their respective genes) in cells from diseased individuals does not tell much about pathogenesis. There are no final answers as to why the Hsps are augmented in the diseased tissues or in the protein deposits, or what they were doing before they coprecipitated with the abnormal proteins. The problem is compounded by the fact that molecular chaperones function as teams. The team members vary depending on the task to be done, the kind of protein to be assisted in its folding and trafficking endeavors, and the cell locale in which they work–cytosol, ER, mitochondria (see Table 7). Thus, the analysis of the role of molecular chaperones in the pathogenesis of degenerative diseases is extremely difficult *in vivo* as well as *in vitro*. If a particular protein is mutated in a patient and has a tendency to misfold, answering the question, which are the chaperones most relevant to its correct folding, is a complex undertaking. Hence, one has to be cautious, for example, in the interpretation of results from laboratory experiments with isolated chaperones, and of data from analyses of the components of protein deposits, in which one or two molecular chaperones are detected.

An additional complication stems from the fact that protein balance is maintained not only by producing new molecules with the correct configuration as they are needed, but also by eliminating molecules that are damaged beyond repair due to stress or because they are structurally unfit (mutated) (Ciechanover, 1998; Hayes and Dice, 1996; Herman and D'Ari, 1998; Piva et al., 1999; Wagner et al., 1994; Yamao, 1999). Thus proteolysis is also a major player whose role in the development of degenerative diseases with protein deposits deserves serious consideration. It may well be that some degenerative diseases are caused, at least in part, by a defect in the protein elimination system. In fact, components of the proteolytic system represented by the ubiquitin-proteasome pathway are found associated with the pathologic protein deposits (see Table 6; and following sections in this Chapter). Failure of the proteolytic systems may contribute to the accumulation of abnormal proteins and/or their partial digestion products. Protein aggregates inside the cells, or in their surroundings, may cause disease by simply imposing a mechanical burden. In addition, there is information suggesting that the protein aggregates, and the peptide fragments themselves, are toxic, and that the latter fragments increase in number when the proteolytic process is incomplete or incorrect, yielding unusual peptides (Small and McLean, 1999; Wang et al., 1999).

Molecular Chaperones and Disease

Alzheimer's Disease (AD)

The search for manifestations of a stress response and for the presence of stress proteins in post-mortem tissues of individuals with AD started many years ago. For example, it was reported that mRNA for Hsp70 increased in the cortex of AD patients in relation to their condition (specifically fever) at the time of death (Guillemette et al., 1986). More recent work has also dealt with the levels of stress proteins in the diseased CNS. In one study, post-morten tissue samples from 18 patients, 55-90 years of age, and age-matched controls were examined using immunohistochemical and Western blotting methods with mouse monoclonal antibodies against human Hsp of 28, 72, 78, and 90 kDa (Hamos et al., 1991). Hsp72 and Hsp78(Grp78) (see Table 7) were the only two that showed differences in the AD cortex as compared with the controls. Hsp72 immunoreactivity was clearly increased in AD, and appeared only in the fibrillary tangles and neuritic plaques (see Table 6) but never in neurons that had a normal morphology. Grp78 immunoreactivity was also increased somewhat in AD as compared with controls, and it was localized only in neurons with an otherwise normal appearance.

The presence of Hsp72 in fibrillary tangles and neuritic plaques does not tell much about the role of the chaperone against or for disease progression. The reported findings only suggest that Hsp72 recognized and bound abnormal peptides in the tangles and plaques or about to form these deposits, and subsequently coprecipitated with the other proteins. The increase in Grp78 inside otherwise normal-looking neurons indicates perhaps that abnormal peptides were present inside the ER, where Grp78 resides (see Table 7), since it is known that this chaperone augments when there is accumulation of abnormal proteins inside the ER (Kozutsumi et al., 1988).

Further evidence of stress proteins in the pathophysiology of Alzheimer's and related disorders comes from studies showing that insults relevant to the pathogenesis of such disorders increase expression of Hsp 70 and Grp78 in neurons, and that suppression of such increases in these stress protein increases neuronal vulnerability to the insults (Yu et al., 1999; Lee et al., 1999). Thus, exposure of cultured hippocampal neurons to amyloid β-peptide or iron, agents that induce oxidative stress, results in increases in levels of Grp78. Treatment of the cultures with a Grp78 antisense oligonucleotide increases neuron death following exposure to amyloid β-peptide and iron. Measurements of levels of oxidative stress and intracellular calcium in control and Grp78 antisense-treated neurons provided evidence that the mechanism whereby Grp78 protects neurons involves stabilization of cellular calcium homeostasis and suppression of free radical production (Yu et al., 1999). Administration of the metabolic stressor, 2-deoxyglucose, to cultured neurons or adult rats results in increased expression of Hsp70 and Grp78, which is correlated with increased resistance of neurons to excitotoxic insults (Lee et al., 1999). These kinds of experimental data suggest that upregulation of stress proteins in neurodegenerative disorders may represent an excitoprotective response.

Another study, this time based on the application of Western blotting only, also revealed higher levels of Hsp72 in homogenates of AD temporal cortex, as compared with those found in homogenates of cerebellum, and of cortex from normal controls (Perez et al., 1991). The synthesis of hsp72 and Hsp73 by polysomes from AD cortex was increased as compared with that of the controls. These findings were taken as specific for AD in as much as they did not correlate with patient's age at the time of death or with cause of death, or with the interval between death and tissue collection for analysis. Hsp72 was found associated with nascent polypeptides, indicating that the chaperone had become strongly bound to new polypeptides cotranslationally, and for some reason could not effect peptide release and avoid aggregation.

The preceding studies suggest that hsp70 chaperones are attracted to mutated polypeptides in the same way they would be attracted to normal ones distorted by stress when they assist them to regain their native configuration. It would seem that in AD, the chaperones once bound to the abnormal protein somehow fail to help the protein to achieve the proper configuration, and do not release the peptide either. Perhaps the chaperones proceed to direct the polypeptide towards degradation via the ubiquitin-proteasome pathway. This supposition is supported by the fact that ubiquitin is present in plaques and tangles (see Table 6).

Small heat shock proteins (see Table 7) are becoming the target of active investigation in a variety of fields because there is increasing evidence suggesting that these chaperones play important roles in cell physiology, and also in pathology. For example, a recent report describes the effects of αB-crystallin on the formation of amyloid deposits (Stege et al., 1999). *In vitro*, αB-crystallin prevented the formation of fibrils by $A\beta_{1\text{-}40}$. However, this effect did not protect cultured neurons from the toxic action of the peptide. Thus, while the interaction of αB-crystallin with the toxic $A\beta_{1\text{-}40}$ peptide had the expected chaperoning effect, i.e., it prevented aggregation of the conformationally abnormal peptide, it did not abate its toxicity. These findings should not damp our expectations about the use of chaperones in the prevention and treatment of degenerative diseases. Rather, the reported effects of αB-crystallin should encourage more research to understand its mechanism of action and to harness its power for keeping abnormal peptides in solution.

Recently, it was found that cytosolic Hsc70(Hsp73) (see Table 7) interacts with the cytoplasmic domain of the amyloid precursor protein (APP) in the presence of proteasome inhibitors (Kouchi et al., 1999). The binding occurs close to the APP's post-transmembrane region. Interestingly, the binding does not occur in the KFFEQ sequence at the C-terminus of APP, which is assumed to be a targeting signal for proteolysis in the lysosome. APPs from different sources, namely familial AD, and Swedish and Dutch type mutations, were bound equally well by Hsp73, suggesting that APP is cleaved for secretion, or for degradation, in the endosome/lysosomal pathway.

It has been suggested that the ubiquitin-mediated proteolytic pathway might be implicated in the formation of plaques and tangles, since these structures are ubiquitinated (see Table 6). For example, it has been reported that while the extracellular form of APP is degraded via the ubiquitinated pathway, the intracellular form is not (Gregori et al., 1994). On the other hand, a novel ubiquitin-activating enzyme related to E1 was reported to be the effector molecule that interacts with the APP's cytoplasmic domain (Chow et al., 1996).

In summary, it is clear from different studies that Hsps and ubiquitin are present in the aggregates typical of AD, in a consistent manner. However, there is little information on how they got there, why, and what they were doing before precipitation and cell death. Also, a seemingly consistent finding is that Aβ peptides elicit the formation of ROS (reactive oxygen species), characteristics of oxidative stress (Goodman and Mattson, 1994). The authors review briefly what is known about the effects of Aβ peptides and protein deposits. Cells that accumulate the peptides show impairment of glucose metabolism, increased protein oxidation and lipid peroxidation, and other signs of oxidative stress. These manifestations may be caused by age-related mitochondrial dysfunction, alteration in Ca^{++} homeostasis, excitoxicity, and most importantly, accumulation of Aβ peptides (see Mattson, 1997 for review). The latter are 40 or 42 residue-long fragments of APP, and constitute a major component of the neuritic plaques. Their aggregation *in vitro* is catalyzed by oxidation and is inhibited by antioxidants. Aβ peptides would cause neuronal death by inducing oxidative stress. In their work, the authors (Mark et al., 1995, 1997; Pereira et al., 1999) used Aβ25-35 and Aβ1-40 peptides, and induced a concentration-dependent accumulation of ROS accompanied by a decrease in cellular redox and ATP levels. These effects due to inhibition of mitochondrial functions and glycolysis by peptides were blocked by the anti-oxidants vitamin E, idebenone, and GSH ethyl ester. Similar effects due to mitochondrial dysfunction have been observed in Friedreich's ataxia, a finding that also supports the notion that oxidative stress is implicated in the pathogenesis of degenerative diseases (Rustin et al., 1999; Vendemiale et al., 1999). In this regard, it is pertinent to mention that a set of sHsps has been found that protects some mitochondrial components from heat and oxidative stresses (Downs et al., 1999).

Spinocerebellar Ataxia Type 1 (SCA-1)

In SCA-1, an abnormal ataxin-1, a mutant with expansion of the poly-Q tract (see Tables 3c and 5), forms a single intranuclear inclusion (see Table 6). It was found that this inclusion contains antigens recognized by antibodies made against the 20S proteasome, and by antibodies made against human HDJ-2(HSDJ) (Cummings et al., 1998). The latter is a member of the Hsp40 family of molecular chaperones (see Table 7). HeLa cells transfected with a vector containing the mutant ataxin-1 showed aggregates that included 20S proteasome antigens and the chaperone. Overexpression of the latter decreased the frequency of cells with ataxin-1 aggregates. These findings point towards an involvement of the Hsp40 chaperone and ubiquitin in the pathologic aggregates of poly-Q expansion diseases, and suggest that, indeed, the chaperone may actually play a role in preventing aggregate formation.

Bulbospinal Atrophy

Bulbospinal atrophy or Kennedy's disease (see Table 3d) belongs to the poly-Q expansion diseases (see Table 5) and shows intracellular aggregates. The aggregates contain several components (see Table 6). These components were identified using immunologic and microscopic techniques and HeLa cells as an experimental model

(Stenoien et al., 1999). Cells were transfected with a vector bearing the gene for the mutant androgen receptor (AR) protein named ARQ48 (poly-Q expansion with 48 Glu residues). ARQ48 formed aggregates in the cytoplasm and in the nucleus of the transfected cells. Aggregates in the two locations had the same ultrastructure. They contained mitochondria, steroid receptor coactivator 1 (SRC-1), Hsp70, Hsp90, HDJ-2(HSDJ), and the protein NEDD8, which is 60% identical to ubiquitin. The Hsps were detected using antibodies. In this regard, it should be emphasized that there was no immunostaining of the aggregates with antibodies for Hsp25, Hsp27, and Hsp110. Coexpression of HDJ-2(HSDJ) with the poly-Q AR gene reduced aggregate formation since it reduced by 50% the number of transfected HeLa cells that developed this structure. When the cells were transfected with a vector bearing ARQ48 and a mutant HSD-J(HSDJ) gene, prevention of aggregate formation was less efficient than when the wild-type gene was used. However, the decrease in the efficiency of protection shown by the mutant chaperone gene was not as marked as one would expect if HSD-J(HSDJ) were the only chaperone involved in the process. This, of course makes sense, since chaperones tend to work in teams (see Table 7). There might be other chaperones in addition to the Hsp40 homolog that participate in the prevention of aggregate formation in Kennedy's disease, as suggested by the finding of Hsp70 and Hsp90 in the aggregates themselves. It has been reported that DnaJ(Hsp40) activates AR in the nucleus (Caplan et al., 1995) and that AR is stabilized by Hsp70, Hsp90, and DnaJ. It is, therefore, not surprising that these chaperones coprecipitate with AR. Are the chaperones in bulbospinal atrophy bound to AR simply because they are exercising their physiologic stabilizing-activating role, or because they have recognized a mutant peptide in need of assistance to avoid misfolding and aggregation? Interestingly, overexpression of AR induces a heat-shock factor (HSF) (Xiao and DiFranco, 1997). HSF is a positive regulator of the stress response in mammals and other eukaryotes (see Table 2).

Prion Disease

The mechanism of conversion of PrP^{c} to the pathologic form PrP^{sc} does seem to involve a conformational change (Harris 1999; Harrison et al., 1999), Hence, participation of molecular chaperones appears probable. Several studies have addressed the question of what the role of chaperones may be in the conversion process and in the pathogenesis, as well as in what might be their potential therapeutic applications to prevent conversion and promote dissolution of aggregates.

The *Saccharomyces cerevisiae* two-hybrid system was used to identify proteins that would interact specifically with PrP^{c} from the Syrian golden hamster (Edenhofer et al., 1996). A HeLa cDNA library was screened, and the major interactor found was identified as Hsp60. It is unclear which member of the Hsp60 family was referred to by the authors. Was it the bacterial-like Hsp60 present in mitochondria, or was it a subunit of the cytosolic CCT (see Table 7)? This point of clarification is pertinent because while both Hsp60 homologs are molecular chaperones, their physiologic roles and mode of action are different.

The interaction site for Hsp60 was mapped on PrP^{c} between amino acids 180 and 210, which span the region containing parts of the H3 and H4 helical domains (Edenhofer et al., 1996). Bacterial GroEL was also tested *in vitro* and shown to interact with the same PrP^c site. The encoded proteins in three of the Hsp60 cDNAs were 146, 228, and 298 amino acids long, all lacking the N-terminal portion of the normal protein, but containing portions of the peptide-binding domain. On PrP^c, the amino acids involved in Hsp60 binding were at positions 180, 198, 200, and 210. The authors put forward some working hypotheses considering their observations and information in the literature: (*i*) Hsp60 is involved in propagation of PrP^{sc} in a manner similar to that reported for Hsp104 and the prion-like factor (*psi*$^+$) in yeast (Chernoff et al., 1995; and see **Fungal Models of Prion Disease**, below). Would Hsp60 be the "other" protein that according to some reports (Telling et al., 1995) seems to be necessary for prion propagation in addition to PrP^{sc}?; (*ii*) Alternatively, Hsp60 could be involved in preventing aggregation by trapping misfolded forms of PrP^c before the balance is tilted toward conversion into PrP^{sc}; and (*iii*) Hsp60 would be called into action to prevent glycation, which is known to contribute to the pathogenesis of protein deposition.

In any event, the above studies indicate that PrP^c and Hsp60 contact each other and, therefore, they should be within the same cellular compartment. But which? One possibility is the ER (Harris, 1999) but there is no known Hsp60 in the ER (see Table 7). Another possibility is that interaction occurs within the endosome, where PrP^c becomes internalized. However, in this case we must assume that cytosolic, or mitochondrial, hsp60 becomes trapped inside the endocytic compartment–a situation difficult to envisage.

Along the same lines, a recent report suggests also that conversion of PrP^c to a proteinase K resistant form requires a component(s) present in cell lysates, in addition to PrP^{sc} (Saborio et al., 1999).

The ubiquitin-proteasome pathway and chaperones other than Hsp60 have also been investigated in attempts to elucidate their role in prion disease. For example, it was found that expression of the polyubiquitin C gene increased in the brains of mice infected with two different strains of scrapie (Kenward et al., 1994). The increase in the gene's expression was more pronounced in the late stages of the disease. It was paralleled by an increase in the expression of the *hsp70* gene. These data were obtained by Northern blotting, and confirmed those obtained by others using immunohistochemical procedures (Lowe et al., 1990; 1992). However, it remains to be determined which among the Hsp70 family of proteins is the one involved in the phenomena described above, and what its role is in pathogenesis.

Stress Response in Prion-Infected Cells

Another aspect of prion disease that deserves consideration is the possible consequences of infection on the cell's stress-response potential and machinery. One study on this topic (Tatzel et al., 1995) used mouse neuroblastoma cells, and focused on the analysis of the effects of common cell stressors such as heat shock, sodium arsenite, and amino acid analogs (see Table 1). Hsp72 and Hsp28 were not induced in prion-infected cells, although they were in control, non-infected counterparts. Likewise, translocation of Hsp73 (or Hsc70; see Table 7) from the cytoplasm to the nucleus did not occur in infected, stressed cells. In contrast, translocation did occur in the control, stressed cells, as expected from cells that respond to heat shock and other stressors in a normal fashion.

Other parameters of the heat-shock response in eukaryotic cells are trimerization and phosphorylation of heat-shock factor 1 (HSF1; see Table 2) (Morimoto, 1998). These two phenomena occurred in both infected and non-infected cells in response to heat stress.

In summary, while some aspects of the stress response were affected by prion infection, others were not. Future studies should sort out the significance of these findings in terms of their relevance to the development of disease, and explore the possibility of using the stress-response machinery to stop progression of the pathogenetic mechanism unchained by conversion of PrP^c into its pathological form.

Fungal Models of Prion Disease

Advances in the understanding of prion disease are expected from the use of fungal models, in which atypical infectious elements resembling prions have been found (Patino et al., 1996; Schirmer and Lindquist, 1997; Silar and Daboussi, 1999). In the budding yeast *Saccharomyces cerevisiae*, there is a genetic element [PSI^+] whose heritable phenotypic changes correlate with heritable changes in the conformation of the protein Sup35 (Schirmer and Lidquist, 1997). The chaperone Hsp104 seems to play a critical role in the maintenance of Sup35 conformation. In a related work, it was demonstrated that conversion of PrP^c to PrP^{sc} was not affected by chaperones in the absence of PrP^{sc} in the reaction mixture (DebBurman et al., 1997). However, when PrP^{sc} was added along with GroEL (the bacterial Hsp60 homolog) and Hsp104, conversion was catalyzed. No such effects were obtained with the "chemical" chaperones sucrose, trehalose, and dimethyl sulfoxide instead of Hsp104 and GroEL. The "chemical" compounds did in fact inhibit conversion of PrP^c to PrP^{sc}.

Recent work has provided insights into the strict requirements for the maintenance of the [PSI^+] phenotype and the physical state of Sup35 in terms of Hsp104 concentrations (Newman et al., 1999). Likewise, it was established that the yeast chaperone SSA1 (a member of the Hsp70 family; see Table 7) was also needed in precise concentrations, along with Hsp104. In contrast, Hsp82 (Hsp90; see Table 7) had no effect.

Perspectives

Evolution — The Three Domains of Life

A revolution in biology and related disciplines was set in motion in the late 1970s when it was pointed out that prokaryotes can be subdivided into two distinct groups: archaebacteria (nowadays called archaea) and bacteria (or eubacteria) (references in Woese, 1998; Woese et al., 1990). All living cells were grouped into three main phylogentic lines of descent or domains: Archaea, Bacteria, and Eucarya. While these novel ideas have permeated most biologic disciplines, they are still relatively unknown within medical sciences with the exception perhaps of clinical microbiology. It is time to introduce these ideas into the core disciplines of medicine. They may prove useful in planning experiments, interpreting results and, most importantly, in the mining of genomes for information that might lead us to understand the causes and mechanisms of many diseases, including those characterized by age-related degenerative processes with protein abnormalities and deposition.

Archaea are useful models to study a variety of problems directly pertinent to eukaryotes, including humans. These prokaryotic organisms resemble eukaryotes (and thus depart from the bacterial realm) in their mechanisms of gene transcription, and DNA replication and repair, just to name the most striking differences with regard to bacteria. Furthermore, archaea possess a cytosolic chaperonin more closely related to CCT, the eukaryotic cytosolic homolog (see Table 7), than to the bacterial counterpart GroEL/S (Klumpp and Baumeister, 1998; Macario and Conway de Macario, 1999), and also possess homologs of at least two subunits of the eukaryotic prefoldin (Vainberg et al., 1998; personal observations). One can, therefore, envisage the use of the archaeal chaperonin and its partners as a model to investigate folding and refolding of defective proteins from degenerative diseases (see below **Experimental. The Complexity of Chaperone Molecules and Machines**).

Molecular Phylogeny — The Mixed Ancestry of the Eukaryotic Cell

The novel classification of organisms and an understanding of their evolution were greatly facilitated by the development of molecular phylogenetic methods based on computer-assisted comparisons of nucleotide and amino-acid sequences. These comparisons have revealed that the eukaryotic cell is a composite of eucaryal, bacterial, and archaeal structures (Gupta, 1995; Gupta et al., 1994; Gupta and Singh, 1994). The origins of eucaryal Hsp90, Hsp70, and Hsp60/10 have been traced back to bacterial and archaeal ancestors (Bustard and Gupta, 1997; Gribaldo et al., 1999; Gupta, 1995; 1995a; 1998; Gupta et al., 1994; Gupta and Singh, 1994). For example, the mitochondrial Hsp70 originated in ancient α-proteobacteria, and is different from the cytosolic homolog. The functions and mechanisms of action of Hsp70 in the two cell locales must be different. Hence, it is pertinent to ask, which is the most relevant to the development of any particular degenerative disease, the mitochondrial or the cytosolic version of Hsp70, or both? Sequence comparisons have also brought to light functional domains within molecules with distinctive evolutionary trajectories. It will be of interest to apply the methods of molecular phylogeny for unveiling clues to the evolution of pathologic proteins and to that of the stress proteins and chaperones involved in their correct folding and refolding. It may become apparent what functional domain or sequence motif within a chaperone subunit is most pertinent for interacting with a defective protein and promoting its correct folding. This information should be useful in developing preventive and therapeutic strategies.

Genomes

The potential contributions of an evolutionary approach to the study of degenerative diseases described in the preceding sections will be greatly enhanced by the availability of full-genome sequences. There are at the present time at least 7 archaeal, 15 bacterial, and 2 eucaryal genomes that have been completely sequenced, and several others –

including the human genome – that are being sequenced. The amount and type of information in these genome sequences that can be extracted by various means will help in the search for the evolution of molecules pertinent to degenerative diseases, including those of the most relevant stress proteins and molecular chaperones. It will then be possible to infer the functional and/or pathogenetic significance of the molecules, or their functional domains and sequence motifs, in regard to the development and progression of the diseases. Once again these hints provided by comparative analyses of sequences will facilitate the planning of experiments that will focus on the most relevant molecules and domains within them.

Experimental — The Complexity of the Chaperone Molecules and Machines

The data available indicate a clear association of stress proteins and chaperones with the pathologic landmarks (e.g., protein deposits) of degenerative diseases. The nature of the association has not yet been fully elucidated. Nonetheless, some inferences can be made on the basis of what has been observed on protein abnormalities, and on the functions of molecular chaperones. On the one hand, it has been established that gene mutations in degenerative diseases result in the production of defective proteins, probably with a tendency to misfold and aggregate. On the other hand, it is known that molecular chaperones participate in protein biogenesis (folding, assembling, trafficking) and in the maintenance of the physiological set of proteins within the quantitative and qualitative levels required for normal life. On the basis of these two premises, future experiments ought to aim at developing systems for testing the folding of a given abnormal protein in the presence of the chaperone team most pertinent to the cell compartment in which the protein resides – or precipitates in pathologic cases. It has to be borne in mind that chaperones and chaperonins form complexes, and that these are the actual protein-folding machines. Hsp70 interacts with other molecules (see Table 7), and the cytosolic chaperonin CCT is composed of up to 9 different subunits, depending on the species, that associate in a double ring structure (Rommelaere et al., 1999; Willison, 1999). sHsps also form oligo- and polymeric complexes (Kim et al., 1998; 1998a; Munchbach et al., 1998). These features of the chaperones and chaperonins ought to be taken into account in studies of their role in degenerative diseases. For example, folding of an abnormal protein ought to be assessed in the presence of an eucaryal cytosolic chaperoning system (or its archaeal homolog), complete with all its subunits and interacting partners. Furthermore, folding and refolding ought to be assessed *in vivo*, using cells or organisms lacking the gene for a chaperone subunit, followed by the same kind of experiment using cells or organisms lacking the gene for another subunit, and so forth, until the most pertinent subunits have been identified. This approach will be facilitated by the availability of archaeal models. These are simpler than eukaryotes. For example, the archaeal chaperonin complex is formed by one or two different subunits depending on the species (Ditzel et al., 1998; Klumpp and Baumeister, 1998; Nitsch et al., 1998). Likewise, the prefoldin system consists of only two or three subunits instead of the six present in eukaryotes (Vainberg et al., 1998; and personal observations). Moreover, gene knockouts may be easier to obtain in archaeal than in eukaryotic cells, and may be amenable to quicker screening.

There is increasing evidence showing that while the process of aging may depend on genetic predisposition, the role of stress (i.e., environmental factors or stressors) in the process is significant (Geller and Potter, 1999; Lin et al., 1998; Rudolph et al., 1999; Rustin et al., 1999; Tatar et al., 1997). Longevity and improved stress resistance seem to go hand in hand, and *vice versa*, the more advanced the aging process is, the less efficient the response to stress becomes.

Indeed, recent studies aimed at understanding the possible beneficial effects of calorie restriction in the aging brain have provided evidence for enhanced stress response of brain cells. Thus, Bruce-Keller et al. (1999) reported that rats maintained on an alternate day dietary restriction schedule exhibit increased resistance of hippocampal neurons and striatal neurons to excitotoxic and metabolic insults. The beneficial effect of the dietary restriction regimen was also reflected in improved behavioral outcome following the brain insults. In a related study, it was shown that dietary restriction improves behavioral outcome and reduces degeneration of dopaminergic neurons in mouse models of Parkinson's diseases (Duan and Mattson, 1999). In the latter study, dietary restriction and 2-deoxyglucose administration upregulated levels of stress proteins in the dopaminergic neurons. Presumably, dietary restriction enhances the ability of cells to cope with damaged proteins. Important consideration must be given to psychological stressors, since present day competitive societies have put humans in a situation characterized by acute and chronic stresses. The impact of these stresses on the cells and their protein constituents has been noted for many years and is now under active investigation (Liu et al., 1997; McEwen, 1998;1999; Timpl et al., 1998; Weninger et al., 1999). It follows that experiments aiming at determining the role and mechanisms of action of stress proteins and chaperones on the folding and refolding of defective proteins, such as those that are typical of degenerative diseases, should also be performed in such a manner that the effect of stress can be measured. One thing is to assess protein folding under conditions adjusted to match those considered optimal, and another, perhaps more pertinent to learn about degenerative diseases, is to evaluate folding under stressful conditions (e.g., at supraoptimal temperatures resembling fever). What is the role of high-temperature fever in starting the misfolding and precipitation of abnormal protein molecules in an individual with a mutated gene? It could very well be that fever in these individuals is an important factor in the initiation of protein deposition.

If indeed molecular chaperones, and other stress proteins, do play a role in preventing misfolding and aggregation of mutant peptides, and if they do participate in dissolving the aggregates, one must consider using them as preventive and therapeutic agents. It has been demonstrated that heat acclimation increases the basal levels of Hsp72 (Hsp70i; see Table 7) and changes its pattern of production upon subsequent stress (Maloyan et al., 1999). Experiments ought to be done for testing whether or not heat acclimation would delay the appearance of protein deposits, or promote their dissolution, in animal models of degenerative diseases. There is some hope in this regard as demonstrated by experiments in which overexpression of an Hsp40 homolog (see Table 7) decreased the frequency of cells with ataxin-1 (Cummings et al., 1998) or with androgen receptor (Stenoien et al., 1999) aggregates.

Summary

Different degenerative diseases of the central nervous system (CNS) appear after middle age, and increase in frequency in the elderly. They seem to be associated with aging. Data suggest that while aging is partly dictated by genetic determinants, environmental stressors also play a role. Normal aging and a decrease in the capacity to mount a stress response go in parallel. Conversely, genetic conditions that confer longevity are accompanied by improved capacity to respond to stressors. Stress proteins and molecular chaperones are the typical manifestations of the stress response. They play a key role in peptide folding and trafficking, and thus maintain the normal cellular set of protein molecules in their native (functional) configuration. This set of proteins enables the cell to function and to survive stressful situations. Due to gene mutations, defective peptides may appear that have a tendency to misfold and form deposits in vital tissues, e.g., those of the CNS, a condition that is usually associated with degenerative diseases. It remains to be determined how much stress proteins and molecular chaperones participate in the aging process, and in the pathogenesis of the degenerative diseases accompanying this process in many individuals. Because of the known role of chaperones in maintaining the normal set of proteins in a functional configuration, one might assume that chaperones are key players in pathogenesis. Chaperones may prevent protein misfolding, and aggregation, and may mediate dissolution of protein deposits. Likewise, chaperones may become trapped in protein aggregates when they bind defective peptides but cannot assist in their correct folding, or cannot release them, owing to a defect in the pathologic peptide. This entrapment is suggested by the presence of chaperones in pathologic protein deposits. The cell's protein balance also depends on degradation. One might say that the normal set of proteins in a cell is maintained by generation of new molecules and, at the other end of the pathway, by proteolysis. Old, damaged, proteins are eliminated by proteolytic systems, and the same may happen to defective peptides due to gene mutations if they cannot be driven by chaperones into the functional pool. Arguably, molecular chaperones are in the central, most critical, portion of the protein assembly line. They play a crucial role as they assist in the correct folding of new peptides, or present those destined for degradation to proteases. Protein deposits in degenerative diseases may be due not only to a failure of the folding mechanism, but also to a failure of the proteolytic system. Participation of the latter is evidenced by the presence of some of its components (e.g., ubiquitin) in the pathologic protein deposits. Future research aiming at elucidating the role of chaperones in degenerative diseases, whether as pathogenetic factors or preventive-curative tools, ought to explore a number of possibilities. For example, folding of mutant peptides ought to be assessed *in vivo* and *in vitro*. One will have to take into account that chaperones function as complex machines and interact with other molecules and structures. Information from comparative analyses of gene, protein, and genome sequences should help in the selection of experimental systems and molecular targets for investigation. The effect of stressors with which humans are confronted daily, ought to be measured to assess their role on peptide misfolding and aggregation, and in disease onset and progression in individuals genetically susceptible to degenerative diseases. One might argue that the presence of a defective peptide requires the chaperone system constantly. When a stressor hits the cell, the chaperone system is partly diverted to protect and rescue normal peptides damaged by stress and must abandon some copies of the defective peptide. These "unprotected" copies misfold and precipitate. Stress would thus play a critical role in disease onset and progression by creating a functional overload to the chaperone system, already overextended because of the presence of the mutant peptide.

Acknowledgements

Work done in the authors' laboratories was supported in part by grants from the US Department of Energy (DOE), and the National Science Foundation (NSF).

References

Ballester, J.M. & Harchelroad, F.P. (1999). Hyperthermia: How to recognize and prevent heat-related illnesses. Geriatrics 54, 20-24.

Ballinger, C.A., Connel, P., Wu, Y., Hu, Z., Thompson, L.J., Yin, L-Y. & Patterson, C. (1999). Identification of CHIP, a novel tetratricopeptide repeat-containing protein that interacts with heat shock proteins and negatively regulates chaperone functions. Mol. Cellul. Biol. 19, 4535-4545.

Bamborough, P., Willie, H., Telling, G.C., Yehiely, F., Prusiner, S.B. & Cohen F.E. (1996). Prion protein structure and scrapie replication: Theoretical, spectroscopic, and genetic investigations. Cold Spring Harbor Symp. Quant. Biol. 61, 495-509.

Borchelt, D.R., Taraboulos, A. & Prusiner, S.B. (1992). Evidence for synthesis of scrapie prion proteins in the endocytic pathway. J. Biol. Chem. 267, 16188-16199.

Bruce-Keller, A.J., Umberger, G., McFall, R. & Mattson, M.P. (1999). Food restriction reduces brain damage and improves behavioral outcome following excitotoxic and metabolic insults. Ann. Neurol. 45, 8-15.

Bustard, K. & Gupta, R.S. (1997). The sequence of heat shock protein 40 (DnaJ) homologs provide evidence for a close evolutionary relationship between the *Deinococcus-Thermus* group and cyanobacteria. J. Mol. Evol. 45, 193-205.

Caplan, A.J., Langley, E., Wilson, E.M. & Vidal, J. (1995). Hormone-dependent transactivation by the human androgen receptor is regulated by a dnaJ protein. J. Biol. Chem. 270, 5251-5257.

Cheetham, M.E., Brion, J-P. & Anderton, B.H. (1992) Human homologues of the bacterial heat-shock protein DnaJ are preferentially expressed in neurons. Biochem. J. 284, 469-476.

Chellaiah, A., Davis, A. & Mohanakumar, T. (1993). Cloning of a unique human homologue of the *Escherichia coli* DNAJ heat shock protein. Biochim. Biophys. Acta 1174, 111-113.

Chen, S. & Smith, D.F. (1998). Hop as an adaptor in the heat shock protein 70 (Hsp70) and Hsp90 chaperone machinery. J. Biol. Chem. 273, 35194-35200.

Chernoff, Y.O., Lindquist, S.L., Ono, B., Inge-Vechtomov, S.G. & Liebman, S.W. (1995). Role of the chaperone protein Hsp104 in propagation of the yeast prion-like factor (psi$^+$). Science 268, 880-884.

Chow, N., Korenberg, J.R., Chen, X-N. & Neve, R.L. (1996). APP-BP1, a novel protein that binds to the carboxyl-terminal region of the amyloid precursor protein. J. Biol. Chem. 271, 11339-11346.

Ciechanover, A. (1998). The ubiquitin-proteasome pathway: On protein death and cell life. EMBO J. 17,

7151-7160.

Craven, R.A., Tyson, J.R. & Stirling, C.J. (1997). A novel subfamily of Hsp70s in the endoplasmic reticulum. Trends Cell Biol. 7, 277-282.

Crawford, F., Town, T., Freeman, M., Schinka, J., Gold, M., Duara, R. & Mullan, M. (1999). The alpha-2 macroglobulin gene is not associated with Alzheimer's disease in a case-control sample. Neurosci. Lett. 270, 133-136.

Cruts, M. Hendriks, L. & van Broechoven, C. (1996) The presenilin genes: A new gene family involved in Alzheimer disease pathology. Hum. Mol. Genet. 5, 1449-1455.

Csermely, P., Schnaider, T., Sôti, C., Prohászka, Z. & Nardai, G. (1998). The 90 kDa molecular chaperone family: Structure, function and clinical applications. Pharmacol. & Ther. 79, 129-168.

Cummings, C.J., Mancini, M.A., Antalfy, B., DeFranco, D.b., Orr, H.T. & Zoghbi, H. (1998). Chaperone suppression of aggregation and altered subcellular proteasome localization imply protein misfolding in SCA 1. Nature Genet. 19, 148-154.

Cyr, D.M., Langer, T. & Douglas, M.G. (1994). DnaJ-like proteins: Molecular chaperones and specific regulators of Hsp70. Trends Biochem. Sci. 19, 176-181.

Davies, S.W., Turmaine, M., Cozens, B.A., DiFiglia, M., Sharp, A.H., Ross, C.A., Scherzinger, E., Wanker, E.E., Mangiarini, L. & Bates, G.P. (1997). Formation of neuronal intranuclear inclusions underlies the neurological dysfunction in mice transgenic for the HD mutation. Cell 90, 537-548.

DebBurman, S.K., Raymond, G.J., Caughey, B. & Lindquist, S. (1997). Chaperone-supervised conversion of prion protein to its protease-resistant form. Proc. Natl. Acad. Sci. USA 94, 13938-13943.

De Jong, W.W., Caspers, G.J. & Leunissen, J.A. (1998) Genealogy of the alpha-crystallin-small heat shock protein superfamily. Int. J. Biol. Macromol. 22, 151-162.

DiFiglia, M., Sapp, E., Chase, K.O., Davies, S.W., Bates, G.P., Vonsattel, J.P. & Aronin, N. (1997). Aggregation of huntingtin in neuronal intranuclear inclusions and dystrophic neurites in brain. Science 277, 1990-1993.

Ditzel, L., Löwe, J., Stock, D., Stetter, K.O., Huber, Huber, R. & Steinbacher, S. (1998). Crystal structure of the thermosome, the archaeal chaperonin and homolog of CCT. Cell 93, 125-138.

Downs, C.A., Jones, L.R. & Heckathorn S.A. (1999). Evidence for a novel set of small heat-shock proteins that associates with the mitochondria of murine PC12 cells and protects NADH:ubiquinone oxidoreductase from heat and oxidative stress. Arch. Biochem. Biophys. 365, 344-350.

Duan, W. & Mattson, M.P. (1999). Dietary restriction and 2-deoxyglucose administration improve behavioral outcome and reduce degeneration of dopaminergic neurons in models of Parkinson's disease. J. Neurosci. Res. 57, 195-206.

Edenhofer, F., Rieger, R., Famulok, M., Wendler, W., Weiss, S. & Winnacker, E.L. (1996). Prion protein PrP^c interacts with molecular chaperones of the Hsp60 family. J. Virol. 70, 4724-4728.

Favatier, F. & Polla, B. (1996). Protéins de stress et vieillissement. L'Année Gérontologique 10, 217-224.

Feany, M.B., Mattiace, L.A. & Dickson, D.W. (1996). Neuropathologic overlap of progressive supranuclear palsy, Pick's disease and corticobasal degeneration. J. Neuropathol. Exp. Neurol. 55, 53-67.

Ferrari, D.M. & Söling, H-D. (1999). The protein disulphide-isomerase family: Unravelling a string of folds. Biochem. J. 339, 1-10.

Fisher, G., Tradler, T. & Zarnt, T. (1998). The mode of action of peptidyl prolyl *cis/trans* isomerases in vivo: Binding vs. catalysis. FEBS Lett. 426, 17-20.

Frydman, J. & Höhfeld, J. (1997). Chaperones get in touch: The Hip-Hop connection. Trends Biochem. Sci. 22, 87-92.

Gebauer, M., Melki, R. & Gehring, U. (1998). The chaperone cofactor Hop/p60 interacts with the cytosolic chaperonin-containing TCP-1 and affects its nucleotide exchange and protein folding activities. J. Biol. Chem. 273, 29475-29480.

Geissler, S., Siegers, K. & Schiebel, E. (1998). A novel protein complex promoting formation of functional γ- and α-tubulin. EMBO J. 17, 952-966.

Geller, L.N. & Potter, H. Chromosome missegregation and trisomy 21 mosaicism in Alzheimer's disease. (1999). Neurobiol. Dis. 6, 167-179.

Glover, J.R. & Lindquist, S. (1998). Hsp104, Hsp70, and Hsp40: a novel chaperone system that rescues previously aggregated proteins. Cell 94, 73-82.

Goodman, Y. & Mattson, M.P. (1994). Secreted forms of β-amyloid precursor protein protect hippocampal neurons against amyloid β-peptide-induced oxidative injury. Exp. Neurol. 128, 1-12.

Gregori, L., Bhasin, R. & Goldgaber, D. (1994). Ubiquitin-mediated degradative pathway degrades the extracellular but not the intracellular form of amyloid β-protein precursor. Biochem. Biophys. Res. Commun. 203, 1731-1738.

Gribaldo, S., Lumia, V., Creti, R., Conway de Macario, E., Sanangelantoni & Cammarano, P. (1999). Discontinuous occurrence of the *hsp70(dnaK)* gene among archaea and sequence features of HSP70 suggest a novel outlook on phylogenies inferred from this protein. J. Bacteriol. 181, 434-443.

Guillemette, J.G., Wong, L., Crapper, McLachlan, C. & Lewis, P.N. (1986). Characterization of messenger RNA from the cerebral cortex of control and Alzheimer-afflicted brain. J. Neurochem. 47, 987-997.

Gupta, R.S. (1995). Evolution of the chaperonin families (Hsp60, Hsp10 and Tcp-1) of proteins and the origin of eukaryotic cells. Mol. Microbiol. 15,1-11.

Gupta, R.S. (1995a). Phylogenetic analysis of the 90 kD heat shock family of protein sequences and an examination of the relationship among animals, plants, and fungi species. Mol. Biol. Evol. 12, 1063-1073.

Gupta, R.S. (1998). Protein phylogenies and signature sequences: A reappraisal of evolutionary relationships among archaebacteria, eubacteria, and eukaryotes. Microbiol. Mol. Biol. Rev. 62, 1435-1491.

Gupta, R.S., Aitken, K., Falah, M. & Singh, B. (1994). Cloning of *Giardia lamblia* heat shock protein HSP70 homologs: Implications regarding origin of eukaryotic cells and endoplasmic reticulum. Proc. Natl. Acad. Sci. USA 91, 2895-2899.

Gupta, R.S. & Singh, B. (1994). Phylogenetic analysis of 70 kD heat shock protein sequences suggests a chimeric origin for the eukaryotic cell nucleus. Curr. Biol. 4, 1104-1114.

Halliwel, B. & Gutteridge, J.M. (1990). Role of free radicals and catalytic metal ions in human disease: An overview. Methods Enzymol. 86, 1-85.

Hamos, J.E., Oblas, B., Pulaski-Salo, B.S., Welch, W.J., Bole, D.G. & Drachman, D.A.(1991). Expression of heat shock proteins in Alzheimer's disease. Neurol. 41, 345-350.

Hardman, C.D. & Halliday, G.M. (1999). The internal globus pallidus is affected in progressive supranuclear palsy and Parkinson's disease. Exper. Neurol. 158, 135-142.

Harhangi, B.S., Farrer, M.J., Lincoln, S., Bonifati, V., Meco, G., De Michele, G., Brice, A., Dürr, A., Martinez, M., Gasser, T., Bereznai, B., Vaughan, J.R., Wood, N.W., Hardy, J., Oostra, B.A. & Breteler, M.M.B. (1999). The Ile93Met mutation in the ubiquitin carboxy-terminal-hydrolase L1 gene is not observed in European cases with familial Parkinson's disease. Neurosci. Lett. 270, 1-4, 1999.

Harris, D.A. (1999). Cellular biology of prion diseases. Clin. Microbiol. Rev. 12, 429-444.

Harrison, P.M., Chan, H.S., Prusiner, S.B. & Cohen, F.E. (1999). Thermodynamics of model prions and its implications for the problem of prion protein folding. J. Mol. Biol. 286, 593-606.

Hartl, F.U. & Martin, J. (1995). Molecular chaperones in cellular protein folding. Curr. Op. Struct. Biol. 5, 92-102.

Hayes, S.A. & Dice, J.F. (1996). Roles of molecular chaperones in protein degradation. Cell Biol. 132, 255-258.

Heine, H., Delude, R.L., Monks, B.G., Espevik, T. & Golenbock, D.T. (1999). Bacterial lipopolysaccharide induces expression of the stress response genes *hop* and *H411*. J. Biol. Chem. 274, 21049-21055.

Herman, C. & D'Ari, R. (1998). Proteolysis and chaperones: The destruction/ reconstruction dilemma. Curr. Op. Microbiol. 1:204-209.

Höhfeld, J., Minami, Y. & F.U. Hartl, F.U. (1995). Hip, a novel cochaperone involved in the eukaryotic Hsc70/Hsp40 reaction cycle. Cell 83, 589-598.

Hosokawa, N., Hohenadl, C., Satoh, M., Kühn, K. & Nagata, K. (1998). HSP47, a collagen-specific molecular chaperone, delays the secretion of type III procollagen transfected in human embryonic kidney cell line 293; A possible role for HSP47 in collagen modification. J. Biochem. 124, 654-662.

Hourichi, M. & Caughey, B. (1999). Specific binding of normal prion protein to the scrapie form via a localized domain initiates its conversion to the protease-resistant state. EMBO J. 18, 3193-3203.

Hsiao, P-W., Lin, D-L., Nakao, R. & Chang, C. (1999). The linkage of Kennedy's neuron disease to ARA24, the first identified androgen receptor polyglutamine region-associated coactivator. J. Biol. Chem. 274, 20229-20234.

Huang, Z., Prusiner, S.B. & Cohen, F.E. (1996). Structures of prion proteins and conformational models for prion diseases. Curr. Top. Microbiol. Immunol. 207, 49-67.

Hunt, C.R., Parsian, A.J., Goswami, P.C. & Kozak, C.A. (1999). Characterization and expression of the mouse Hsc70 gene. Biochim. Byophys. Acta 1444, 315-325.

Iwaki, T., Kime-Iwaki, A., Liem, R.K.H. & Goldman, J.E. (1989). αB-crystallin is expressed in non-lenticular tissues and accumulates in Alexander's disease brain. Cell 57, 71-78.

Iwaki, T., Iwaki, A., Tateishi, J., Sakaki, Y & Goldman, (1993). Alpha-B-crystallin and 27-kd heat-shock protein are regulated by stress conditions in the central nervous system and accumulate in Rosenthal fibers. Am. J. Pathol. 143, 487-495.

Jackson, G.S., Hill, A.F., Joseph, C., Hosszu, L., Power, A., Waltho, J.P., Clarke, A.R. & Collinge, J. (1999). Multiple folding pathways for heterologously expressed human prion protein. Biochem. Biophys. Acta 1431, 1-13.

Jaikaran, E.T.A.S., Marcon, G., Levesque, L., St George-Hyslop, P., Fraser, P.E. & Clark, A. (1999). Localisation of presenilin 2 in human and rodent pancreatic islet β-cells; Met239Val presenilin 2 variant is not associated with diabetes in man. J. Cell Sci. 112, 2137-2144.

Kaemmerer, W.F. & Low, W.C. (1999). Cerebellar allografts survive and transiently alleviate ataxia in transgenic model of spinocerebellar ataxia type-1. Exp. Neurol. 158, 301-311.

Kato, S., Hirano, A., Umahara, T., Llena, J.F., Herz, F. & Ohama, E. (1992). Ultrastructural and immunohistochemical studies on ballooned cortical neurons in Creutzfeldt-Jakob disease: Expression of alpha-B-crystallin, ubiquitin and stress-response protein 27. Acta Neuropathol. Berl. 84, 443-448.

Kenward, N., Hope, J., Landon, M. & Mayer, R.J. (1994). Expression of polyubiquitin and heat-shock protein 70 genes increases in the later stages of disease progression in scrapie-infected mouse brain. J. Neurochem. 62, 1870-1877.

Kim, K.K., Kim, R. & Kim, S-H. (1998). Crystal structure of a small heat-shock protein. Nature 394, 595-599.

Kim, R., Kim, K.K., Yokota, H. & Kim, S-H. (1998a). Small heat shock protein of *Methanococcus jannaschii*, a hyperthermophile. Proc. Natl. Acad. Sci. USA 95, 9129-9133.

Klumpp, M. & Baumeister, W. (1998). The thermosome: Archetype of group II chaperonins. FEBS Lett. 430, 73-77.

Kouchi, Z., Sorimachi, H., Suzuki, K. & Ishiura, S. (1999). Proteasome inhibitors induced the association of Alzheimer's amyloid precursor protein with Hsc73. Biochem. Biophys. Res. Commun. 254, 804-810.

Kozutsumi, Y., Segal, M., Normington, K., Gething, M-J. & Sambrook, J. (1988). The presence of malfolded proteins in the endoplasmic reticulum signals the induction of glucose-regulated proteins. Nature 332, 462-464.

Kurt, M.A., Davies, D.C. & Kidd, M. (1999). β-amyloid immunoreactivity in astrocytes in Alzheimer's disease brain biopsies: An electron microscope study. Exper. Neurol. 158, 221-228.

Lavin, M.F. (1999). ATM: The product of the gene mutated in ataxia-telangiectasia. Intl. J. Biochem. Cell Biol. 31, 735-740.

Lee, J., Bruce-Keller, A.J., Kruman, Y., Chan, S.L. & Mattson, M.P. (1999). 2-Deoxy-D-Glucose protects hippocampal neurons against excitotoxic and oxidative injury: Evidence for the involvment of stress proteins. J. Neurosci. Res. 57, 48-61.

Li, M., Miwa, S., Kobayashi, Y., Merry, D.E., Yamamoto, M., Tanaka, F., Doyu, M., Hashizume, Y., Fischbeck, K.H. & Sobue, G. (1998). Nuclear inclusions of the androgen receptor protein in spinal and bulbar muscular atrophy. Ann. Neurol. 44, 149-254.

Li, H., Li, S-H., Cheng, A.L., Mangiarini, L., Bates, G.P. & Li, X-J. (1999). Ultrastructural localization and progressive formation of neuropil aggregates in Huntington's disease transgenic mice. Hum. Mol. Genet. 8, 1227-1236.

Licastro, F., Pedrini, S., Govoni, M., Pession, A., Ferri, C., Annoni, G., Casadei, V., Veglia, F., Bertolini, S. & Grimaldi, L.M. (1999). Apolipoprotein E. and α-1-antichymotrypsin allele polymorphism in sporadic and familial Alzheimer's disease. Neurosci. Lett. 270, 129-132.

Lin, Y-J., Seroude, L. & Benzer, S. (1998). Extended life-span and stress resistance in the *Drosophila* mutant *methuselah*. Science 282, 943-946.

Lindner, R.A., Kapur, A., Mariani, M., Titmus, S.J. & Carver, J.A. (1998). Structural alterations of α-crystallin during its chaperone action. Eur. J. Biochem. 258, 170-183.

Linnik, K.M. & Herscovitz, H. (1998). Multiple molecular chaperones interact with apolipoprotein B during its maturation. J. Biol. Chem. 273, 21368-21373.

Litt, M., Kramer, P., LaMorticella, D.M., Murphey, W., Lovrien, E.W. & Weleber, R.G. (1998). Autosomal dominant congenital cataract associated with a missense mutation in the human alpha crystallin gene CRYAA. Hum. Mol. Genet. 7, 471-474.

Liu, D., Diorio, J., Tannenbaum, B., Caldji, C., Francis, D., Freedman, A., Sharma, S., Pearson, D., Plotsky, P.M. & Meaney, M.J. (1997). Maternal care, hippocampal glucocorticoid receptors, and hypothalamic-pituitary-adrenal response to stress. Science 277, 1659-1662.

Love, S., Saitoh, T., Quijada, S., Cole, G.M. & Terry, R.D. (1988). Alz-50, ubiquitin and tau immunoreactivity of neurofibrillary tangles, Pick bodies and Lewy bodies. J. Neuropathol. Exp. Neurol. 47, 393-405.

Lowe, J., Blanchard, A., Morrell, K. Lennox, G., Reynolds, L., Bilette, M., Landon, M. & Mayer, R.J. (1988). Ubiquitin is a common factor in intermediate filament inclusion bodies of diverse type in man, including those of Parkinson's disease, Pick's disease, and Alzheimer's disease, as well as Rosenthal fibres in cerebella astrocytomas, cytoplasmic bodies in muscle, and Mallory bodies in alcoholic liver disease. J. Pathol. 155, 9-15.

Lowe, J., McDermott, H., Kenward, N., Landon, M., Mayer, R.J., Bruce, M., McBride, P., Somerville, R. & Hope, J. (1990). Ubiquitin conjugate immunoreactivity in the brains of scrapie infected mice. J. Pathol. 162, 61-66.

Lowe, J., Fergusson, J., Kenward, N., Landon, M., Brown, J., Hope, J. & Mayer, R.J. (1992). Immunoreactivity to ubiquitin-protein conjugates is present early in the disease process in the brains of scrapie-infected mice. J. Pathol. 168, 169-177.

Lowe, J., Mayer, R.J. & Landon, M. (1993). Ubiquitin in neurodegenerative diseases. Brain Pathol. 3,

55-65.

Macario, A.J.L. (1995). Heat-shock proteins and molecular chaperones: Implications for pathogenesis, diagnostics, and therapeutics. Intl. J. Clin. Lab. Res. 25, 59-70.

Macario, A.J.L. & Conway de Macario, E. (1999). The archaeal molecular chaperone machine: Peculiarities and paradoxes. Genetics 152, 1277-1283.

Macario, A.J.L. &. Conway de Macario, E. (2000) Heat shock response. Overview. In: The Encyclopedia of Stress (Fink, G., ed.), in press. Academic Press, San Diego, CA.

Maloyan, A., Palmon, A. & Horowitz, M. (1999). Heat acclimation increases the basal HSP72 level and alters its production dynamics during heat stress. Am. J. Physiol. 276 (5 Pt 2): R1506-R1515

Mark, R.J., Hensley, K., Butterfield, D.A. & Mattson, M.P. (1995). Amyloid β-peptide impairs ion-motive ATPase activities: Evidence for a role in loss of neuronal Ca^{2+} homeostasis and cell death. J. Neurosci. 15, 6239-6249.

Mark, R.J., Pang, Z., Geddes, J.W., Uchida, K. & Mattson, M.P. (1997). Amyloid β-peptide impairs glucose transport in hippocampal and cortical neurons: Involvement of membrane lipid peroxidation. J. Neurosci. 17, 1046-1054.

Mattson, M.P. (1997). Cellular action of β-amyloid precursor protein and its soluble and fibrillogenic derivatives. Physiol. Rev. 77, 1081-1132.

Mayer, M.P. & Bukau, B. (1999). Molecular chaperones: The busy life of Hsp90. Curr. Biol. 9. R322-R325.

McEwen, B.S. (1998). Protective and damaging effects of stress mediators. New Eng. J. Med. 338, 171-179.

McEwen, B.S. (1999). Stress and hippocampal plasticity. Annu. Rev. Neurosci. 22, 105-122.

Meacham. G.C., Lu, Z., King, S., Sorscher, E., Tousson, A. & Cyr, D.M. (1999). The Hdj-2/Hsc70 chaperone pair facilitates early steps in CFTR biogenesis. EMBO J. 18, 1492-1505.

Misselwitz, B., Staeck, O., Matlack, K.E.S. and Rapoport, T.A. (1999). Interaction of BiP with the J-domain of the Sec63p component of the endoplasmic reticulum protein translocation complex. J. Biol. Chem. 274, 20110-20115.

Mori, H., Kondo, J. & Ihara, Y. (1987). Ubiquitin is a component of paired helical filaments in Alzheimer's disease. Science 235, 1641-1644.

Morimoto, R.I. (1998). Regulation of the heat shock transcriptional response: Cross talk between a family of heat shock factors, molecular chaperones, and negative regulators. Genes & Dev. 12, 3788-3796.

Muchowski, P.J., Valdez, M.M. & Clark, J.L. (1999). αB-crystallin selectively targets intermediate filament proteins during thermal stress. Invest. Ophtalmol. Vis. Sci. 40, 951-958.

Munchbach, M., Nocker, A. & Narberhaus, F. (1998). Multiple small heat shock proteins in rhizobia. J. Bacteriol. 181, 83-90.

Munro, S. &. Pelham, H.R.B. (1986). An Hsp70-like protein in the ER: Identity with 78 kd glucose-regulated protein and immunoglobulin heavy chain binding protein. Cell 46, 291-300.

Naylor, D.J., Hoogenraad, N.J. & Hoj, P.B. (1996). Isolation and characterization of cDNA encoding rat mitochondrial GrpE, a stress-inducible nucleotide-exchange factor of ubiquitous appearance in mammalian organs. FEBS Lett. 396, 181-188.

Naylor, D.J., Stines, A.P., Hoogenraad, N.J. & Hoj, P.B. (1998). Evidence for the existence of distinct mammalian cytosolic, microsomal, and two mitochondrial GrpE-like proteins, the co-chaperones of specific Hsp70 members. J. Biol. Chem. 273, 21169-21177.

Netzer, W.J. & Hartl, F.U. (1998). Protein folding in the cytosol: Chaperonin-dependent and –independent mechanisms. Trends Biochem. Sci. 23, 68-73.

Newnam, G.P., Wegrzyn, R.D., Lindquist, S.L. & Chernoff, Y.O. (1999). Antagonistic interactions between yeast chaperones Hsp104 and Hsp70 in prion curing Mol. Cell. Biol. 19, 1325-1333.

Nishikawa, S. & Endo, T. (1997). The yeast JEM1p is a DnaJ-like protein of the endoplasmic reticulum membrane required for nuclear fusion. J. Biol. Chem. 272, 12889-12892.

Nishiyama, K., Trapp, B.D., Ikezu, T., Ransohoff, R.M., Tomita, T., Iwatsubo, I., Hsiao, K.K., Lisanti, M.P. & Okamoto, T. (1999). Caveolin-3 upregulation activates β-secretase-mediated cleavage of the amyloid precursor protein in Alzheimer's disease. J. Neurosci. 19, 6538-6548.

Nitsch, M., Walz, J., Typke, D., Klumpp, M., Essen, L.O. & Baumeister, W. (1998). Group II chaperonin in an open conformation examined by electron tomography. Nature Struc. Biol. 5, 855-857.

Oh, S., Iwahori, A. & Kato, S. (1993). Human cDNA encoding DnaJ protein homologue. Biochim. Biophys. Acta 1174, 114-116.

Oh, H.J., Easton, D., Murawski, M., Kaneko, Y. & Subjeck, J.R. (1999). The chaperoning activity of hsp110. J. Biol. Chem. 274, 15712-15718.

Parsell, D.A., Kowal, A.S., Singer, M.A. & Lindquist, S. (1994). Protein disaggregation mediated by heat-shock protein Hsp104. Nature 372, 475-478.

Patino, M.M., Liu, J-J., Glover, J.R. & Lindquist, S. (1996). Support for the prion hypothesis for inheritance of a phenotypic trait in yeast. Science 273, 622-626.

Paulson, H.L., Perez, M.K., Trottier, Y., Trojanowski, J.Q., Subramony, S.H., Das, S.S., Vig, P., Mandel, J-L., Fischbeck, K.H. & Pittman, R.N. (1997). Intranuclear inclusions of expanded polyglutamine protein in spinocerebellar ataxia type 3. Neuron 19, 333-344.

Penco, S., Schenone, A., Bordo, D., Bolognesi, M., Abbruzzese, M., Bugiani, O., Ajmar, F. & Garre, C. (1999). A *SOD1* gene mutation in a patient with slowly progressing familial ALS. Neurol. 53, 404-406.

Pereira, C., Santos, M.S. & Oliveira, C. (1999). Involvement of oxidative stress on the impairment of energy metabolism induced by Aβ peptides on PC12 cells: Protection by antioxidants. Neurobiol. Dis. 6, 209-219.

Perez, N., Sugar, J., Charya, S., Johnson, G., Merril, C., Bierer, L., Perl, D., Haroutunian, V. & Wallace, W. (1991). Increased synthesis and accumulation of heat shock 70 proteins in Alzheimer's disease. Mol. Brain Res. 11, 249-254.

Perng, M.D., Cairns, L., van den Ijssel, P., Prescott, A., Hutcheson, A.M. & Quinlan, R.A. (1999). Intermediate filament interactions can be altered by HSP27 and αB-crystallin. J. Cell Sci. 112, 2099-2112.

Perry, G., Friedman, R., Shaw, G. & Chau, V. (1987). Ubiquitin is detected in neurofibrillary tangles and senile plaque neurites of Alzheimer disease brains. Proc. Natl. Acad. Sci. USA 84, 3033-3036.

Pfanner, N. & Meijer, M. (1997). Mitochondrial biogenesis: The Tom and Tim machine. Curr. Biol. 7, R100-R103.

Piva, R., Cancelli, I., Cavalla, P., Bortolotto, S., Dominguez, J., Draetta, G.F. & Schiffer, D. (1999). Proteasome-dependent degradation of p27/kip1 in gliomas. J. Neuropathol. Exp., Neurol. 58, 691-696.

Poduslo, J.F., Curran, G.L., Sanyal, B. & Selkoe, D.J. (1999). Receptor-mediated transport of human amyloid β-protein 1-40 and 1-42 at the blood-brain barrier. Neurobiol. Dis. 6, 190-199.

Prusiner, S.B. (1982). Novel proteinaceous infectious particles cause scrapie. Science 216, 136-144.

Ranson, N.A., White, H.E. & Saibil, H.R. (1998). Chaperonins. Biochem. J. 333, 233-242.

Ranum, L.P.W., Lundgren, J.K., Schut, L.J., Ahrens, M.J., Perlman, S., Aita, J., Bird, T.D., Gomez, C. & Orr, H.T. (1995) Spinocerebellar ataxia type I and Machado-Joseph disease: Incidence of CAG expansions among adult-onset ataxia patients from 311 families with dominant, recessive, or sporadic ataxia. Am. J. Hum. Gen. 57, 603-608.

Rassow, J., von Ahsen, O., Bömer, U. & Pfanner, N. (1997). Molecular chaperones: Towards a characterization of the heat-shock protein 70 family. Trends Cell Biol. 7, 129-133.

Rogalla, T., Ehrnsperger, M., Preville, X., Kotlyarov, A., Lutsch, G., Ducasse, C., Paul, C., Wieske, M.,

Arrigo, A-P., Buchner, J. & Gaestel, M. (1999). Regulation of Hsp27 oligomerization, chaperone function, and protective activity against oxidative stress/tumor necosis factor α by phosphorylation. J. Biol. Chem. 274, 18947-18956.

Rommelaere, H., De Neve, M., Melki, R., Vandekerckhove, J. & Ampe, C. (1999). The cytosolic class II chaperonin CCT recognizes delineated hydrophobic sequences in its target proteins. Biochemistry 38, 3246-3257.

Rosenberg, R.N. (1995). Spinocerebellar ataxias and ataxins. New Eng. J. Med. 333, 1351-1353.

Ross, C.A., Margolis, R.L., Becher, M.W., Wood, J.D., Engelender, S. & Sharp, A.H. (1998) Pathogenesis of polyglutamine neurodegenerative diseases: Toward a unifying mechanism. In: Genetic Instabilities and Hereditary Neurological Diseases (Wells, R.D. and Warren, S.T., eds.), pp.761-776, Academic Press, San Diego, CA.

Rudolph, K.L., Chang, S., Lee, H-W., Blasco, M., Gottlieb, G.J., Greider, C. & DePinho, R.A. (1999). Longevity, stress response, and cancer in aging telomerase-deficient mice. Cell 96, 701-712.

Rustin, P., von Kleitst-Retzow, J-C., Chantrel-Groussard, K., Sidi, D., Munnich, A. & Rötig, A. (1999). Effect of idebenone on cardiomyopathy in Friedreich's ataxia: A preliminary study. The Lancet 354, 477-479.

Ryan, M.T., Herd, S.M., Sberna, G., Samuel, M.M., Hoogenraad, N.J. & Hoj, P.B. (1997) The genes encoding mammalian chaperonin 60 and chaperonin 10 are linked head-to-head and share a bidirectional promoter. Gene 196, 9-17.

Ryan, M.T., Naylor, D.J., Hoj, P.B., Clark, M.S. & Hoogenraad, N.J. (1997a). The role of molecular chaperones in mitochondrial protein import and folding. Intl. Rev. Cytol. 174, 127-193.

Saborio, G.P., Soto, C., Kascsak, R.J., Levy, E., Kascsak, R., Harris, D.A. & Frangione, B. (1999). Cell-lysate conversion of prion protein into its protease-resistant isoform suggests the participation of a cellular chaperone. Biochem. Biophys. Res. Commun. 258, 470-475.

Saris, N., Holkeri, H., Craven, R.A., Stirling, C.J. & Makarow, M. (1997). The Hsp70 homologue Lhslp is involved in a novel function of the yeast endoplasmic reticulum, refolding and stabilization of heat-denatured protein aggregates. J. Cell Biol. 137, 813-821.

Schenk, D., Barbour, R., Dunn, W., Gordon, G., Grajeda, H., Guido, T., Hu, K., Huang, J., Johnson-Wood, K., Khan K., Kholodenko, D., Lee, M., Kiao, Z., Liebergurg, I., Motter, R., Mutter, L., Soriano, F., Shopp, G., Vasquez, N., Vandevert, C., Walker, S., Wogulis, M., Yednock, T., Games, D. & Seubert, P. (1999). Immunization with amyloid-β attenuates Alzheimer-disease-like pathology in the PDAPP mouse. Nature 400, 173-177.

Scherzinger, E., Lurz, R., Turmaine, M., Mangiarini, L., Hollenbach, B., Hasenbank, R., Bates, G.P., Davies, S.W., Lehrach, H. & Wanker, E.E. (1997). Huntingtin-encoded polyglutamine expansions form amyloid-like protein aggregates in vitro and in vivo. Cell 90, 549-558.

Schirmer, E.C. & Lindquist, S. (1997). Interactions of the chaperone Hsp104 with yeast Sup35 and mammalian PrP. Proc. Natl. Acad. Sci. USA 94, 13932-13937.

Schneider, J.A., Watts, R.L., Gearing, M., Brewer, R.P. & Mirra, S.S. (1997) Corticobasal degeneration: Neuropathologic and clinical heterogeneity. Neurol. 48, 959-969.

Schweisguth, F. (1999). Presenilins in their infancy. Chem. Biol. 6, R187-R190.

Silar, P. & Daboussi, M-J. (1999). Non-conventional infectious elements in filamentous fungi. Trends Genet. 15, 141-145.

Skinner, P.L., Koshy, B.T., Cummings, C.J., Klement, I.A., Helin, K., Servadios, A., Zoghbi, H.Y. & Orr, H.T. (1997). Ataxin-1 with an expanded glutamine tract alters nuclear matrix-associated structures. Nature 389, 971-974.

Small, D.H. & McLean, C.A. (1999). Alzheimer's disease and the amyloid β protein: What is the role of amyloid? J. Neurochem. 73, 443-449.

Stege, G.J.J., Renkawek, K., Overkamp, P.S.G., Verschuure, P., van Rijk, A.F., Reijnen-Aalbers, A., Boelens, W.C., Bosman, G.J.C.G.M. & de Jong, W.W. (1999). The molecular chaperone αB-crystallin enhances amyloid β neurotoxicity. Biochem. Biophys. Res. Commun. 262, 152-156.

Stenoien, D.L., Cummings, C.J., Adams, H.P., Mancini, M.G., Patel, K., DeMartino, G.N., Marcelli, M., Weigel, N.L. & Mancini, M.A. (1999). Polyglutamine-expanded androgen receptors form aggregates that sequester heat shock proteins, proteasome components and SRC-1, and are suppressed by the HDJ-2 chaperone. Hum. Mol. Gen. 8, 731-741.

Tatar, M., Khazaeli, A.A. & Curtsinger, J.W. (1997). Chaperoning extended life. Nature 390, 30.

Tatzelt, J., Zuo, J., Voelly, R., Scott, M., Harti, U., Prusiner, S.B. & Welch, W.J. (1995). Scrapie prions selectively modify the stress response in neuroblastoma cells. Proc. Natl. Acad. Sci. USA 92, 2944-2948.

Telling, G.C., Scott, M., Mastrianni, J., Gabizon, R., Torchia, M., Cohen, F.E., DeArmond, S.J. & Prusiner, S.B. (1995). Prion propagation in mice expressing human and chimeric PrP transgenes implicates the interaction of cellular PrP with another protein. Cell 83, 79-90.

The Huntington's Disease Collaborative Research Group. (1993). A novel gene containing a trinucleotide repeat that is expanded and unstable on Huntington's disease chromosome. Cell 72, 971-983.

Timpl, P., Spanagel, R., Sillaber, I., Kresse, A., Reul, J.M.H.M., Stalla, G.K., Blanquet, V., Steckler, T., Holsboer, F. & Wurst, W. (1998). Impaired stress response and reduced anxiety in mice lacking a functional corticotropin-releasing hormone receptor 1 Nature Genet. 19, 162-166.

Tran, P.B. & Miller, R.J. (1999). Aggregates in neurodegenerative disease: Crowds and power? Trends Neurol. Sci. 22, 194-197.

Trombetta, E.S. & Helenius, A. (1998). Lectins as chaperones in glycoprotein folding. Curr. Op. Struct. Biol. 8, 587-592.

Vainberg, I.E., Lewis, S.A., Remmelaere, H., Ampe, C.,Vandekerckhove, J., Klein, H.L. & Cowan, N.J. (1998). Prefoldin, a chaperone that delivers unfolded proteins to cytosolic chaperonin. Cell 93, 863-873.

Vendemiale, G., Grattagliano, I. & Altomare, E. (1999). An update on the role of free radicals and antioxidant defense in human disease. Intl. J. Clin. Lab. Res. 29, 49-55.

Vicart, P., Caron, A., Guicheney, P., Li, Z., Prvost, M-C., Faure, A., Chateau, D., Chapon, F., Tomé, F., Dupret, J-M., Paulin, D. & Farceau. (1998). A missense mutation in the αB-crystallin chaperone gene causes a desmin-related myopathy. Nature Genet. 20, 92-95.

Wagner, I., Arlt, H., van Dyck, L., Langer, T. & Neupert, W. (1994). Molecular chaperones cooperate with PIM1 protease in the degradation of misfolded proteins in mitochondria. EMBO J. 13, 5135-5145.

Wang, J., Dickson, D.W., Trojanowsi, J.Q. & Lee, V. M-Y. (1999) The levels of soluble versus insoluble Aβ distinguish Alzheimer's disease from normal and pathologic aging. Exp. Neurol. 158, 328-337.

Weninger, S.C., Dunn, A.J., Muglia, L.J., Dikkes, P., Miczek, K.A., Swiergiel, A.H., Berridge, C.W. & Majzoub, J.A. (1999). Stress-induced behaviors require the corticotropin-releasing hormone (CRH) receptor, but not CRH. Proc. Natl. Acad. Sci. USA 96, 8283-8288.

Westermann, B. & Neupert, W. (1997). Mdj2p, a novel DnaJ homolog in the mitochondrial inner membrane of the yeast *Saccharomyces cerevisiae*. J. Mol. Biol. 272, 477-483.

Wilhelmsen, K.C. (1999). The tangled biology of tau. Proc. Natl. Acad. Sci. USA 96, 7120-7121.

Will, R.G., Ironside, J.W., Zeidler, M., Cousens, S.N., Estibeiro, K., Alperovitch, A., Poser, S., Pocchiari, M., Hofman, A. & Smith, P.G. (1996). A new variant of Creutzfeldt-Jakob disease in the UK. The Lancet 347, 921-925.

Willison, K.R. (1999). Composition and function of the eukaryotic cytosolic chaperonin-containing TCP-1.

In: Molecular Chaperones and Folding Catalysts (Bukau, B., ed.) pp. 555-571. Harwood Academic Publishers, Australia.

Woese, C.R. (1998). A manifesto for microbial genomics. Curr. Biol. 8, R781-R783.

Woese, C.R., Kandler, O. & Wheelis, M.L. (1990). Towards a natural system of organisms: Proposal for the domains Archaea, Bacteria, and Eucarya. Proc. Natl. Acad. Sci. USA 87, 4576-4579.

Won, K-A., Schumacher, R.J., Farr, G.,W., Horwich, A.L. & Reed, S.I. (1998). Maturation of human cyclin E requires the function of eukaryotic chaperonin CCT. Mol. Cellul. Biol. 18, 7584-7589.

Wood, S.J., Wypych, J., Steavenson, S., Louis, J-C., Citron, M. & Biere, A.L. (1999). α-synuclein fibrillogenesis is nucleation-dependent. J. Biol. Chem. 274, 14509-14512.

Xia, W., Zhang, J., Perez, R., Koo, E.H. & Selkoe, D.J. (1997). Interaction between amyloid precursor protein and presenilins in mammalian cells: Implications for the pathogenesis of Alzheimer disease. Proc. Natl. Acad. Sci. USA 94, 8208-8213.

Xiao, N. & DeFranco, D.B. (1997) Overexpression of unliganded steroid receptors activates endogenous heat shock factor. Mol. Endocrinol. 11, 1365-1374.

Yamao, F. (1999) Ubiqutin system: Selectivity and timing of protein destruction. (1999). J. Biochem. 125, 223-229.

Yang, A.J., Chandswangbhuvana, D., Shu, T., Henschen, A. & Glabe, C.G. (1999). Intracellular accumulation of insoluble, newly synthesized Aβn-42 in amyloid precursor protein-transfected cells that have been treated with Aβ1-42. J. Biol. Chem. 274, 20650-20656.

Yasojima, K., McGeer, E.G. & McGeer, P.L. (1999). Complement regulators C1 inhibitor and CD59 do not significantly inhibit complement activation in Alzheimer disease. Brain Res. 833, 297-301.

Yu, Z., Luo, H., Fu, W. & Mattson, M.P. (1999). The endoplasmic reticulum stress-responsive protein GRP78 protects neurons against excitotoxicity and apoptosis: Suppression of oxidative stress and stabilization of calcium homeostasis. Exp. Neurol. 155, 302-314.

Zapun, A., Jakob, C.A., Thomas, D.Y. & Bergeron, J.J.M. (1999). Protein folding in a specialized compartment: The endoplasmic reticulum. Structure 7, R173-R182.

Note: After this Chapter was submitted for publication, pertinent work was published. A few of the most relevant and recent publications are listed below, grouped by main topic.

Composition of protein deposits.

Kokubo, H., Lemere, C.A. & Yamaguchi, H. (2000). Localization of flotillins in human brain and their accumulation with the progression of Alzheimer's disease pathology. Neurosc. Lett. 290, 93-96.

Shashidharan, P., Good, P.F., Hsu, A., Perl, D.P., Brin, M.F. & Olanow, C.W. (2000). Torsin A accumulation in Lewy bodies in sporadic Parkinson's disease. Brain Res. 877, 379-381.

Role of Hsp in controlling protein aggregation and aggregate dissolution.

Carmichael, J., Chatellier , J., Woolfson , A., Milstein, C., Fehrst, A.R. & Rubinsztein, D.C. (2000). Bacterial and yeast chaperones reduce both aggregate formation and cell death in mammalian cell models of Huntington's disease. Proc. Natl. Acad. Sci. USA 97, 9701-9705.

Jana, N.R., Tanaka, M., Wang, G. & Nukina, N. (2000). Polyglutamine length-dependent interaction of Hsp40 and Hsp70 family chaperones with truncated N-terminal huntingtin: Their role in suppression of aggregation and cellular toxicity. Hum. Mol. Genet. 9, 2009-2018.

Macario, A.J.L. & Conway de Macario, E. (2000). Stress and molecular chaperones in disease. Intl. J. Clin. Lab. Res. 30, 49-66.

Muchowski, P.J., Schaffar, G., Sittler, A., Wanker, E.E., Hayer-Hartl, M.K. & Hartl, F.U. (2000). Hsp70 and Hsp40 chaperones can inhibit self-assembly of polyglutamine proteins into amyloid-like fibrils. Proc. Natl. Acad. Sci. USA 97, 7841-7846.

Chaperones and prion.

Moriyama, H., Edskes, H.K. & Wickner, R.B. (2000). [URE3] prion propagation in *Saccharamocyces cerevisiae*: Requirement for chaperone Hsp104 and curing by overexpressed chaperone Ydj1p. Mol. Cellul. Biol. 20, 8916-8922.

Jin, T., Gu, Y., Zanuso, G., Sy, M., Kumar, A., Cohen, M., Gambetti, P. and Singh, N. (2000). The chaperone protein BiP binds to a mutant prion protein and mediates its degradation by the proteasome. J. Biol. Chem. 275, 38699-38704.

Stress, oxidative stress, mitochondria, calcium, glucose and glutamate transport, in pathogenesis and protection.

Finkel, T. & Holbrook, N.J. (2000). Oxidants, oxidative stress and the biology of aging. Nature 408, 239-247.

Giasson, B.I., Duda, J.E., Murray, I.V.J., Chen, Q., Souza, J.M., Hurtig, H.I., Ischiropoulos, H., Trojanowski, J.Q. & Lee, V.M-Y. (2000). Oxidative damage linked to neurodegeneration by selective alpha-synucleinopathy lesions. Science 290, 985-989.

Guo Z.H., Ersoz, A., Butterfield, D.A. & Mattson, M.P. (2000). Beneficial effects of dietary restriction on cerebral cortical synaptic terminals: Preservation of glucose and glutamate transport and mitochondrial function after exposure to amyloid beta-peptide, iron, and 3-nitropropionic acid. J. Neurochem. 75, 314-320.

Guo Z.H., Kindy, M.S., Kruman, I. & Mattson, M.P. (2000). ALS-linked Cu/Zn-SOD mutation impairs cerebral synaptic glucose and glutamate transport and exacerbates ischemic brain injury. J. Cer. Blood Flow Metab. 20, 463-468.

Guo, Z.H. & Mattson, M.P. (2000). Neurotrophic factors protect cortical synaptic terminals against amyloid- and oxidative stress-induced impairment of glucose transport, glutamate transport and mitochondrial function. Cer. Cortex 10, 50-57.

Guo Z.H. & Mattson, M.P. (2000). In vivo 2-deoxyglucose administration preserves glucose and glutamate transport and mitochondrial function in cortical synaptic terminals after exposure to amyloid beta-peptide and iron: Evidence for a stress response. Exp. Neurol. 166, 173-179.

Malecki, A., Garrido, R., Mattson, M.P., Hennig, B. & Toborek, M. (2000). 4-Hydroxynonenal induces oxidative stress and death of cultured spinal cord neurons. J. Neurochem. 74, 2278-2287.

Pedersen,W.A., Culmsee, C. Ziegler, D., Herman, J.P. & Mattson, M.P. (1999). Aberrant stress response associated with severe hypoglycemia in a transgenic mouse model of Alzheimer's disease. J. Mol. Neurosc. 13, 159-165.

Role of the ubiquitin-proteasome system in protein degradation and pathogenesis.

Lam, Y.A., Pickart, C.M., Alban, A., Landon, M., Jamieson, C., Ramage, R., Mayer, R.J. & Layfield, R. (2000). Inhibition of the ubiquitin-proteasome system in Alzeimer's disease. Proc. Natl. Acad. Sci. USA 97, 9902-9906.

Shimura, H., Hattori, N., Kubo, S., Mizuno, Y., Asakawa, S., Minoshima, S., Shimizu, N., Iwai, K., Chiba, T., Tanaka, K. & Suzuki, T. (2000). Familial Parkinson disease gene product, parkin, is a ubiquitin-protein ligase. Nature Genet. 25, 302-305.

Wyttenbach, A., Carmichael, J., Swartz, J., Furlong, R.A., Narain, Y., Rankin, J. & Rubinsztein, D.C. (2000). Effects of heat shock, heat shock protein 40 (HDJ-2), and proteasome inhibition on protein aggregation in cellular models of Huntington's disease. Proc. Natl. Acad. Sci. USA 97, 2898-2903.

Zhang, Y., Gao, J., Chung, K.K.K., Huang, H., Dawson, V.L. & Dawson, T.M. (2000). Parkin functions as an E2-dependent ubiquitin-protein ligase and promotes the degradation of the synaptic vesicle-associated protein, CDCrel-1. Proc. Natl. Acad. Sci. USA 97, 13354-13359.

THE ROLE OF TELOMERASE IN AGE-RELATED DEGENERATIVE DISEASE AND CANCER

MICHAEL FOSSEL

Michigan State University, Box 630, Ada, Michigan 49301

Table of Contents

Recent work in cell biology and genetics has suggested that at the most fundamental level, age-related diseases result from cell senescence. Our growing understanding of the process of cell senescence, senescence associated gene expression, and telomerase biology will soon allow us to prevent, ameliorate, or to a significant degree, even reverse such pathology. This potential includes an effective control of cancer. The dual assumptions that aging is merely the passive accumulation of "wear and tear" and age-related diseases merely the outgrowth of such passively accumulated dysfunction are both erroneous. Aging occurs as the result of a predictably altered pattern of gene expression that is specific to cell type and it culminates in dysfunction and disease. *In vitro*, the process can be prevented or (within predictable limits) reversed. The implications for clinical medicine are broad, startling, and almost without precedent.

Interorganellar Signaling in Age-Related Disease, Edited by Mark P. Mattson, 163—204

Introduction

Age-related diseases have generally been considered a disparate and motley collection of pathologies, united only conceptually (but not causally) by their increasing incidence with increasing age. This assumption is widespread and generally unstated, but equally unsupported: to date it has hinged merely on the lack of any sufficient contrary hypothesis or contradicting data. In the same sense that many diseases were discovered to have microbial etiologies a century ago, recent evidence strongly suggests that age-related diseases have their etiology in cell senescence. This does not undercut, but rather extends, our current understanding of many different diseases. In the case of atherogenesis, for example, the usual suspects (hypertension, diabetes, hypercholesterolemia, tobacco use, etc.) remain indicted, but we begin to discern the ringleader that orchestrates and underlies the entire process. This same extension not only serves to explain the anomalies (such as atherogenesis in progeric children who lack the usual risk factors), but more importantly, offers us a new and potentially profoundly more effective avenue for clinical intervention in such diseases.

In the same sense that tetanus was once "prevented" with hygiene and "treated" with wound cauterization, we now prevent heart disease with diet, drugs, and behavioral change and treat heart disease with angioplasty, transplantation, etc. An understanding of immunization allowed us to prevent tetanus; an understanding of cell senescence will, *pari passu*, allow us to treat age-related diseases in a radically more effective way.

History

Although a knowledge of microbes was widespread in scientific and medical circles since the early 1800's, by the mid 1870's, infectious disease was still not broadly accepted as a disease etiology. To the contrary, a great many intelligent and well-respected scientists felt that scientists such as Pasteur and Koch were at best premature, but at worst (and more likely), simply wrong. Cellulitis was the product of "devitalization" of healthy tissues, felt to be due to some subtle "toxic factor"; bacterial growth from a wound was purely opportunistic, and not strictly related to causation, morbidity, or mortality. Immediately prior to Koch's epochal discoveries, there was little to contradict such a view, summed up in Satterthwaite's classic remark that "we have no reliable evidence that the definite forms described [bacteria] are capable of producing the diseases in question" (Starr, 1982), and E. P. Hurd's equally remarkable statement that, "the inoculation of a healthy person with bacteria ... is not necessarily dangerous" (Starr, 1982). Such views were not irrational in their historical context. To the contrary, and like similar views today, they were entirely consistent with the known data, yet insufficient and an obstruction to clinical progress. This view was disproved not by a more elegant theory or even by pure laboratory data, but rather by the overwhelming efficacy of clinical therapy based on a microbial etiology of disease, viz., the success of antisepsis, inoculations, and antibiotics.

Our knowledge of aging stands at a similar juncture. As occurred in the case of microbial theory, the triumph of the new concepts will lie not in theory or in the increasingly remarkable *in vitro* data, but rather in clinical trials. The success will (and must) be measured in diseases prevented and cured, rather than in cell models or conceptual fit. The concepts that we have employed to date have been adequate but byzantine, having allowed us to merely classify and discuss the data obtained. In a strict sense, a unifying concept of aging has eluded us; while in a loose sense, the notion of "wear and tear" has served in its place. This is usually expressed in the more modern equivalent of accrual of damage caused by free radicals to DNA, proteins, and lipid structures within the cell.

Aging

Discussions of aging depend on context. Practically and clinically, aging is an increasing susceptibility to a panoply of diseases as age increases. Most of us would prefer not to define aging as a disease, and appropriately so: it is an all-but universal process found in single cells and organisms, culminating in apparently irrevocable, dysfunctional changes at all levels. Yet, as aging is the cause of these changes and the force behind age-related disease, it is culpable and therefore a justifiable, fundamental target for clinical research to cure such disease. As Ralph Waldo Emerson wrote, "Old age seems the only disease, all others run into this one".

If aging is unavoidable, if it is wear-and-tear, if it is simply the "cost of doing business" as an aerobic organism in an entropic world, then, ultimately, clinical therapy is futile. The evidence, however, contradicts such a view. Not only are there immortal multicellular organisms (e.g., the hydra), but all organisms are dependent upon cell lines that demonstrate no evidence whatsoever of aging. Germ cell lines, from which each organism's somatic cells are the necessary and supportive "sidetrack", have been extant for three and a half billion years. The eukaryotes (such as ourselves) have had symbiotic mitochondria for perhaps a billion and a half years. Although most free radicals (which drive a substantial portion of degenerative changes) originate in mitochondria, they have not caused senescence, aging, or any cumulative dysfunction within germ cell lines. For more than a billion years eukaryotic germ cells have withstood aging, yet senescent changes occur within a matter of decades when such cells differentiate into human somatic cells. Any attempt to understand aging and the fundamental etiology of age-related disease, must not only explain why aging occurs, but why it *does not* occur for more that a billion years in such cell lines.

To explain aging and understand age-related diseases, we must first understand chromosomal changes and then the changes in gene expression which succeed them.

Telomeres

In 1888, Waldeyer described the dark clumps of material in the cell nucleus and named them chromosomes. In the 1930's, Muller, Darlington, and Haldane describe the terminal portions of the chromosomes and in 1938, Muller named them telomeres (Muller, 1938). Quickly thereafter, Barbara McClintock (who won the Nobel Prize for her work on the chromosomes of maize) proved that telomeres were essential for chromosomal integrity: without them, the chromosomes become unstable, "sticky", and prone to significant mutation (McClintock, 1942). The strictly inaccurate – but poetic and common – concept that the telomeres "keep the chromosomes from unraveling" has a certain utility in understanding the dangers that occur if telomeric maintenance is faulty, but the concept hides the far more important aspect of telomere function, that of the regulation of gene expression.

In 1972, James Watson (Watson, 1972) pointed out that a corollary of the accepted mechanism of linear chromosome replication was that a small portion of each telomere would be lost during cell division. The process would necessarily result in progressive telomeric erosion until, ultimately, there would be a risk to chromosomal integrity and, hence, cell survival. The implications were that either 1) all cells with linear chromosomes, including those of James Watson, had died and life was extinct, or 2) his understanding of chromosomal replication was substantially flawed, or 3) there was an unknown mechanism which was responsible for relengthening telomeres (at the very least, in the germ cell line). The first possibility has been discounted. The second has proven to be partially true: we probably do not fully understand either chromosomal replication or the (several) mechanisms of telomere shortening. The third mechanism is entirely true: the enzyme "telomere terminal transferase" or telomerase is responsible for relengthening telomeres in eukaryotic cells.

Curiously, however, it was not Watson who first suggested the danger of telomere loss, but Alexey Olovnikov, a Russian, who published the theory in 1971 (Olovnikov, 1971). Olovnikov, however, went even further. He suggested that telomere shortening might be responsible for cell senescence.

Cell Senescence

The fact that cells senesce was at first, as with the mere existence of microbes, a heresy. By the late 19th century, technical advances had allowed us to culture cells in the laboratory. Based on their laboratory culture of chicken cells, the two major figures in cell culture work, Alexis Carrel and Albert Ebeling, believed that cells would divide and grow indefinitely under appropriate laboratory conditions: somatic cells were inherently immortal. If Ebeling had allowed his cells to divide (and saved them) for thirty-four years, "the mass of chicken cells ... would have been greater than the sun" (Hayflick, 1994). In 1961 (Hayflick & Moorhead, 1961), it took remarkable courage, perhaps stubbornness and a blithe disregard for their professional prospects, for Leonard Hayflick and Paul Moorhead, then at the Wistar Institute, to suggest that cells had "life spans" and that earlier results of indefinite cell life spans were due to technical errors. Yet today, the "Hayflick limit" is one of the major pillars of cell biology. Cancer cells may be immortal, but somatic cells have replicative limits. All non-cancerous somatic cells have a defined replicative life span characteristic of the species, tissue cell type, and age of the donor. On the average, human embryonic cells are limited to fifty successive doublings (Hayflick, 1994).

Two questions remained: 1) what was the clock that governed this replicative limit, and 2) what was the relevance (if any) of this replicative limit to aging in the organism? The answer to the first of these two questions was (as above) suggested by Alexey Olovnikov, but it was not until the last decade that we have proven him correct. In 1990,

Calvin Harley, Carol Greider, and Bruce Futcher published a landmark paper (Harley et al., 1990) showing that telomere lengths correlate precisely with replicative limits. A series of similarly correlative papers followed, culminating in the 1998 publication of a landmark paper (Bodnar et al., 1998) which proved causation, thereby "ending the debate" (Lange, 1998).

The critical experiment (Bodnar et al., 1998) was done by Harley's group (largely at Geron Corporation) and his colleagues at the University of Texas (Jerry Shay and Woody Wright) and has been replicated many times since (Vaziri and Benchimol, 1998; Vaziri, 1998; Caddle et al., 1998). Transfecting the gene for the catalytic component for human telomerase into three lines of human somatic cells (lung fibroblasts, skin fibroblasts, and pigmented retinal epithelial cells), they showed that the resultant cells became immortal. These cells have yet to demonstrate either a replicative limit or any alteration to a senescent pattern of gene expression, typical of senescent cells approaching such a replicative limit. In the four decades since it was first hypothesized, telomerase has changed from being merely Watson's theoretical "loose end" of chromosomal replication (and a potent mechanism for repairing chromosomal loose ends) to a method of immortalizing human somatic cells with enormous potential for human disease. What is this enzyme and how does it function?

Telomerase

Structure of Telomerase

Telomerase is a reverse transcriptase. Capable of adding bases to human telomeric DNA using an RNA template (Greider and Blackburn, 1985; Yu et al., 1990; Singer and Gottschling, 1994), human telomerase (hTERT) is composed of two moieties: the RNA template and a catalytic component.

The RNA template is specific for the telomere sequence but has additional bases which may have catalytic activity as well. In humans (and, indeed, all known vertebrates and a great many invertebrates), the telomere sequence is a series of tandem repeats of TTAGGG (thymine, thymine, adenine, guanine, guanine, guanine), often for several thousand base pairs. The corresponding RNA template of telomerase matches this sequence but also has an overlapping portion and is capable of "climbing" along the telomere as it adds bases (Greider, 1996). In human somatic cells, the RNA component is usually down-regulated although not uniformly so: in some cases it is constitutively expressed (Feng et al., 1995; Blasco et al. 1995; Blasco et al., 1997; Avilion, 1996). It is, therefore, not necessarily the limiting factor for telomerase activity in somatic cells and in some cases it is regulated independently of the catalytic component.

The catalytic component (human telomerase reverse transcriptase or hTERT) is a protein and the "active" portion of telomerase. The amino acid sequence is known, as is its corresponding gene sequence (Nakamura et al., 1997; Nakayama et al., 1998). As a general rule (see below for exceptions), human somatic cells do not express the catalytic component constitutively and hTERT is therefore the "controlling element" of telomerase activity in many normal human somatic cells (Lingner et al., 1997; Nakamura et al., 1997; Nakayama et al., 1998; Harrington et al., 1997; Kilian et al., 1997).

Since normal somatic cells generally express the RNA component but not the catalytic component (Greenberg et al., 1998), there are three ways to test the effect of telomere length on cell senescence and gene expression. The experimenter can:

1) transfect the gene for the catalytic component of telomerase into a somatic cell which normally lacks it (as was done in Bodnar et al., 1998),
2) "knockout" the telomerase gene altogether in a cell or organism (as in mice; see below), or
3) reversibly induce expression of the catalytic component of telomerase in somatic cells which do not routinely express it (as occurs normally in hematopoetic stem cells; see below).

Currently, we are technically capable of transfecting the telomerase gene *in vitro* or *ex vivo*, and potentially (though with extremely limited success) *in vivo* as well. The knockout technique is currently a standard in many genetics laboratories and the example of a knockout mouse will be discussed. A great many researchers are trying to understand how to induce "native" telomerase expression in cells. Reversible induction occurs routinely in some somatic stem cells and, theoretically, it should be feasible to reversibly induce telomerase expression both *in vitro* and *in vivo*, but we are currently incapable of doing so. This goal has become the Holy Grail of research on the effects of cell senescence on aging.

Is Telomerase Good for Anything?

In the absence of telomerase, organ renewal *in vivo* or serial passage of somatic cells *in vitro* leads to telomere shortening and (as we shall see) subsequent changes in gene expression, cell morphology, and the rate of cell cycling. This is the current "rule of thumb" and remains firmly supported by literally hundreds of studies over the past decade. Nonetheless, is telomerase necessary for an organism?

There are a number of studies that bring the utility of telomerase into question. Mouse embryonic fibroblasts (MEF's) for example, nullizygous for telomerase RNA, can spontaneously immortalize without any apparent telomerase activity at all (Blasco et al., 1997). Although the majority (perhaps 92%; see below) of human malignancies express telomerase, there *are* immortal human tumor lines which have long telomeres yet lack any detectable telomerase activity. Despite such exceptions, the rule is generally followed and has become accepted: the all but consistent repression of telomerase in normal somatic cells and its overwhelming expression in malignancy suggests that telomerase is a general requirement for cell immortalization.

The early results of the telomerase knockout mouse were therefore greeted with mixed surprise and uncertainty. Mice are a useful laboratory animal for aging research: they are cheap, well understood, and have short life spans. Curiously, mice (unlike humans) have measurable baseline telomerase activity in their somatic cells. This fact, coupled with the (perhaps causally) associated high incidence of cancer in mice (Prowse and Greider, 1995), made a telomerase knockout mouse an understandable priority in aging research. The initial results (Blasco et al., 1997) suggested that such mice suffered no ill effects of lacking telomerase.

Consider what such initial results implied. Cells lacking telomerase are seldom a problem: most normal human somatic cells lack telomerase and humans generally do quite well. However, the germ cell line (and as we shall see, some stem cell lines) depends upon being able to relengthen its telomeres in order to maintain the ability to divide: a necessity in the long run for species survival and perhaps even in the short run for fetal development. Many were therefore startled (and some disappointed) when the initial results showed that telomerase deficient mice grew and reproduced normally. Lifespan, motor activity, weight gains, hematopoetic histology, peripheral blood counts, immune response, and *in vitro* cell growth were within normal limits for the strain (Blasco et al., 1997; Lee et al., 1998). In summary, "telomerase activity *per se* is not essential in organogenesis or postnatal organ function or maintenance" in the mouse (Lee et al., 1998).

Subsequent results, however, have put the initial data in a new light (Lee et al., 1998; Rudolph et al., 1999). Some first generation mice had erosive dermatitis. Fertility problems became apparent in both sexes by the fifth generation and there were no sixth generation offspring. This was at least partially attributable to cell losses and organ changes seen in the reproductive systems (spermatogenesis in the seminiferous tubules contrasted with the less proliferative Sertoli and Leydig cells, uterine horn tissues, ovaries), but other organ systems with highly proliferative cell populations were similarly affected. The long-term renewal capabilities of hematopoetic stem cells was clearly diminished by telomere shortening when tested *in vitro*, although this was not grossly apparent in either peripheral smears or marrow pathology. Similar cryptic dysfunction was found in the mitogenic response of lymphocytes, whose cells showed a marked increase in the percentage of late apoptotic cells. Abnormal chromosomal fusions and other evidence of genomic instability also increased over generation times in telomerase negative mice.

The long delay in such potentially critical defects has been attributed to the long telomeres found in mice (at least in the particular strain used). The mouse has a longer telomeric "buffer", as it were, but nevertheless suffers the consequences once this buffer erodes (Lee et al., 1998).

What, if any, is the message from such studies? The primary message is that *telomerase expression per se is not the issue*. The fundamental aspect of aging at the cellular level is not telomerase expression. Perhaps, as might be inferred from the work cited above, the fundamental aspect is telomere length. But in fact, as we shall see, even this is not fundamental.

Defining Cell Senescence

What is cell senescence? In any conversation on cell senescence, indeed in any review of the literature, it quickly becomes apparent that senescence is a concept without a common usage. This problem occurs at two levels.

The first issue is one of "how senescent is senescent?" Assume for a moment that we have a well-accepted measuring stick for senescence (we clearly don't, which is the second issue). Our scale might run from a maximum of one hundred (young) down to a minimum of zero (fully senescent). At the extreme high end of the scale, one hundred represents an extremely "young" cell (perhaps one with long telomeres?), in which there is no indication whatsoever of senescence under anyone's definition or by any measure. At the extreme low end of the scale, zero represents a cell which is not merely senescent, but in fact so irrevocably senescent that it is perhaps dead (we will forego discussing necrotic versus apoptotic cell death). Most researchers, forced into using such as scale, might agree that quite low numbers – for example, cells with an average "senescence score" below five – are certainly senescent and cannot be rescued by telomerization or other intervention. With the advent of telomerization, the scale is no longer merely academic because we can now ask at what number we can still rescue cells from the "cliff" of irrevocable senescence. In this sense, using this scale, is a cell senescent at twenty? What about fifty? At what number does "true senescence" set in?

Once we can telomerize cells, there is disagreement about whether we can reverse senescence or not and the confusion hinges on your cutoff for defining senescence. If any number below 99 represents senescence, then cell senescence can be reversed in almost all cells. If only scores below 3 are truly senescent, then cell senescence is never reversible. While no one would suggest that we can reset senescence in cells at zero (after all, such cells are dead by our definition), there is considerable disagreement whether or not cell senescence can be reversed at all. This is largely attributable to disagreement as to the cutoff for senescence. How senescent? We lack a universal "cutoff".

Non-senescent cells can be immortalized (Bodnar et al., 1998). Although there is a publicly unstated (although privately expressed) and almost universal acceptance that there is a "point of no return" beyond which cell senescence is irrevocable, there is no currently accepted method for defining and measuring such a point. Researchers therefore disagree about whether we can immortalize senescent cells or perhaps only "presenescent", early senescent, or non-senescent cells. Being unaware of our assumptions and the discrepancy in our definitions, postpones any possibility of resolution.

The second issue is one of achieving a consensus of what is being discussed when we refer to senescence in a cell or a population of cells in the first place. One common, historically sanctioned measure has been population doublings. The early literature is replete with references to the "PD's" of cell populations. Such a definition and measure of cell senescence has two problems. The first is that it measures populations, not cells. We are now capable of discussing mechanisms of senescence at the intracellular level and require a term that reflects our increasing technical sophistication. The second problem is that the number of population doublings or even (with reference to particular cells) the replicative limit varies with species (human versus mouse), pathology (e.g., progeria), cell phenotype (fibroblast versus hematopoetic stem cell), and donor age (although to a arguable and complex extent; see below). Under these circumstances, it is misleading and bootless to discuss whether two cells are equally "senescent" if we rely merely on the number of previous or anticipated population doublings or even if we rely on the exact number of cell divisions for a specific cell. Cell divisions *per se* does not and cannot define cell senescence any more than telomere length does.

How then can we define cell senescence in a meaningful, useful, and predictive way?

Senescent Gene Expression

As the intelligent reader will have surmised, gene expression itself is becoming the universal standard in discussions of cell senescence, although the transformation is an informal one and by no means complete. This is the result of two factors: 1) we require such an objective and useful measure in order to compare different studies, cell phenotypes, and interventions, and 2) we are now capable of doing so.

Using the pattern of gene expression as the basis for definition is neither a smooth nor a simple process. With regards to senescence of skin fibroblasts, for example, a common marker is beta galactosidase or beta-gal (Dimri and Campisi, 1994; Dimri et al., 1995; Watanabe et al., 1997; Katakura et al., 1997), yet the marker has been heavily criticized (Yegorov et al., 1998). It is certainly neither one hundred percent specific nor one hundred per cent sensitive.

The primary benefit of using gene expression or "Senescence Associated Gene Expression" (SAGE), to define cell senescence is not merely that it is objective and measurable, nor that it accounts for cell phenotypes (and in fact their genotypes), but that it references the cell changes that we are most concerned with practically and even clinically. We care little whether a cell expresses telomerase or not, has long or short telomeres, or how many divisions it has undergone (and at what point do we start such a replicative count, by the way?). To the contrary, however, we care *a great deal* how a cell expresses its genes: that pattern determines cell function and, in the context of other cells, determines organ function and clinical disease.

For a practical analogy, consider that every cell is a time bomb. This was a common analogy a decade ago (Harley, 1991; I made a model of such a bomb that still sits on the bookshelf of one of the field's most prominent researchers). Suppose that the fuse is the telomere and that the bomb itself the altered pattern of gene expression (SAGE) which ensues when the telomere "burns down" sufficiently. In predicting cell senescence, telomere is insufficient. Mice have long fuses, but how fast do their cells divide? The "burn rate" of those telomeric fuses is equally important. But even knowing the length of the telomeric "fuse" and the "burn rate" of cell division is still insufficient. The outcome depends on the "explosive" in our time bomb. If gene expression changes dramatically (perhaps even with the earliest changes in the length of a very long telomere), the outcome may be catastrophic, whereas in the obverse case, if gene expression changes only minimally and then only when the telomere has all but disappeared, the outcome may be inconsequential. We need to understand exactly what genes change expression and at what telomeric length, not simply the telomere length and rate of telomere base loss.

Given a choice of two time bombs, the question is not the length of the fuse, nor how fast the fuse burns, but, more importantly, what the bomb does. Is the explosion dramatic and life threatening or a mere dud in its effects on cell function?

In cell senescence, the fundamental issue remains neither telomerase, nor telomeres, nor cell division, but rather the practical outcome in cell function, i.e., changes in gene expression. It is gene expression that defines cell senescence. If cell senescence has implications for age-related pathology, it will be understood through our increasing knowledge of senescence associated gene expression.

The Linkage: Telomeres and Senescent Gene Expression

Even granting that gene expression is the fundamental orchestra to which we should listen if we are to understand cell senescence, it is the telomere that appears to do most of the conducting. To obtain a beneficial clinical outcome through resetting gene expression, we must concentrate not on each orchestra member as an individual, but on assuring that the final symphony is the one we prefer to hear. It is the conductor that determines the choice of symphony, but how does this process work? How does the telomere modulate gene expression? The question is important: we have convincing evidence that not only does the telomere direct gene expression, but that gene expression may be reset and maintained indefinitely (Bodnar et al., 1998). How does the conductor tell the orchestra what to play?

Considerable work has gone into this question, predominantly in yeast (Saccharomyces), an organism that may bear little resemblance to human cells in this regard. The theoretical models of telomere/gene expression linkage have been summarized (Fossel, 1996), but the question remains open.

A recent paper by Olovnikov has suggested that ionic "fountains" might be involved (**Olovnikov, 1999**). In this model, telomeres bind to the inner nuclear membrane and modulate local ion concentrations and their availability to chromosomes, thereby in turn modulating transcription. Although consistent with known data, there is nothing to substantiate such a model. Most models of telomere/gene expression linkage assume a heterochromatin "hood" which covers the telomere and a variable length of the subtelomeric chromosome. As the telomere shortens, the hood either "slides" further down the chromosome (the heterochromatin hood remains invariant in size and simply moves with the shortening terminus) or the hood itself shortens (as the telomere is less capable of retaining heterochromatin). In either case, the result is an alteration of transcription from portion of the chromosome immediately adjacent to the telomeric complex (either up-regulating or down-regulating such genes). These genes, in turn, modulate other more distant genes (and sets of genes) elsewhere on the chromosomes. There is some direct evidence for such modulation of genes in the subtelomere, but overall, while there is clearly a linkage, the mechanism(s) of such a linkage remain unclear.

An additional possibility, and one almost certain to play a significant role in human cell senescence, is that of the cell cycle braking system. The last decade has seen an explosive growth in this field, particularly as it bears on mechanisms of carcinogenesis. In general, the cell "monitors" gene damage using Damaged-DNA-Binding-Proteins (DDBP's), which signal the cell that the DNA is damaged and trigger a cascade of events involving a minimum of several dozen factors, including p53, p21, D cyclins, cdk's, etc. These stop the cell cycle, but also up-regulate DNA repair mechanisms. The cell either repairs the errors or remains unable to divide. Such a mechanism need not be strictly all-or-nothing and, to the extent that telomere shortening is interpreted by the cell as DNA damage, may account for the gradual slowing of the cell cycle that occurs as cells senesce. In this regard, cell senescence shares common pathways with cell mechanisms which prevent carcinogenesis and with apoptosis.

Although important if we are to understand cell senescence, its role in age- related disease, and effective points for therapeutic intervention, the linkage between telomere shortening and senescent gene expression remains enigmatic and doubtless full of unexpected quirks in the years to come.

Effects on Cell Activity

If the genome is an orchestra, then with aging, one might suggest that it has gone out of tune. The alteration is not one of genetic damage, at least in marginally senescent cells (see discussion above regarding the problems defining senescence). There is no evidence for accruing genetic damage with age, rather gene expression changes. To return to the orchestra analogy, the instruments are neither missing nor out of tune: they are merely playing discordantly because the conductor has chosen a discordant score.

In the same vein, there is a natural, but erroneous, tendency to equate aging, even at the cellular level, with wear-and-tear. The cell does not suffer wear-and-tear blindly and helplessly. It does so because gene expression changes and therefore cell function changes. Within the cell, these changes include a decrease in DNA repair, a decrease in protein turnover, a decrease in lipid turnover, a decrease in sequestration and trapping of free radicals, and an increase in free radical production (see discussion in Fossel, 1996). As cells age, it is quite likely SAGE that allows mitochondria to produce increasing numbers of reactive oxygen species (ROS) for every mole of ATP produced. SAGE allows mitochondrial membranes to become "leakier" as lipid turnover slows. SAGE prevents the cell from trapping ROS that make it into the cytoplasm (or worse yet the nucleoplasm). And it is SAGE that allows damage to cellular structures (e.g., structural proteins, ribosomes, etc.) to accumulate as turnover rates decrease (Fossel, 1996).

Wear-and-tear is not unavoidable: to the contrary, every cell in the germ cell line has managed to maintain itself in excess of a billion years, albeit in the face of mutational changes, massive cell loss, and selection. Cells senesce because (and as) they alter their pattern of gene expression. But if we look at cell senescence as a facet of aging, then we would conclude that cells do not age because of wear-and-tear, rather they age because they *permit* wear-and-tear. Cell senescence is (experimentally, *in vitro*) entirely and indefinitely preventable. Depending on where we draw the line (and understanding that there is a point beyond which reversal is at least unlikely and in the final analysis even conceptually impossible), cell senescence can also be reversed.

To what extent does this imply that clinical aging and age-related diseases may also be prevented or (again recalling our caveats) even reversed? The answer pivots precisely on the extent to which cellular senescence underlies clinical aging and those same diseases.

Current Theory: Aging and Cell Senescence

The weak model of the cell senescence theory of aging has been stated above and is well supported by current data: telomere shortening modulates gene expression and the consequent changes in gene expression define and control cell senescence. Furthermore, they are entirely preventable and, within limits, reversible.

The strong model of the cell senescence theory of aging goes further, suggesting that cell senescence underlies aging and age-related diseases. To the extent that we can prevent (or reverse) cell senescence, we will also prevent (or reverse) clinical aging and age-related disease.

Before we consider what data there is in support of the strong form of the theory, let us pause and make two immediate observations. The first is that the notion that aging is preventable (or reversible) has no substantiated precedent in human, scientific, or medical history. To the contrary, every such claim to date has proven to be at best misguided and at worst an unconscionable sham. To state that we might fundamentally affect human aging and thereby obviate current (basically palliative) approaches to age-related diseases is suspect, remarkable, and stunning in its hubris. The second observation is that the arguments raised to date to falsify such a theory are, in every case, straw men. The data is complex and most of the mechanisms that underlie aging remain still unknown and largely unexplored, but current criticisms do not undercut the theory that human aging and age-related diseases are preventable or reversible via modulation of cell senescence. Before addressing such criticism further, note that the ultimate question is not "is the theory elegant, comprehensive, or consistent?" but rather "what happens when we try it?"

What the Theory is Not

Straw man arguments are inappropriate attempts to falsify an hypothesis through the use of faulty logic, rather than data and are, unfortunately, common. They consist of a faulty syllogism: a major premise, a related minor premise, and an inappropriate conclusion that the original statement is therefore falsified. For example, in the context of microbial theory of disease in the 1870's, such a straw man argument would be:

1) If bacteria cause infections, bacteria should be present in every infection.
2) Bacteria cannot be recovered from all infections.
3) Therefore, bacteria cannot be the cause of infections.

In reality, only a small percentage of aspirate cultures from cellulitis are positive under optimal circumstances. Nonetheless, this result does not constitute a disproof of the bacterial etiology of cellulitis. To the contrary, we give antibiotics and most patients improve. The "proof" of a bacterial etiology for cellulitis lies not in unreliable aspirate cultures, but in a remarkably good clinical response to appropriate antibiotic treatment. In the case of cell senescence, however, the issues are far more subtle and complex than a mere lack of technical ability to measure a single variable, such as the presence of bacteria or the length of telomeres. Aging is not so simple and falsification of the model is not simple either. Nonetheless, several straw men arguments have been cited as people misunderstand and attempt to falsify the strong form of the theory.

Among the most common straw man is that "telomeres don't cause aging." This is analogous to stating that "cholesterol doesn't cause atherosclerosis", i.e., strictly true and yet quite misleading. Cholesterol doesn't "cause" atherosclerosis, yet is both a major factor in the atherogenesis that occurs in most patients and, moreover, a convenient pathological, conceptual, and laboratory marker for the disease process. The lack of strict causation is scarcely a recommendation for ignoring either serum or dietary cholesterol. In a similar vein, no one actually claims that telomeres "cause" aging (hence, the straw man designation). Rather, telomeres are considered a rough cellular "clock", functioning as both a modulator of gene expression and a marker of some (but limited) utility in measuring cell senescence in the laboratory. Telomeres don't cause aging; they serve to time it within a cell and are a potentially useful target for therapeutic intervention. Whether telomerase will serve as a means to prevent organ aging (as it does cell senescence; Bodnar et al., 1998) remains to be seen. Arguing that telomeres "don't cause aging" is specious and disingenuous; it abuses the ambiguity of the concept of "cause" while ignoring both the well-supported data on telomeres and cell senescence and the potential testable clinical targets implied by such data.

A second straw man is that not all cells in older individuals are senescent. This is, of course, true and quite necessarily so. If all such cells *were* senescent (bearing in mind the problems in defining such cells, discussed above), then the donor patient would almost certainly have died already. Senescent cells are not effective cells. Clinical studies show that every elderly patient has cells of varying degrees of senescence, even within the population of cells which divide. The notion that all cells in all tissues divide at constant rates and therefore senesce in tandem is naive. Not only do cells in some organs (e.g., hematopoetic stem cells in the marrow) consistently divide more frequently than cells in other organs (e.g., neurons in the CNS), but even within a single organ or tissue, there is significant variance in cell division rates (see below) and this is borne out when we measure cell senescence from living donors. In a recent reevaluation (Cristofalo et al., 1998a), Cristofalo has even argued that in skin fibroblasts there is *no correlation* between cell senescence and donor age in postnatal samples. Although this study looked only at population doublings, work which focuses on more useful and appropriate measures of cell senescence, gene expression, morphology, and cell function, found that while there was a predictable increase in the proportion of senescent cells with donor age, as we should expect, not all cells were senescent at any donor age.

A related straw man argument is that not all cells from older donors have the same telomere lengths. Again, how could it be otherwise? Cells do not divide in lock-step, even within a specific, apparently homogenous tissue. The classic graph of telomere lengths versus time suggests that germ cells maintain their lengths while the length of telomeres in somatic cells falls continuously and uniformly as expressed schematically in Figure 1. The line representing somatic cell telomere lengths is a useful heuristic tool, but misleading. It forms the basis of some of the criticism (both straw man arguments and those based on actual data) of the theory that cell senescence underlies aging in the organism. For somatic cells DO NOT show a simple, linear, and uniform decrease in telomere lengths. The truth is that cells, even within a tissue (*in vivo* or *in vitro*), vary substantially in telomere lengths as time passes. By a similar argument, it is extraordinarily unlikely that any two daughter cells will maintain the same vector of telomere shortening, as they are unlikely to receive precisely the same integration of signals favoring and restraining cell division, bear precisely the same environmental stresses, or have precisely the same direct cell-cell interactions. This is certainly the case for stem cells (see below) where the daughter cells are genetically distinct and pursue intentionally divergent courses vis-a-vis cell division, but is almost equally true of most dividing somatic cells in which the daughter cells are often considered to be exactly equivalent. A more accurate representation of telomere shortening over time, (see Figure 2) would suggest that every cell pursues a separate vector of telomere loss and an even more accurate figure (which would be so complex as to be neither readable nor printable) would show that at every cell division, the two subsequent lines diverged, curved, overlapped, tangled, paralleled, and all but knotted as the cells continued their gradual, discontinuous, and unsteady course into a complex and interactive senescence.

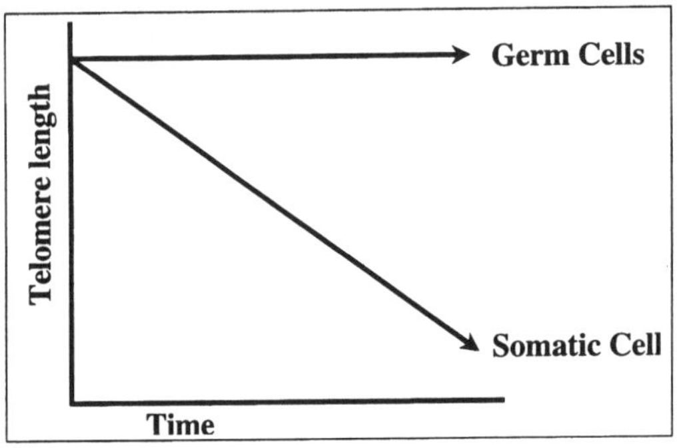

Figure 1. Classic model of telomere shortening.

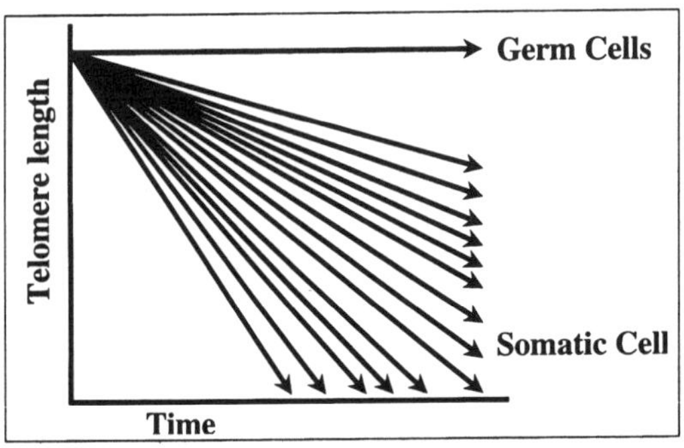

Figure 2. The reality of telomere shortening.

Even in the most controlled and uniform of environments (which in this respect, the human body is not), cells will vary not only in telomere erosion, but, far more importantly and far more interestingly, in the onset and effects of senescence itself.

The argument that all cells should be senescent (or have equivalent telomere lengths) in older donors is specious, belying an ignorance of cell function at the tissue level and of the "demographics" of cell division in living tissue over time. Aging tissues show a broadening range of senescence as measured by gene expression, telomere lengths, morphology, and function. Not all cells in older patients are senescent, nor should they be so. As tissues age, a range of senescence should be, and is, typical, both theoretically and in actuality. Cells do not senesce uniformly: any individual, no matter their age, will have a fair percentage of remarkably "young" (i.e., non-senescent) cells extant in any tissue sampled.

The "trump card" of critics remains remarkably persistent, while being remarkably credulous: cell senescence cannot play a role in aging and age-related disease because "not all cells divide". The observation is (approximately) accurate, the conclusion (entirely) unwarranted. Classical histology maintains that certain cells never divide in the mature (or even the postnatal) individual. Until recently, neurons and myocytes have been considered examples of such post-replicative, differentiated cells. In reality, there are exceptions to this dictum: some neurons not only divide in the adult, but may even play a standard role in learning and memory; satellite cells might likewise routinely serve to replace myocyte loss in adult muscle. But assume otherwise. The exceptions are academic in this context and are orthogonal to the main point: *non-dividing cells are critically dependent upon dividing cells.* Non-senescing cells are not immune to the problems caused by senescing cells.

The most obvious example is cardiac muscle. There is little to suggest that aging occurs significantly in such tissue (Lakatta, 1994; Fossel, 1996). To the contrary, "heart disease", the most common killer in most developed nations is, in actuality, primarily vessel disease, and vessel disease, primarily atherosclerotic. A paucity of evidence that cardiac myocytes divide or senesce notwithstanding, these same cells are critically dependent upon patent coronary arteries for their survival (and that of the patient), and there is excellent evidence (Chang and Harley, 1995) that the endothelial cells lining arterial walls divide, their telomeres shorten, and they senesce. Furthermore, a clear argument can be made that their senescence plays the major role in atherogenesis (see Fossel, 1996; also Cooper et al., 1994; Kumazaki, 1993; Fossel, 1998a, b; Bodnar et al., 1998).

Nor should this surprise us. The plethora of "causes" in atherogenesis includes hypertension, hypercholesterolemia, hyperglycemia, tobacco use (though at least three mechanisms), viral or bacterial infection in the vessels, dietary effects (both lipid amount and type, as well as other mineral and vitamin intake), and the ubiquitous and interactive etiology of genetics. Curiously, in the case of Hutchinson-Gilford progerics, the usual cause of death is atherosclerosis, yet they lack any of the commonly identified risk factors for atherogenesis. These children, although appearing to "age rapidly", do so segmentally, i.e., they show changes in their skin and vessels typical of aging, yet have no correlated increase in cancer rate, Alzheimer's or other dementia, CNS dysfunction, immune senescence, etc. The average age at death is 12.7 years, overwhelmingly of atherosclerotic causes.

Surely our standard model of (and unspoken assumptions about) atherogenesis is not so much wrong, as uncomfortably incomplete. Could it be that just as skin fibroblasts in progeric children demonstrate early senescence, so too do their endothelial cells, prompting, even forcing rapid atherogenesis? Certainly we do see cell senescent changes prior to the onset of and correlated with the process of atherogenesis in "normal" patients (Chang and Harley, 1995). As we shall see below, the implication is that non-dividing cells are in all cases, directly or indirectly, at the mercy of dividing cells.

This same observation is pertinent in the case of the central nervous system. Neurons may not (generally) divide, but they are critically dependent not only upon endothelial cells that line the cerebral arteries, but equally upon the glial cells that make up the preponderance of cells in the brain. And although most glial cells (particularly oligodendrocytes) themselves may not divide in the mature brain, many glial cells (particularly microglia) clearly do (Sturrock, 1976). Such dependence may include maintenance of the ionic milieu, guidance to arborization, insulation from inappropriate exposure to excitatory transmitters, metabolic support, and, perhaps most critically in this context, trophic factor maintenance (Nishi, 1994). Senescing cells have an altered pattern of gene expression and, as we shall see, an altered pattern of trophic factor production and secretion. In the case of the glial-neural interaction, this may yet be found to be the crucial factor in the etiology of neural apoptosis.

The argument that cell (division or) senescence cannot play a role in aging or age-related disease because not all cells (divide or) become senescent is ill-conceived. To maintain it is to display an ignorance of normal tissue function, disease processes, and logic.

On the other hand, although each of these straw man arguments is unsophisticated, that does not lend support to the argument that cell senescence plays a role in aging and disease. The fact that there is no evidence or logical argument to disprove a theory is entirely different from proving that a theory is correct or even marginally useful. The question remains: what is the evidence in favor of such a theory? To answer this question, let us consider what the theory actually says.

Understanding What SAGE is

The theory maintains that cell senescence underlies and explains at least part (perhaps all) of human aging and age-related disease. It further maintains and predicts that to the degree that we prevent or reverse cell senescence, we could have a beneficial outcome for aging and age-related diseases. The outline of the theory is as follows.

In the aging human (and many other species as well), some (though not all) cells divide. Cell division is accompanied by telomere shortening, generally (though not exclusively) as a result of the chromosomal end-replication problem first described by Olovnikov (Olovnikov, 1971) and Watson (Watson, 1972) in the early 1970's. Cells may, in some cases, re-extend their telomeres through the use of telomerase (although other mechanisms may occasionally play a role). The majority of normal somatic cells do not express telomerase and have no measurable telomerase activity. Cells (including cancer cells) expressing telomerase are immortal.

Although most somatic cells never express telomerase, a restricted set of somatic cells can express it selectively upon appropriate stimulation. Such cells include hematopoetic, basal skin, GI crypt, hepatic, hair follicle, and potentially several other stem cell lines.

Typically, such a stem cell responds to local and specific cytokine signals by transiently expressing telomerase, re-extending its telomere, and dividing. One daughter cell then differentiates and is thereafter unable to re-express telomerase even in the face of further cytokine stimulation. Its telomeres shorten progressively with all further divisions. The other daughter cell remains in the stem cell line with a marginally shorter telomere and is both capable of re-expressing telomerase to appropriate cytokine stimulation and capable of dividing. The result is that stem cell telomeres shorten but at a demonstrably slower rate than other cells with a similar rate and number of cell divisions. This is, presumptively and teleologically, to allow stem cells to function over the entire lifetime of the organism. This is especially true if the organism (a human, for example) has an extended lifetime and will require cell replacement (in the erythrocyte line, for example) to occur in astronomically high numbers over the span of a century.

Somatic cells which divide but are not in the stem cell line show a predictable and steady erosion of telomere lengths. Somatic cells which do not divide, generally maintain their telomere lengths intact.

As the telomere shortens in any cell, there is a change in gene expression. This "Senescence Associated Gene Expression" is typical of a species and each cell phenotype. SAGE has two classes of effects: intracellular and extracellular.

Intracellular effects consist of slower protein elongation and therefore protein turnover, resulting in a gradual accretion of dysfunctional enzymatic and structural proteins within a cell (for the mathematical analysis of such accretion, see Fossel, 1996). Additionally, DNA repair becomes both slower and less effective, resulting in genomic instability in senescent cells. Mitochondria produce a higher frequency of ROS per mole of ATP produced, resulting in both diminished cell energy economy and an increased risk of ROS damage. Furthermore, the mitochondrial sequestration in increasingly faulty, allowing such ROS to diffuse through the cytoplasm in higher concentrations. ROS trapping also decreases as trapping compounds of intracellular origin are expressed and available in decreasing amounts. These four ROS problems (increased production, poorer sequestration, decreased trapping, and decreased turnover of damaged macromolecules) result in a sharp increase in cell dysfunction as senescence ensues. Additionally, the cell becomes increasingly resistant to apoptotic signals, the cell cycle slows, and the cell becomes less capable of dividing and replacing local cell losses.

Extracellular effects also occur and are paramount in explaining disease and age-related tissue and organ changes. Such changes begin locally, but progress systemically. Not only do senescent cells change their pattern of gene expression, but the result is a change in the production of secretory products, such as trophic factors (Cristofalo et al., 1998b; Funk, personal communication). The same is true of structural or enzymatic secretory products, such as collagen, elastin, collagenase, elastase, or beta galactosidase (Dimri and Campisi, 1994). Inflammatory factors, and cytokines in general, also change as the cell senesces. Some cells (e.g., skin fibroblasts) show a decreased pattern of inflammatory cytokines (Funk, personal communication). Other cells (e.g., endothelial cells) show an increasingly inflammatory pattern. Cell senescence is a generic process, but its expression differs between cells and is specific to each cell type.

The senescence of one cell accelerates the senescence of surrounding cells. If one set of cells is highly stressed and prone to cell loss, the remaining cells divide and therefore senesce. They become less capable of meeting the local tissue needs as they do so. As they senesce, they divide more slowly and become less able to replace lost cells. This increases the stress and the rate of cell division in yet other, nearby, previously less senescent cells. The outcome is that senescing cells "drag down" other cells and increase the overall rate of senescence. Such intercellular dependencies may be significant, yet are unlikely to be "tight"; the senescent variance of cells within a tissue will therefore remain substantial. There will be very few unaffected cells and likewise few totally senescent cells, but rather a normal distribution, rather than a bimodal or flat distribution of senescence within a cell population as an organism ages. The degree of interdependence of the cells (in terms of cell replacement, cell stress, and cell division) determines the degree of leptokurtosis of the senescence distribution curve.

The effects do not remain local. To paraphrase John Donne, "no cell is an island, entire of itself; every cell is a piece of the organism, a part of the body." Extracellular effects may start as local as a subtle change in the function of a neighboring cell or a subtle change in the extracellular matrix, but the effects also progress to neighboring tissues planes, to entire organs, to downstream parts of the vascular bed in which they lie, and, ultimately, throughout the organism. What starts locally, ends systemically. What begins as cell senescence, ends as age-related disease.

The theory that cell senescence underlies aging in the organism and predominates as the fundamental etiology in age-related disease, must not only survive well-intended but naive criticisms, it must also do two other things:
1) it must explain what else we know of aging in a broad context, and
2) it must make predictions that are testable and survive such testing.

With regards to the first requirement, how does this theory fit within the framework of other, previous aging models?

In the Context of Previous Theories

The theory that cell senescence is fundamental to aging and age-related diseases is remarkably robust, predictive and testable. Although this is not the appropriate venue to explore the historical background of aging research and prior theory, a few observations will highlight its consistency with other models and its greater consistency with what we currently know of biology and pathology.

The belief that free radicals (or reactive oxygen species, ROS) underlie much of the damage in aging organisms has a firm theoretical foundation and, despite a surprising lack of proof, free radicals have an almost unquestioned acceptance as being pivotal in aging at any level and in almost any organism. They are likewise, as discussed above, a major force in cell senescence. Yet they are not an explanation of aging. They do not explain why the germ cell line, which in eukaryotes has been subject to mitochondrial ROS damage since the intracellular symbiosis began, has NOT senesced (retrospectively) for more than a billion years of germ cell lineage, yet does so rapidly in the post natal organism. The question that lies at the heart of free radical theory should not be "how *does* it usually cause aging or senescence?" but rather, "why *doesn't* it usually cause aging or senescence?" for the equivalent source of free radicals is present in both germ and somatic lines, yet one is immortal, the other senescent. Free radicals may underlie much of aging, but there is something even more fundamental that turns the process of free radical damage on or off. This more fundamental process is SAGE, generally modulated by telomere length, regulated in turn by telomerase. The same breach of underlying causation and myopic look at evolutionary cell history is inherent in the entire panoply of "wear-and-tear" based theories of aging: they may explain why some organisms age, but they do not explain why others do not.

One might be tempted to attribute this division to unicellularity versus multicellularity, suggesting that germ cells are immortal, but once multicellularity commences, the somatic lines are doomed to degradation by some variant of simple wear-and-tear. This assumption is often made, but the data contradicts it. Some multicellular organisms age, some unicellular organisms do not. Some tissue cultures age, some do not. The correlation is not made along the division of unicellularity versus multicellularity, but along the division of telomere maintenance versus telomere loss and this correlation (while still brooking some interesting exceptions) is all but universal. Wear and tear is almost certainly part of the aging process, it does not explain aging and purely "free radical theories of aging" are insufficient to account for aging.

Many theories would attribute aging to hormonal or organ-level clocks, e.g., DHEA, growth hormone, melatonin, etc. Not only is there no data that aging per se can be altered by these approaches (Wolfe, 1998; Kala et al., 1998; Fossel, 1996; Fossel, 1998a, b), but each begs the more basic question: if the declining value of hormone X is fundamental to aging, then what clock determines the decline in hormone X? Note that this is not to say that any of these substances might not have beneficial, even lifespan increasing effects. Estrogen has exactly such effects. An adequate diet likewise is beneficial and increases the mean lifespan. Good hygiene, immunizations, seatbelts, regular exercise, and screening exams may all have these attributes. Yet none of these is a reasonable explanation for either aging or the fundamental processes which underlie age-related diseases generically. They have value, but that value is neither universal nor is it sufficient to prevent age-related disease.

Is there any intervention which has been shown to postpone the fundamental process of aging and extend not merely the mean life span of a species, but the maximum life span as well? There are three such approaches: caloric restriction, breeding, and genetic alteration.

The first approach is to cut the caloric intake of a group of animals (usually by about 30% of *ad libitum* intake) while required nutrients are carefully maintained. The outcome is, in every species so far studied, a substantial increase in the mean and maximum life span and a substantial postponement of age-related disease processes including (in relevant species) atherosclerosis and cancer. Initial results suggest (Lane, 1998) that this is also true in Rhesus macaques. There is every reason to suspect that it would be equally effective in humans, although no clear research data confirms this. It is certainly consistent with the fall in death rates in the Netherlands during food shortages in World War II and likewise with studies on dramatic diet changes in clinical studies of those at high risk for heart disease (Ornish, 1998). Nonetheless, there is no accepted theoretical understanding of this process. Most discussion of the mechanism focuses on insulin and glucose metabolism, metabolic efficiency, free radical generation, etc, but there is no current consensus. Certainly it remains possible that any of these are consistent with cell senescence, which would still offer a broader explanation for aging phenomena and disease in all contexts.

Both of the last two approaches, breeding studies (reviewed in Rose and Charlesworth, 1980; Rose, 1998) and genetic alteration (e.g., transfection of genes into Drosophila, Jazwinski, 1996; or into C. elegans, Larsen et al., 1995), have focused not only on changes in gene expression (via breeding or laboratory alteration), but usually on specific gene expression relevant to energy and free radical metabolism. Once again, these two approaches are entirely consistent (and perhaps predictable) within the context of cell senescence as a theory of aging and age-related disease. The heart of cell senescence is an alteration of gene expression and whether fundamentally positive effects are accomplished by telomerase (which resets or postpones SAGE), by breeding (which may select for specific genes involved in senescence or its postponement), or transfection of genes (which affect energy and free radical metabolism and thereby delay the cascade of damage described already above), the outcomes would be the same. So-called "genetic theories of aging" are precisely the stage upon which we should be discussing cell senescence.

The cell senescence theory of aging is consistent with what we know of aging and with the theories that frame what we know of aging. But what of age-related disease? How does human clinical pathology occur in the aging individual and how might we intervene in such diseases?

Age-Related Diseases

Overview of Each Disease System: Theory and Data

To an extent Emerson remains correct that "aging is the only disease". And yet – however fundamental aging may be, whatever it is, and however it plays a role in the cascade of causation that define any pathology – we do not die of aging; we die of disease. We die of heart attacks, aneurysms, strokes, cancer, and infection. We suffer from arthritis, decubitus ulcers, and dementia. Ultimately the practical (indeed the only serious) question is not "is the theory correct", but rather "does it help treat human disease?" In short, can we alleviate human suffering? All else is, at best, a tool to achieve such an end. If the tool works, we are gratified by the outcome. If the tool is the optimal one to do the job, so much the better, as we are more likely to succeed and to do so quickly. Current therapy, for atherosclerosis, for example, is effective to a degree, but is it optimal or is there a more effective, fundamental, and universal therapy? To this end, what role does cell senescence play in age-related disease?

Vessel Senescence — Atherosclerosis

Atherosclerosis underlies most adult deaths worldwide and not only in developed nations. It usually expresses itself as death due to cardiac or central nervous system failure. Although organ specific etiologies exist within both the heart and the brain, the overwhelming cause of organ failure in the aged is the arterial blood supply and the overwhelming cause of this failure of the arterial blood supply is, in turn, atherosclerosis. The major outline of atherosclerotic pathology has been understood since the late 19th century and is well characterized at the gross and microscopic levels. During the late twentieth century, we have advanced in understanding the pathology at both the broad epidemiologic and narrow cellular levels.

In a general way, there are two major contributors to the pathology: extrinsic and intrinsic factors. The former factors include diet, exercise, alcohol, tobacco, and viral and bacterial infections. They will not be addressed here except tangentially. The latter include not only broadly genetic processes such as diabetes, immune function, hypertension, and disorders of lipid metabolism, but also "intrinsic aging processes" (Bierman, 1994) and at least one extremely rare, but illuminating genetic disorder, Hutchinson-Gilford syndrome (early onset progeria).

The cascade was once felt to begin within the smooth muscle layers of the arterial wall, but more recently the endothelial cell layer has been implicated as lying "upstream" in the etiology (Cooper et al., 1994). This has been extended by the observations of Harley's group (Chang and Harley, 1995), who found that cells in areas of high vascular stress showed telomere shortening prior to onset of vascular disease. The circumstantial evidence of death due to atherosclerosis in progeric children is likewise consistent with the model that cell senescence not only underlies atherosclerosis, but would also be the most effective interventional target.

An overview of the model best begins in the endothelial cell wall. Cells of the endothelial layer are exposed not only to circulating infectious agents and toxins (tobacco products, for example), but also to the direct effects of shear stress during systole. As a consequence of the several stresses to cell in the intima, endothelial cells frequently die and the transient denudation is subsequently repaired by the division of surviving, neighboring cells. The precise "turnover rate" of *in vivo* endothelial cells is unknown, but presumed to be high relative to similar cells elsewhere (e.g., in veins) and to other types of cells immediately underlying the intima. This is borne out in measures of telomere lengths in arterial endothelial cells compared to venous wall cells (Chang and Harley, 1995). Although it is clear that there are changes in gene expression both *In vitro* and *in vivo* (Cooper et al., 1994), and that the model sketched out here is more consistent and broader than previous models, the precise chain of causation still requires elucidation.

What is relatively clear is that events then progress "downstream" much as in the classical model of atherogenesis, with the rapid onset of inflammatory changes, cholesterol deposition, foam cells, smooth muscle proliferation, thrombus formation, and general atheromatous change. However, instead of these changes being merely passive (e.g., the passive accumulation of lipid products), they are the active response of cells in the medial layer to an altered pattern of trophic factors (from intimal cells) on which they depend for their normal function.

In this same purview, the reason that progerics have heart disease becomes explicable. Although lacking the standard four major risk factors – hypercholesterolemia, smoking, hypertension, and diabetes – their endothelial cells probably partake of the same abnormally early cell senescence found in their skin fibroblasts (Counter et al., 1995), viz., shortened telomeres. Very likely, their senescent endothelial cells have SAGE at birth, hence an altered pattern of trophic factor production, upon which cells in the media depend, hence the otherwise remarkable onset of atherosclerosis in their first decade of life, resulting in rapid death due to stroke or myocardial infarct. Whether this explanation, while consistent, is entirely accurate still remains to be seen. To date, we have no progeric donation of arterial endothelial tissue and remain unlikely to obtain any in the near future. Assays of telomere length and cell senescence in progerics have thus far been restricted to skin fibroblasts and leukocytes (see below).

None of this would be pertinent were it not for the possibility of intervention. Current interventions in atherosclerosis do not affect fundamental cell function. Although heart transplants, coronary artery bypass grafting, endarterectomy, and vascular stenting are effective to a degree, they by no means interrupt the onset of pathology. And although drug alteration of lipid metabolism, dietary changes, exercise, and control of contributory risks (diabetes, hypertension, and smoking) lower the vector of disease progression and do so at a presumptively more fundamental level than the repair or replacement of vessels, they still do not strike deeply enough. An optimum intervention would have four desirable characteristics:

1) it would be cheap (compared to coronary artery bypass grafting, for example)
2) it would affect the disease uniformly, irrespective of genetics, environment, etc
3) it would be preventative (as immunization is in the case of infection)
4) it would be curative (as antibiotics are in the case of infection)

Contrast each of these characteristics with our current and historical treatment of disease.

Coronary artery bypass grafting, endarterectomy, and cardiac transplantation all have a degree of efficacy, yet are extremely expensive contrasted to the costs of immunization and prevention. Most health plans, most underdeveloped nations, and most uninsured patients cannot afford unlimited use of such technology in treating heart disease.

Current preventative measures for atherogenesis are also more patient-specific than is optimal: antihypertensives, drugs to alter lipid absorption or metabolism, tobacco cessation, platelet inhibitors, etc vary in their efficacy with genetic background and patient behavior. Again, contrast this specific efficacy with the more uniform applicability of immunization in microbial disease. A century ago, we had no thoroughly effective means of preventing tetanus; now the rate of tetanus death is limited only to our ability to ensure universal immunization. In a like vein, we can avoid much, but not all atherosclerotic deaths through diet, drugs, and behavioral intervention, but could certainly do far better by preventing cell senescence. The survival rate from tetanus is still only 50% with the best of medical care. The survival rate for those with known significant atherosclerosis is also poor and we are extremely limited in our ability to reverse the disease process. If cell senescence underlies atherogenesis, then we might well be able to reverse much of the process prior to end-organ damage.

Overall, consider the concept of the "most efficient point of intervention". Even if we could universally afford to transplant damaged organs (whether hearts, knees, or livers), the most effective (and cost effective) point of intervention would be "upstream". There is a current tendency to attribute a degree of the sacred to any medical approach which relies on prevention rather that treatment. Such approaches are often more effective, yet the assumption that prevention is always better is seldom analyzed. The value of any intervention which is "upstream", or "at a fundamental level", or "preventative" lies NOT in any inherent superiority, but in their probable relative efficacy: they are usually cheaper, more effective, less painful, and easier for all concerned. In the same vein, an intervention which employs the prevention or reversal of cell senescence to treat atherosclerosis and its clinical outcomes will be much preferable to current therapies not because it is somehow more "fundamental", but only because it is demonstrably more effective. Clinical therapy aimed at cell senescence will be aimed at the most efficient point on intervention in such disease.

Skin Senescence — Beyond the Cosmetic

Not merely cosmetic. Skin is not simply "the bag of tissue that keeps us from falling apart," nor are concerns for skin aging vested solely in cosmetic concerns. To the contrary, skin aging is a significant problem in the aging population and often manifested in poor (or non) healing of lacerations, infected (and occasionally fatal) decubitus ulcers, an increasing risk of infection, and poorer thermal homeostasis. Skin becomes a less effective barrier to trauma, heat loss, microbes, and other assaults. Cell senescence plays a demonstrated role in skin aging and we can, in limited laboratory models, already reverse this process and reconstitute at least some aspects of young human skin. There is the potential to prevent and reverse human skin aging through the use of transfected or induced telomerase expression. This potential has important clinical implications far beyond the commercial potential of preventing wrinkles, liver spots, or dryness. The process of skin aging is not only more important clinically than many realize, it is also far more complex.

The Cascade of Processes in Aging Skin

Skin aging, like atherosclerosis, is "caused" by a number of factors. These include exposure to ultraviolet light and repetitive trauma of all sorts: shear forces, pressure, heat, infectious, etc. Other factors also include tobacco use, diabetes, and a host of genetic diseases. It is difficult (and probably unnecessary) to strictly distinguish "local" factors from vascular causes; very likely there is a enormous interactional component between vascular aging *per se* and aging which occurs in skin *per se*. In fact, the cell senescence view of aging would not only regard strict apportionment of such local and distant factors as unrealistically exact, but would further regard such interactive effects as lying at the heart of the process. Aging, in any organ, is dependent upon not only local factors which drive cell senescence, but upon bidirectional interactional effects between local and more distant cells and tissues.

However complex the sources driving cell senescence in the skin, the result is a gradual, lifetime history of cell division which can be accelerated through UV exposure, trauma, systemic disease, or tobacco exposure. The outcome is increased cell division as underlying stem cells replace cell losses. Certain stem cells, among them basilar stem cells in the skin, may transiently express telomerase at the time of division (Ramirez et al., 1997; Shay and Bacchetti, 1997; and see Fossel, 1998a, b), thus slowing telomere loss and consequent changes in gene expression. Nonetheless, the pattern of gene expression does change and does so in a cell-type-specific manner as the tissue ages.

There are several labs that have characterized cell and tissue specific SAGE patterns. In one case, Beta galactosidase, a specific marker has come to represent cell senescence, for better or for worse (Dimri et al. 1995; Dimri & Campisi, 1994; Campisi, 1998). Although Beta galactosidase is an appropriate marker for senescence in human skin fibroblasts, there are two naive mistakes that are often made with regard to its use.

The first mistake is to assume that is functions with 100% specificity and sensitivity, that is, that Beta galactosidase is only and always found in senescent human skin fibroblasts. In fact, part of the problem is that it is such a good marker that we can identify Beta galactosidase positive cells among a population of cells, some of which are senescent and some of which are not. The fact that senescent tissue is a mosaic of cells of different degrees of senescence leads some to suggest that beta galactosidase is not a good marker because not all skin fibroblasts in aging skin are positive. But not all skin fibroblasts in aging skin are senescent either. In fact, even if beta galactosidase were a flawless, all-or-nothing marker (which it is not; Yegorov et al., 1998) which validly and reliably identified only "fully senescent cells" (however defined), it would still show a mosaic of positive cells in aging skin and at no time would we expect to find either no staining at all (perhaps fetal skin?) or 100% staining of all cells. Results are consistent with a more sophisticated understanding of the role of beta galactosidase as a marker for cell senescence: it is a fairly accurate marker of cell senescence in specific human skin fibroblasts as they senesce and, equally, it is a useful marker of cell senescence in human skin fibroblast populations as well.

The second error is to attempt to use beta galactosidase as a universal marker of cell senescence in all cell phenotypes. Cells differ and, almost by definition, their differences are both caused by and associated with differences in gene expression. Beta galactosidase is no exception. Even in other types of cells which may express beta galactosidase constitutively (Yegorov et al., 1998), we cannot expect that its expression should necessarily parallel the changes which occur in senescing skin fibroblasts. That beta galactosidase does not function as a marker for senescence in all cell phenotypes does not detract from its usefulness as a marker within skin fibroblasts.

An extension of these observations is that there are other markers that may work equally well (or far better) in other cell phenotypes. In fact, as technical advances allow us to portray cells in terms of gene arrays, the appropriate question is no longer how a *marker* changes with senescence, but how the *pattern* of gene expression changes with senescence. In the same sense that the array varies between cells, so too does the array vary as cells senesce and can be used to adequately characterize SAGE. Work done by Funk is typical of this approach. Using a broad array of genes, he has shown that not only does gene expression change with cell senescence, but that telomerization "resets" gene expression to the non-senescent pattern (Funk, personal communication). The array becomes no longer the marker of senescence, but the definition itself. Although this is not only useful but probably appropriate for *in vivo* work, at the clinical level we must continue to ask a more important question, viz., does tissue function reflect these changes? So far, the answer is that it does precisely this.

Using telomerization of human skin cells, not only do cells revert to a non-senescent pattern, but morphology and function do likewise (see below; Shelton et al., 1999). At issue here is not merely the astonishing result, but the notion that we must continue to follow function at the tissue level and not merely expression at the cell level.

Remaining with the issue of SAGE for the moment, however, how does intracellular function change as cells senesce? Within human skin, there is a measurable increase in the production of inflammatory factors (Shelton et al., 1999a). This pattern is typical of human skin fibroblasts (as distinct from other human cell phenotypes) and is senescence dependent. Additional changes, such as the predictable changes in elongation factors, energy metabolism, lipid turnover, etc are also found, but, again, vary between cell phenotypes. The overall pattern for aging cells is one of diminished function and efficiency, but the specific *pattern* of SAGE varies between cells.

As we move from the intracellular level to the intercellular level, cell senescence causes additional changes. SAGE causes slower cell division as the cell cycle slows down markedly with cell senescence. The result is a slower rate of cell replacement. Although cell "production" slows, cell losses do not. There is a constant, indeed perhaps accelerated, loss of cells with age as dermal tissues become less functional. In humans, for example, sensory losses alone result in more accidental burns, abrasions, and pressure losses as aging progresses. As the balance of cell production and cell loss shifts, the result is a smaller population of cells at any one moment. Cells grow fewer and tissues layers grow thinner as aging occurs. This is evident in human skin and reflected in the long observed thinning of the epidermal, dermal, and subcutaneous fat layers in aging human skin.

But the changes are more complex than mere thinning of major identifiable skin layers. At the microscopic and functional levels, there is a parallel reduction in rete ridges, vascularity, mast cell number and function, vitamin D production, dermal epidermal junction convolutions, Langerhans cells, healing rates, dermal epidermal adhesion, sensory receptors, heat conservation, and tear resistance. The dermal epidermal junction not only becomes less convoluted and less well bound, but the result is an increasing propensity toward bullae formation and sloughing along tissue planes with age. Nor are these the only changes. There is poorer control of melanocyte division and localization, and they diminish by 10 to 20% per decade, accelerating cell damage attributable to UV penetration.

This last process is typical of aging tissues: the diminution of one function causes an acceleration of damage and cell turnover, resulting in accelerating cell senescence, in turn accelerating the very diminution of function drives the process. One might almost say that "cell senescence itself causes more cell senescence". It is certainly true that at both the intracellular and the intercellular levels, senescing cells feed the very process of senescence. Cells depend upon one another: when one cell senesces, it increases the rate of senescence of other cells, interdependent upon the first.

We have considered the result of senescence at the genetic, intracellular, and intercellular levels. The process does not stop there. The culmination of the process is clinical dysfunction. At the gross level, the result is wrinkles, poor healing, decubiti, heat loss. Although the result is universal and obvious, the cause of skin aging is subtle and poorly understood. It is certainly not simply a matter of photoaging through

UV exposure, nor is it mere trauma, toxic damage, ROS production. It is not even solely the shortening of telomeres, the slowing of cell division, the alteration of gene expression, or any other single factor. In a nutshell, it is all of these things, summed as cell senescence. However defined or understood, though, the practical question remains: can it be altered? In a sense, there are already a multitude of therapies to slow skin aging or hide its outcomes. Entire commercial ventures are founded on no more than retinoids, face lifts, moisturizers, sun blocks, and similar approaches to avoiding the tread of time. But is there any evidence that we can actually reverse the process?

Interventions

We know that cell senescence can be prevented or (within limits) reversed *in vitro*. Can we do the same with *in vivo* skin? Initial results confirm that we can do precisely that. Using the model of an immune deficient mouse onto which we can plant human skin cells without rejection, Funk showed that skin aging can be reversed in the organism (Shelton et al., 1999; Funk et al., 2000).

Using human keratinocytes and fibroblasts, they showed that young cells layer out and form typical young human skin on the mouse. Microscopically, grossly, and genetically, the skin resembles young skin. The dermal epidermal junction is appropriately convoluted, the layers are typically thick, the cells well differentiated and normal in their pattern of gene expression. Repeating the same process with senescent cells results in tissue that appears to be old human skin, replete with thin tissue layers, simplified dermal epidermal junction, easy microbullae formation, and senescent gene expression.

Using transfection, the cells had their telomeres relengthened. Not only was gene expression reset, but replanting these cells onto the mouse model resulted in tissue which was again typical of young human skin. The dermal epidermal junction was, once again, suitably complex; the skin layers were, once again, thicker; the cells were functionally, morphologically, and genetically typical of young skin.

This experiment by no means proves that we can actually rejuvenate human skin at will. For one thing, we are only marginally capable of transfecting the telomerase (or any) gene into an *in vivo* model. For another, we can scarcely expect to perform a clinical "rejuvenation" on an experimental subject by removing all their skin before replanting transfected autologous or homologous skin cells. Nor are these the only technical or ethical obstacles. Nonetheless, the result is startling and it firmly supports the theory outlined here that not only does cell senescence underlie aging, but (despite our assumptions to the contrary) aging is modifiable at the most fundamental level. This work suggests that our knowledge of the role of cell senescence in aging can be translated into clinical interventions for aging and age-related disease.

Immune System Senescence

The immune system is much like the brain: it starts young and vigorous, but ignorant, and ends old and slow, but knowledgeable. In a sense, the senescent immune system has a broad experience of antigens and pathogens, but a contrasting tendency to be slower and less specific than we might wish. Initial work already suggests that this process is the result of cell senescence. Older leukocytes have shorter telomeres. The same correlation is found in trisomy 21 patients, in which shortened telomere lengths in leukocytes are correlated with immune dysfunction (Vaziri et al., 1993).

The role of telomerase in stem cells was first identified in the immune system (Chiu et al., 1996). The stem cell does not express telomerase constitutively, but only under limited circumstances. Upon stimulation with lymphokines, the stem cell transiently expresses telomerase, reextends its telomere, shuts down telomerase expression, then goes through one cell cycle. Typically, one daughter cell differentiates and is thereafter insensitive to lymphokine stimulation of telomerase expression: it is no longer capable of resetting telomere length, which shortens progressively as in other somatic cells. The other daughter cell remains in the reserve, stem cell population and remains capable of responding to lymphokine stimulation by reexpression of telomerase, although is does not express it without such stimulation. Telomerase activity in this daughter cell is sufficient to slow, but not halt, telomerase expression; telomere length in this stem cell population therefore continues to degrade, albeit at a far slower rate than is typical of somatic cells. Curiously, even circulating lymphocytes can be shown to express telomerase to stimulation (Counter et al., 1995), although this ability falls with age (Vaziri et al., 1993). The rate of fall off with age has not yet been determined. Parenthetically, unpublished data on two progeric children show that their telomerase expression is normal as determined by an age-matched volunteer (Shay and Fossel, unpublished data).

The overall effect is that the immune system demonstrates reserve capacity to replace cell loss. Although most immune system cells senesce as do other somatic cells, the presence of stem cells which are capable of expressing telomerase upon stimulation results in a slower overall "rate of aging" in the immune system than might be expected from consideration of either 1) the number of divisions required to produce circulating lymphocytes over the human life span or 2) the telomere lengths found in the human hematopoetic stem cell population at birth.

Nonetheless, immune senescence does occur, and is well known clinically on the basis of recurrent herpes zoster infections, poor response to immunization, high morbidity to influenza infection, etc. Additionally, cancer surveillance has long been considered impaired in the elderly and is potentially attributable to immune senescence (see below; Effros and Valenzuela, 1998). If we look at lymphocytes directly, the most profound effects of senescence are seen within the T cell population, profoundly diminishing its ability to defend against not only infections, but cancer. T cell senescence is associated with the loss of expression of critical signaling molecules and has been reviewed by Effros (Effros and Valenzuela, 1998). Although a great many markers do remain invariant with T cell senescence (surface markers reflecting lineage, adhesion antigens, and receptor structures), CD28, responsible for an essential activation signal in these cells, shows a profound decline. In neonates, only 1% of T cells lack CD28, in centenarians, nearly 40% lack the marker (Boucher et al., 1998), making immune senescence one of "the most dramatic cases of examples of *in vivo* replicative senescence" (Effros and Valenzuela, 1998). CD28 has a multitude of critical functions, but in this context, it is worth noting that it is crucial for stimulated expression of telomerase in lymphocytes. Senescing cells lose the ability to express telomerase; senescence accelerates in aging lymphocytes, as they can no longer slow telomere loss.

The same effect occurs when senescence is accelerated iatrogenically, as in the case of bone marrow transplantation. In these cases, the marrow is forced to undergo cell division on a massive scale to reconstitute the entire lymphatic system, with resulting rapid senescence. Estimates suggest that marrow transplantation is (by estimates of telomere length) equivalent to 40 years of *in vivo* aging (Shay, 1998; Moore, 1997; Wynn et al., 1998). This places (and explains the) severe limits upon serial autologous marrow transplantation.

Additional clinical relevance is found in HIV disease. Estimates suggest that lymphocytes are being destroyed at the rate of 1-2 billion cells per day and replaced at essentially the same rate (in the asymptomatic patient). Although there is no clear senescent effect in the CD4+ lymphocyte population (Wolthers et al., 1996), senescence is accelerated within the CD28+ population, particularly within the CD8+ subset, whose average telomere restriction fragment lengths approximate those of centenarians) (Effros et al., 1996). The proportion of CD28 negative lymphocytes correlates directly with disease progression (Effros et al., 1996; Wolthers et al., 1996), suggesting that stem cell exhaustion may play a role in progression from HIV+ status to clinical AIDS, although there remains substantial disagreement in this issue (Zeichner et al., 1999).

Interventions

In each of the above cases, normal immunosenescence, marrow transplantation, and HIV, the senescence of the lymphocyte population might be ameliorated by use of transfection. The Bodnar work (Bodnar et al., 1998) clearly establishes our ability to reset cell senescence. Technically, we are capable of identifying stem cells in peripheral circulation, transfecting them, and replacing them *ex vivo*. The obstacles to such clinical experimentation on any of these three groups (and additionally on trisomy 21 patients for reasons discussed above) are largely ethical, not technical. To what extent do the risks of these conditions justify an experimental genetic alteration of circulating lymphocytes, with the attendant risks of malignancy, immune dysfunction, etc? In the case of normal immune senescence, the risk is substantial if the patient is currently healthy and if the patient has a significant illness in which immune reconstitution might have value, there would be insufficient time to perform the transfection and reconstitution. The same is true in trisomy 21 patients. In the case of HIV or marrow transplantation, however, the risks of the disease itself are acutely higher and may justify exactly this sort of immune reconstitution under certain conditions. Immune "rejuvenation" experiments, via lymphocyte telomerization, are not only technically doable (Effros and Valenzuela, 1998), but, in certain cases, currently justifiable on ethical and clinical grounds. We are at the point where we could, particularly in the case of immune senescence, prove the therapeutic utility of the theory that cell senescence underlies age-related disease.

Senescence in the CNS: Alzheimer's and Parkinson's Diseases

Why it is Hypothetical?

Alzheimer's dementia is a disease in which much is known and little is understood. Clearly we have a good grasp of the descriptive pathology at almost all levels. What is not clearly understood is how various etiologies (genetics, viral and bacterial infections, diet, tobacco use, diabetes, vascular disease, toxin exposure, behavior, etc) interact to produce (or avoid) Alzheimer's dementia. As discussed in the section on "straw man" arguments, many naively suggest that cell senescence cannot play a role in Alzheimer's dementia because neurons don't divide in the adult. Although this is technically inaccurate, it is essentially true. Nonetheless, neurons are the minority of cells in the CNS and glial cells, particularly microglia, do divide in the mature organism. Additionally, microglia have been implicated in the inflammation which occurs early in Alzheimer's pathology (Lemke et al., 1999). A case can be made (Fossel, 1996) that, as in vascular senescence, a host of risk factors contribute to senescence of cells which do divide (microglia) and these senescing cells result in pathology in cells which do not divide (cortical neurons). Certainly such neurons are dependent upon glia for their normal function in a host of ways, including trophic support. Evidence in other cell systems is uniform in suggesting that trophic factor expression changes in senescing cells. Although this might be precisely the case in Alzheimer's dementia – as microglia senesce and the neurons therefore suffer the consequences – we have thus far been technically unable to measure senescence in microglia *in vivo*, as distinct from surrounding cells in the matrix. The theory that cell senescence lies at the heart of the pathological cascade of events that results in Alzheimer's is entirely consistent with known data, parallel with what we know of age-related diseases in other tissues, and, indeed, elegant, but the entire concept remains unproven. Proof will be technically difficult, but worthwhile if it allows us to intervene therapeutically and prevent the disease. Our experimental abilities are not yet up to the importance of the outcome in exploring the role of cell senescence in Alzheimer's disease.

Interestingly, recent studies have demonstrated the presence of telomerase activity in postmitotic neurons of embryonic and early postnatal mice and rats (Fu et al., 1999a). Suppression of telomerase activity by treatment of cultured hippocampal and cortical neurons with TERT antisense DNA results in increased vulnerability of the neurons to cell death induced by insults relevant to the pathogenesis of Alzheimer's disease and other age-related neurodegenerative disorders. For example, neurons with reduced telomerase activity exhibit increased vulnerability to glutamate, an excitatory transmitter implicated in a number of neurological disorders, and iron, an agent which induces oxidative stress and is implicated in Alzheimer's and Parkinson's diseases. Additional data, in the latter study and a previous study (Fu et al., 1999b), suggests that overexpression of telomerase in neurons can increase their resistance to various insults. These recent findings have important implications for the potential therapeutic potential of manipulations that increase telomerase activity in the treatment of age-related neurodegenerative disorders.

Other Tissues, Other Organs

There is support for the role of cell senescence in age-related disease in many other systems. Chondrocytes show telomere shortening, lung fibroblasts senesce, and vascular changes may lie behind a host of other age-related changes such as sarcopenia. In every case, there is no contradiction of what we already know of the pathology and risk factors involved, but rather an extension of the theoretical underpinnings along with a tantalizing suggestion for novel approaches to therapeutic intervention. In the case of osteoarthritis, for example, the accumulated pressure stress upon cells within the joint would be expected to result in increased cell loss, cell turnover, and hence cell senescence. The fact that chondrocytes do, indeed, demonstrate senescence begs the immediate question of what would happen if we could reset gene expression within such cells via telomerization? Would we give the joint "a new lease on life", allowing the cells to meet the needs imposed upon them? As in the case of Alzheimer's disease, technically we are not quite up to answering this question yet. Yet the possibility remains that we might undercut the known causes of such diseases, and prevent or reverse the pathology at the most fundamental cellular level with a profound clinical benefit.

Cancer

Given the role that telomeres, telomerase, and cell senescence plays in human disease, it is appropriate to ask what evolutionary benefit derives from such a costly mechanism. The outcome of cell senescence is to limit cell division, precisely the aim of an organism in defending itself against the risk of malignant transformation. Teleologically, we age to lower our risk of cancer. This is at first sight not only peculiar, but ironic, since cell senescence increases genomic instability and decreases immune surveillance, in both cases increasing the risk of clinical malignancy. Nonetheless, within the cell, cell senescence acts to limit inappropriate cell division.

Cancer is the inappropriate division of cells and/or lack of appropriate apoptosis. Quite commonly, cancer cells are also dysfunctional, i.e., they do not perform the tasks typically required in cells of their location, lineage, and phenotype. The organism's primary defense against inappropriate cell division is a complex signaling system constantly giving information to each cell: ordering this cell to enter the cell cycle and divide and (more critically in this context) that cell not to do so. Received signals are integrated within each cell and their summed effect either prevents the cell from progressing or initiates cell cycling. When a cancer cell divides despite strong intercellular signals to the contrary, then from the perspective of the community of cells (the organism) and its ultimate survival, the division is inappropriate and by definition, the cell malignant.

Overview of Cancer to Date

The clinical goal of cancer therapy is "the magic bullet," a clinical therapy which kills cancer cells only and does not affect normal cells in any way. This goal has never been met, and, indeed, it has become an article of faith in much of the cancer research community that such a goal is, strictly speaking, unattainable. This admittedly pessimistic, but historically warranted, view was based upon the assumption that there was no way to reliably distinguish cancer cells from normal cells.

Consider where we currently stand in regards to oncotherapy. Killing cancer cells is trivially simple: a few milligrams of potassium cyanide is almost instantly effective in killing all cancer cells in all patients, yet the side effects (simultaneous and equally certain death of all normal cells) are costly. Flippancy aside, this is precisely the problem inherent in current chemotherapy: the problem of killing cancer cells without significant morbidity for the patient. Too often, we are unable to do so without significant collateral damage to the immune system, hematopoesis, gastrointestinal function, hair growth, etc. Until approximately 1989, oncology research was, even theoretically, caught in the horns of this dilemma. We sought to employ superficial differences between normal and cancerous cells to our clinical advantage, all the while suspecting that there were no fundamental differences distinguishing the two that might be leveraged into a therapeutic "magic bullet". Furthermore, there was a feeling that by its very nature, cancer was not a single disease, but rather that it was a mere conceptual collection of diseases that resulted from idiosyncratic damage to each cancer cell. Every cancer cell was, potentially, the result of a distinct DNA error and therefore every cancer cell would require a slightly different "bullet". In the most extreme view, there was no "cancer", only cancers in the plural sense, each different, each a separate therapeutic dilemma.

This view changed as we recognized the importance of the cell cycle braking system in preventing cancer (Hartwell and Kastan, 1994; Jacks and Weinberg, 1996). Essentially, the cell recognizes DNA damage and unless it can be repaired, refuses to enter the cell cycle and divide. When the cell cycle braking system is activated, two simultaneous events occur: 1) the cell up regulates DNA damage repair and 2) it refuses to enter the cell cycle and replicate the damaged DNA. The cell accomplishes both of these objectives through DDBP's (damaged DNA binding proteins) which, in binding to damaged DNA, release p53 molecules and invoke both arms of the entire complex cascade that is the braking system.

The implication was that a cancer cell must not only have damaged DNA (thereby ignoring extrinsic signals telling it not to divide), but it must also have a damaged cell cycle braking system (thereby ignoring intrinsic signals telling it not to divide). More to the point for clinical therapy, however, the cell cycle braking system was identical in all cells and if it malfunctioned, then there were only a limited number of possible parts that could have gone wrong and *these constituted a limited number of therapeutic targets*. Prior to this observation, a cancer cell might differ from a normal cell in any of a thousand ways; now a cancer cell also had to differ from a normal cell in only one of a handful of quite specific parts of the cell cycle braking system. For example, 60% (Rao, 1996; Sherr, 1996) of all cancers had an aberrant form of p53 and hence shared a common therapeutic target. Suddenly, cancer cells were not only fundamentally different from normal cells, but this difference was identifiable and might constitute a therapeutic target (Fossel, 1996; Banks and Fossel, 1997; Fossel, 1998a). The concept of a "magic bullet" remained mythic only to a degree: we might potentially learn to distinguish cancer cells from normal cells, killing the former without harming the latter.

Context of Cell Senescence

But DNA repair mechanisms and the cell cycle braking system are not the cell's only defenses against cancerous transformation and malignancy. The cell also contains a "dead man switch": the telomere. In old locomotives, the train would stop unless the engineer held open the "dead man switch"; release it and the train stopped regardless. Telomeres function as exactly such a switch. Even if the cell cannot repair its DNA, and even if has a damaged cell braking system, it will still be limited to a defined number of cell divisions. Cell senescence will set in as the telomere erodes. Furthermore, even if the cell could evade senescence, it would destroy its own chromosomal integrity.

A useful current analogy is the accelerator, the brake, and the gas tank (de Lange & DePinho, 1999). In the absence of external cellular signals for division (the accelerator), the cell will not divide significantly (nor would a car move very fast). If the cell has damaged DNA and divides inappropriately (as when the car's accelerator is stuck in the down position), the cell detects the problem and applies the brakes (as does the car's driver). In the final resort, if the cell divides despite problems in both mechanisms (as when the accelerator is stuck down and the brakes also fail), then sooner or later, the cell will senesce, slow, and finally stop (as the car would run out of gas). The analogy is apt: we have a limited gas tank in each cell, lowering our risk of unlimited cell division (and malignancy), but the cost is that the entire organism also finally runs out of gas and ages. In the evolutionary balance, we do so after ensuring survival of our offspring and the species cost is small compared to the benefit of avoiding death by malignancy prior to reproduction. The species cost is appropriate, the personal cost remains high in terms of age-related diseases.

But telomeres offer us far more than a familiar transportation analogy: they offer us a unified target distinguishing normal cells from cancer cells. The prospect is not precisely a magic bullet, but a bullet that is far more magical than we though we could ever attain. The prospect is for a therapy which distinguishes cancer cells with a greater than 90% sensitivity (Shay and Gazdar, 1997; Lee and Blasco, 1998). Conceptually, cancer cells must maintain their telomeres to continue dividing. They can do so either by expressing telomerase (Harley et al., 1994) or through some alternate pathway independent of telomerase (Bryan et al., 1995, 1997, Blasco et al., 1997). Failing this, they continue to lose telomere length, crisis ensues, genetic instability becomes overwhelming, and cell death occurs (Lee et al., 1998). Even if they circumvent the "prominent genetic roadblock" of telomerase by oncogene activation (e.g., Myc) or loss of tumor suppressor function (e.g., RB/p53), they will still suffer the consequences of genomic instability and cell death unless the telomere is reextended (de Lange and DePinho, 1999). Like Myc (Greenberg et al., 1998; Wang, et al. 1998), telomerase is found in virtually all human cancers, but inhibition of telomerase activity arrests and ultimately kills malignant cells (de Lange & DePinho, 1999; Counter et al., 1994).

In reality, not only do the majority (perhaps 92%) of human malignancies express telomerase (Shay and Gazdar, 1997), but *all* metastatic malignant cells apparently do so. With regard to cells versus telomerase activity, there are four possibilities:

1) Cancer cells which express telomerase,
2) cancer cells which do not express telomerase,

3) normal cells which express telomerase, and
4) normal cells which do not express telomerase.

Most cells fall into the first and last categories, i.e., telomerase positive cancer cells or telomerase negative normal somatic cells. Assuming that telomerase inhibition were effective, then it would be the second and third categories that might limit such therapeutic efficacy.

Of telomerase negative cancer cells, many will actually be senescing and will soon slow and finally stop dividing. These should be "benign tumors", which is precisely what some of these cells are. Initial work by Shay suggests that in the case of neuroblastomas, we can divide the ultimately malignant from the ultimately benign cases by knowing the telomerase status of the neuroblastoma cells (Maitra et al., 1999; Hiyama et al., 1995, 1997). In this example, telomerase activity functions as a diagnostic marker of malignancy, as it may in a myriad of other cancers under current investigation. There remains the possibility that there are either malignant cancers which re-extend their telomeres through other means not involving telomerase (Bryan et al., 1995, 1997; Blasco et al., 1997) or benign cancers (in the cell senescence sense) that despite almost being senescent will kill their host because of their location. A lipoma of very limited size, for example, might technically be benign, but would certainly be malignant were it to occur in the mid medulla. Even if a cell has a finite potential for division, it might kill the host before reaching that limit. In that sense, the question is not simply "does the cancer express telomerase?" but "will it reach senescence before death of the host ensues?" Nonetheless, telomerase negative cancers are probably of limited concern due to the fact that 1) they are a small percentage (perhaps 8%) of human malignancies and 2) they are often telomerase negative only because it is still early in what will ultimately be found to be a benign clinical course.

The second category of concern is that of somatic cells which express telomerase and therefore might potentially be affected by the clinical use of a telomerase inhibitor. This category comprises germ cells (or their supporting cells) and stem cells. Stem cells in this category include basilar skin cells, hematopoetic cells, GI crypt cells, hair follicle cells, and perhaps several others (Ramirez et al., 1997; Shay and Bacchetti, 1997; Fossel, 1998a, b). In all cases, however, such cells 1) have relatively long telomeres compared to cancer cells and 2) insofar as we know, they do not express telomerase constitutively but transiently upon adequate stimulation. In either case, however, telomerase inhibitors – which will be used therapeutically against telomerase positive cancer cells – would be administered over a limited span of time (hours to weeks?) and would have limited impact upon such telomerase positive somatic cells. Such somatic cells might lose more telomere than otherwise, bearing in mind the two factors above, should suffer minimally compared to cancer cells which have marginal telomere lengths to begin with (Shay and Gazdar, 1997).

While excessive uncontrolled cell proliferation has long been implicated in aberrant regulation of the cell cycle in cancer, mounting evidence supports a major role for lack of appropriate cell death in the formation of many cancers. Indeed, the data showing that mutations in tumor suppressor proteins cause many cancers strongly suggest that a major mechanism of cancer is the increased resistance of cells to apoptosis induced by signals that normally promote appropriate turnover of cells. Recent findings suggest that telomerase can suppress apoptosis of cancer cells (Fu et al., 1999b). Exposure of pheochromocytoma cells to telomerase inhibitors increases their vulnerability to apoptosis induced by a variety of stimuli. Telomerase appears to suppress an apoptotic signal that triggers mitochondrial dysfunction and caspase activation, two key steps in apoptosis (Fu et al., 1999b). This action of telomerase may be relatively independent of its ability to inhibit telemere shortening. Thus, telomerase inhibitors may be particularly valuable in the treatment of cancers, as they would selectively induce apoptosis of telomerase-expressing cancer cells, while sparing somatic cells that have little or no telomerase activity.

These issues are precisely the pending issues which need addressing in clinical trials with the advent of telomerase inhibitors as a therapeutic modality. We must understand the risks and trade offs inherent in using telomerase inhibitors, despite theoretical considerations which suggest that such risks are minimal. The context here is also important: the question is not simply one of the risks and benefits involved, but of the risks and benefits compared to current oncotherapeutic approaches, which are significant and daunting.

The Value of Lengthening Telomeres to Prevent Cancer

There is a distinct value (evolutionary and otherwise) to the default that telomeres shorten, thereby placing a finite limit upon somatic cell division. In light of this general assumption that shortening telomeres have a specific value in arresting cancer cell division, what is the value, if any, of having longer telomeres? This issue will soon become particularly pressing in light of the issue raised throughout the bulk of this paper: telomere reextension is an appropriate therapeutic intervention for age-related diseases and might come into general therapeutic use. If we extend telomere length in order to prevent cell senescence and its associated age-related diseases, do we not increase the risk of malignant transformation?

Certainly this is an appropriate *a priori* concern. We could scarcely condone the prospect of creating tumors with long telomeres and recognize that, in some cells, this mechanism may be all that remains between their initial defects (in signal response and cell cycle braking) and indefinite divisions with consequent mortality to the host. Although we might reasonably prefer to pretreat a patient with telomerase inhibitors (for cancer) before treating the same patient with telomerase inducers (for age-related diseases), thereby obviating some of this risk, the concern still lingers. To what extent does telomerase induction increase our risk of cancer?

Initial result suggest that the opposite is, in fact, the case. Telomerase expression not only does not increase the risk of malignant transformation, but it may additionally serve to protect us against cancer by at least two (independent) mechanisms.

The first issue, that of malignant transformation as a by-product of telomerase expression, has been addressed in a series of papers looking at telomerase transfection in normal human cells. The initial landmark paper by Bodnar and her colleagues (Bodnar et al., 1998) not only showed that telomerase transfection into human foreskin fibroblasts, lung fibroblasts, and retinal pigmented epithelial cells was effective at extending the cell life span (now in excess of 400 divisions and still counting), but that this did not show any evidence of causing malignant transformation as evidenced by karyotype, loss of contact inhibition, growth in low serum, or other gross phenotypic or morphologic changes. This same point has been addressed repeatedly since that time (e.g., Harley, 1998; Jiang et al., 1999) and the results are uniformly negative. There is no evidence, nor should we expect any on theoretical grounds, that telomerase expression *per se* causes malignant transformation.

We should, in fact, expect that telomerase expression would lower the risk of malignant transformation by stabilizing the genome (Counter et al., 1992). This is precisely what is found: telomerized cells have a lower incidence of genomic instability than do equivalent nontelomerized cells. The use of telomerase therapeutically may actually lower the incidence of malignant transformation.

Additionally, transformation is not the only issue involved. Once a cancer survives the intracellular defenses, it must still survive the extracellular defenses, notably immune surveillance. Immune surveillance declines with cell senescence (Effros and Valenzuela, 1998). The therapeutic use of telomerase induction may well provide us with additional protection against cancer by improving T cell performance.

The Prospects for Telomerase Inhibitors to Treat Cancer

Telomerase inhibitors are clearly effective (Wang et al., 1998; van Steensel et al., 1998; Karsleder et al., 1999). The issues remaining fall into two categories: technical and clinical. The first issue is that of finding an effective candidate to take to human trials. Several candidate families have been identified, including RNA anti-sense compounds, nucleopeptides, and others which interfere with telomerase binding or activity, competitively or otherwise. While these are effective *In vitro*, the issue is one of finding a candidate that is stable, safe, and effective *in vivo*. It must survive for a sufficiently long period in the face of enzymatic degradation and liver metabolism, not interfere with other cellular and intercellular activity, and cross both the plasma membrane and the nuclear membrane with good solubility. The technical aspects of identifying such a candidate will doubtless be overcome soon, leaving us with the second issue: clinical performance.

Clinically, the issue will be that of the "magic bullet" question or, put more pragmatically, what is the therapeutic ratio? This will involve many of the issues already discussed above, such as the incidence of side effects on germ line and stem cells. Additional considerations also pertain: what is the time course of action? The rapidity with which such a candidate can kill tumor cells is crucial. The time course is likely to be that of weeks to months rather than days to hours, based on an understanding of the mechanism involved. The cancer cell must divide, shorten its telomeres further, and senesce sufficiently to stop dividing (and perhaps die). This is not an immediate mechanism.

This raises the issues of coordinate therapy: surgery, radiation, and standard chemotherapy. At its best, telomerase inhibition might function without adjunct, but it is more likely that it will first be used as itself an adjunct, to "mop up" cells which escape surgical excision or debulking procedures, for example. Nonetheless, the potential of telomerase inhibition is remarkable and, while conservative prediction for its place in therapy is appropriate, it may well become the single most effective oncotherapy to date. Cancer is the lack of cellular control on cell division; telomerase inhibition reconstitutes one very effective mechanism of such control and as such will have no precedent in mechanism or efficacy.

Clinical Implications

We are about to achieve the ability to treat age-related diseases, including cancer, in a way not previously possible. The initial question was to what extent does cell senescence play a role in age-related disease. Our tentative answer is that it is the lead character and the most reasonable single target for therapeutic intervention. To what extent can we apply this understanding clinically? At this moment we are capable of three sorts of intervention:

1) *In vitro*, we can modulate senescence at the cellular level. Prevention of cell senescence has been achieved without qualification. To the extent that a cell is "not completely senescent", we can reverse the process.
2) *Ex vivo*, we are capable of doing the same for human lymphocytes or any other stem cell or even non-stem somatic cells population that can be identified, extracted, and replaced. This procedure would be appropriate in serial marrow transplantation, perhaps in HIV patients, and potentially in normal immune senescence or in trisomy 21 patients. In regard to human skin (having the example of the mouse in mind), we should be technically capable of reconstituting young human skin in limited patches on an intact elderly patient. In these regards, we could reverse aging *ex vivo*. Though technically within our ability, this has not yet been done in humans.
3) *In vivo*, we are currently incapable of re-extending telomeres as a means of resetting gene expression and reversing cell senescence. In limited cases, we could transfect hTERT genes into intact human skin using a gene gun, but although the possibility is under current consideration (in the case of Hutchinson Gilford progeria, for example, by my laboratory), this has not been done in humans. Transfection, whether directly by gene gun or indirectly through vectors, is not the most likely method to achieve efficient modulation of cell senescence. Telomerase induction, by means of a small molecular weight, gene specific molecule, would be the preferred approach and is actively being pursued by several groups (confidential communication). Whether any of these approaches will succeed remains unknown.

The clinical potential is enormous. If we can achieve control of telomere length (and thereby of senescence associated gene expression and cell senescence), we stand to cure a multitude of diseases at one technological blow. The analogy of curing countless bacterial infections through a knowledge of microbiology and immunity is tempting, but historically insufficient. Telomerase therapy promises to break upon us more suddenly and achieve far more clinically than was the case with infectious disease.

In addition, there are profound fiscal and social implications that while far reaching (Fossel, 1996; Banks and Fossel, 1997; Fossel, 1998a, b), cannot be pursued here. We can, however, reasonably expect that telomerase therapy will not only be more effective, but far less costly than current medical approaches to the diseases of aging.

In terms of health care costs alone, we can expect that, again without any significant social precedent, we will see a drop in the percentage of the GNP spend on medical care.

Summary and Conclusions

Aging is caused by cell senescence and an altered pattern of gene expression in the context of complex interdependencies of the multitude of tissues and organs that constitute the human body. All cells do not senesce. Those that do are responsible for the pathology that we associate with aging, even in cells that do not themselves senesce. Cell senescence is, without intervention at the genetic level, unavoidable, but both cell senescence and its clinical outcome (age-related disease) can be slowed or hastened through non genetic means including, but not limited to, behavioral changes, environmental exposures, diet, drugs, infections, genetic inheritance, and concurrent diseases. In the same manner, such interventions can slow or hasten the overall process of aging in the body.

Although it can be slowed or hastened, aging cannot be prevented, stopped, or reversed by any intervention that does not affect cell senescence and specifically senescence associated gene expression. Although many other approaches might be feasible, the most efficient means of either treating age-related diseases, cancer, or the fundamental process of aging itself will almost certainly be the reversible control of native telomerase activity by induction of expression and, in the case of cancer, by telomerase inhibitors.

Such intervention is currently, although within strict technical limits, possible. Within the next decade, we will likely be capable of therapeutic modulation of telomerase; with telomerase therapy will come not only effective control for age-related diseases and cancer, but an unprecedented alteration of the genetic controls on life span. The fiscal, social, and personal implications of such therapy are almost impossible to predict, but likely to be stunning.

Acknowledgment

The author acknowledges the help of Walter Funk, Cal Harley, Alexey Olovnikov, and Len Hayflick.

References

Avilion, A.A. (1996). Human telomerase RNA and telomerase activity in immortal cell lines and tumor tissues. Cancer Res. 56, 645-650.

Banks, D.A. & Fossel, M. (1997). Telomeres, Cancer, and Aging; Altering the Human Lifespan. J.A.M.A. 278, No. 16, 1345-1348.

Benchimol, S. & Vaziri, H. (1998). Reconstruction of telomerase activity in normal cells leads to elongation of telomeres and extended replicative life span. Current Biology 8, 279-282.

Bierman, E.L. (1994). Aging and Atherosclerosis. Principles of Geriatric Medicine and Gerontology, Third Edition. pp.509-516.

Blasco, M.A., DePinho, R.A., Greider, C.W., Hande, M.P., Lansdorp, P.M., Lee, H.W. & Samper, E. (1997). Telomere shortening and tumor formation by mouse cells lacking telomerase RNA. Cell 91, 25-34.

Blasco, M.A., Funk, W., Greider, C.W. & Villeponteau, B. (1995). Functional characterization and developmental regulation of mouse telomerase RNA. Science, Vol. 269, 1267-1270.

Bodnar, A.G., Chiu, C., Frolkis M., Harley, C.B., Holt, S.E., Lichtsteiner, S., Morin, G.B., Ouellette, M., Shay, J.W. & Wright, W.E. (1998). Extension of Life-Span by Introduction of Telomerase into Normal Human Cells. Science 279, 349-352.

Boucher, N., Dufeu-Duchesne, T., Vicaut, E., Farge, D., Effros, R.B. & Schachter, F. (1998). CD28 expression in T cell aging and human longevity. Exp. Gerontol. 33, 267-282.

Bryan, T.M., Englezou, A., Gupta, J., Bacchetti, S. & Reddel, R.R. (1995). Telomere elongation in immortal human cells without detectable telomerase activity. EMBO J. 14, 4240-4248.

Bryan, T.M., Marusic, L., Bacchetti, S., Namba, M. & Reddel, R.R. (1997). The telomere lengthening mechanism in telomerase-negative immortal human cells does not involve the telomerase RNA subunit. Hum. Mol. Genet. 6, 921-926.

Campisi, J. (1998). The role of cellular senescence in skin aging. J. Investig. Dermatol. Symp. Proc. 3, 1-5.

Chang, E. & Harley, C.B. (1995). Telomere length as a measure of replicative histories in human vascular tissues. Proc. Natl. Acad. Sci. USA 92, 1190-1194.

Chiu, C.P., Dragowska, W., Kim, N.W., Vaziri, H. Yui, J. Thomas, T.E., Harley C.B. & Lansdorp, P.M. (1996). Differential expression of telomerase activity in hematopoietic progenitors from adult human bone marrow. Stem Cells 14, 239-248.

Cooper, L.T., Cooke, J.P. & Dzau, V.J. (1994). The vasculopathy of aging. J. Gerontol. 49, 191-196.

Counter C.M., Hirte, H.W., Bacchetti, S. & Harley, C.B. (1994). Telomerase activity in human ovarian carcinoma. Proc. Natl. Acad. Sci. USA 91, 2900-2904.

Counter, C.M., Avilion, A.A., LeFeuvre, C.E., Stewart, N.G., Greider, C.W., Harley, C.B. & Bacchetti, S. (1992). Telomere shortening associated with chromosome instability is arrested in immortal cells which express telomerase activity. EMBO J. 11, 1921-1929.

Counter, C.M., Gupta, J., Harley, C.B., Leber, B. & Bachetti, S. (1995). Telomerase activity in normal leukocytes and in hematologic malignancies. Blood 85, 2315-2320.

Counter, C.M., Meyerson, M., Eaton, E.N., Ellisen, L.W., Caddle, S.D., Haber, D.A. & Weinberg, R.A. (1998). Telomerase activity is restored in human cells by ectopic expression of hTERT (hEST2), the catalytic sub-unit of telomerase. Oncogene 16, 1217-1222.

Cristofalo, V.J., Allen, R.G., Pignolo, R.J., Martin, B.G. & Beck, J.C. (1998a). Relationship between donor age and the replicative lifespan of human cells in culture: A reevaluation. Proc. Natl. Acad. Sci. USA 95, 10614-10619.

Cristofalo, V.J., Volker, C., Francis, M.K. & Tresini, M. (1998b). Age-Dependent Modifications of Gene Expression in Human Fibroblasts. Critical Reviews in Eukaryotic Gene Expression 8, 43-80.

DePinho, R.A., de Lange, T. (1999). Unlimited Mileage from Telomerase? Science 283, 947-950.

Dimri, G.P. & Campisi, J. (1994). Molecular and cell biology of replicative senescence. Cold Spring Harbor Symp. Quant. Biol.54, 67-73.

Dimri, G.P., Lee, X., Basile, G., Acosta, M., Scott, G., Roskelley, C., Medrano, E., Linskens, M., Rubelj, I., Pereira-Smith, O., Peacock, M. & Campisi, J. (1995). A biomarker that identifies senescent human cells in culture and in aging skin in vivo. Proc. Natl. Acad. Sci. USA 92, 9363-9367.

Effros, R.B. & Valenzuela, H.F. (1998). Immunosenescence: analysis and genetic modulation of replicative senescence in T cells. J. Anti-Aging Med.1, No.4, 305-313.

Effros, R.B., Allsopp, R., Chiu, C.P., Hausner, M.A., Hirji, K., Wang, L., Harley, C.B., Villepontaeu, B., West, M.D., Giorgi, J.V. (1996). Shortened telomeres in the expanded CD28-CD8+ cell subset in HIV disease implicate replicative senescence in HIV pathogenesis. AIDS 10, F17-22.

Feng, J., Funk, W.D., Wang, S.S., Weinrich, S.L., Avilion, A.A., Chiu, C.P., Adams, R.R., Chang, E., Allsopp, R.C., Yu J, Le S, West, M.D., Harley, C.B., Andrews, W.E., Greider, C.W. & Villeponteau, B. (1995). The RNA component of human telomerase. Science 269, 1236-1241.

Fossel, M. (1996). Reversing Human Aging. William Morrow and Company, New York.

Fossel, M. (1998a). Telomerase and the Aging Cell; Implications for Human Health. JAMA 279, 1732-1735.

Fossel, M. (1998b). Implications of Recent Work in Telomeres and Cell Senescence. J. Anti-Aging Med. 1, No.1, 39-43.

Fu, W., Killen, M., Pandita, T. & Mattson, M.P. (1999a). Telomerase protects developing neurons against apoptosis. Submitted for publication.

Fu, W., Begley, J.G., Killen, M.W. & Mattson, M.P. (1999b). Anti-apoptotic role of telomerase in pheochromocytoma cells. J. Biol. Chem. 274, 7264-7271.

Funk, W.D., Wang, C.K., Shelton, D.N., Harley, C.B., Pagon, G.D., Hoeffler, W.K. (2000). Telomerase expression restores dermal integrity to in vitro-aged fibroblasts in a reconstituted skin model. Exp Cell Res 25, 270-278.

Greenberg, R.A., Allsopp, R.C., Chin, L., Morin, G.B. & DePinho, R.A. (1998). Expression of mouse telomerase reverse transcriptase during development, differentiation and proliferation. Oncogene 16, 1723-1730.

Greider, C.W. & Blackburn, E.H. (1985). Identification of a specific telomere terminal transferase activity in Tetrahymena extracts. Cell 43, 405-413.

Greider, C.W. (1996). Telomere length regulation. Ann. Rev. Biochem. 65, 337-365.

Harley, C. (1998). Telomerase and cell immortality in humans. Program and abstracts of the 89th Annual Meeting of the American Association for Cancer Research. March 28-April 1, 1998; New Orleans, La.

Harley, C.B. (1991). Telomere loss: mitotic clock or genetic time bomb? Mutat. Res. 256, 271-282.

Harley, C.B., Futcher, A.B. & Greider, C.W. (1990). Telomeres shorten during aging of human fibroblasts. Nature 345, 458-460.

Harley, C.B., Kim, N.W., Prowse, K.R. & et al. (1994). Telomerase, cell immortality, and cancer. Cold Spring Harb. Symp. Quant. Biol. 59, 307-315.

Harrington, L., Zhou, W., McPhail, T., Oultion, R., Yeung, D.S., Mar, D., Bass, M.D. & Robinson, M.D. (1997). Human telomerase contains evolutionarily conserved catalytic and structural subunits. Genes Dev. 11, 3109-3115.

Hartwell, L.H. & Kastan, M.B. (1994). Cell cycle control and cancer. Science 266, 1821-1828.

Hayflick, L. & Moorhead, P.S. (1961). The limited in vitro lifetime of human diploid cell strains. Exp. Aging Res. 25, 585-621.

Hayflick, L. (1994). *How and Why We Age*. Ballantine Books, New York.

Hiyama, E., Hiyama, K., Ohtsu, K., Yamaoka, H., Ichikawa, T., Shay, J.W. & Yokoyama, T. (1997). Telomerase activity in neuroblastoma: is it a prognostic indicator of clinical behaviour? Eur. J. Cancer 33, 1932-1936.

Hiyama, E., Hiyama, K., Yokoyama, T., Matsuura, Y., Piatyszek, M.A. & Shay, J.W. (1995). Correlating telomerase activity levels with human neuroblastoma outcomes. Nat. Med.1, 249-255.

Jacks, T. & Weinberg, R.A. (1996). Cell-cycle control and its watchman. Nature 381, 643-644.

Jazwinski, S.M. (1996). Longevity, genes, and aging. Science 273, 54-59.

Jiang, X., Jimenez, G., Chang, E., Frolkis, M., Kusler, B., Sage, M., Beeche, M., Bodnar, A.G., Wahl, G.M., Tlsty, T.D. & Chiu, C. (1999). Telomerase expression in human somatic cells does not induce changes associated with a transformed phenotype. Nat. Genet. 21, 111-114.

Kala, D.N., Orhii, P.B., Chen, C., Lee, D.Y., Hubbard, G.B., Lee, S. & Olatunji-Bello, Y. (1998). Aged-rodent models of long-term growth hormone therapy: lack of deleterious effect on longevity. J. Gerontol. Biol. Sci. 53, 452-463.

Karsleder, J., Broccoli, D., Dai, Y., Hardy, S. & de Lange, T. (1999). p53- and ATM- dependent apoptosis induced by telomeres lacking TRF2. Science 283(5406):1321- 5.

Katakura, Y., Yamamoto, K., Miyake, O., Yasuda, T., Uehara, N., Nakata, E., Kawamoto, S. & Shirahata, S. (1997). Bidirectional regulation of telomerase activity in a subline derived from human lung adenocarcinoma. Biochem. Biophys. Res. Commun. 237, 313-317.

Kilian, A., Bowtell, D.D., Abud, H.E., Hime, G.R., Venter, D.J., Keese, P.K., Duncan, E.L., Reddel R.R. & Jefferson, R.A. (1997).Isolation of a candidate human telomerase catalytic subunit gene, which reveals complex splicing patterns in different cell types. Hum. Mol. Genet. 6, 2011-2019.

Kumazaki, T. (1993). Modulation of gene expression during aging of human vascular endothelial cells. Hiroshima J. Med. Sci 42, 97.

Lakatta, E.G. (1994). Alterations in Circulatory Function. Principles of Geriatric Medicine and Gerontology, Third Edition. pp 493-508.

Lane, M.A., Black, D.K., Ingram, D.K. & Roth, G.S. (1998). Calorie restriction in nonhuman primates: implications for age-related disease risk. J. Anti-Aging Med. 1, No. 4, 315-326.

Lange, T. (1998). Telomeres and senescence: ending the debate. Science 279, 334-335.

Larsen, P.L., Albert, P.S. & Riddle, D.L. (1995). Genes that regulate both development and longevity in Caenorhabditis elegans. Genetics 139, 1567-1583.

Lee, H.W., Blasco, M.A., Gottlieb, G.J., Horner, J.W. 2nd, Greider, C.W. & DePinho, R.A. (1998). Essential role of mouse telomerase in highly proliferative organs. Nature 392, 569-574.

Lemke, M.R., Glatzel, M. & Henneberg, A.E. (1999). Antimicroglia antibodies in sera of Alzheimer's disease patients. Biol. Psychiatry 45, 508-511.

Lingner, J., Hughes, T.R., Mann, M., Lundblad, V. & Cech, T.R. (1997). Reverse transcriptase motifs in the catalytic subunit of telomerase. Science 276, 561-567.

Maitra, A., Yashima, K., Rathi, A., Timmons, C.F., Rogers, B.B., Shay, J.W. & Gazdar, A.F. (1999). The RNA component of telomerase as a marker of biologic potential and clinical outcome in childhood neuroblastic tumors. Cancer 85, 741-749.

McClintock, B. (1942). The stability of broken ends of chromosomes in Zea mays. Genetics 26:234-82.

Moore, M.A. (1997). Stem cell proliferation: ex vivo and in vivo observations. Stem. Cells 15, 239-248.

Muller, H.J. (1938). The remaking of chromosomes. Collecting Net. 13, 182-195, 198.

Nakamura, T.M., Morin, G.B., Chapman, K.B., Weinrich, S.L., Andrews, W.H., Lingner, J., Harley, C.B. & Cech, T.R. (1997). Telomerase catalytic subunit homologs from fission yeast and human. Science 277, 955-959.

Nakayama, J., Tahara, H., Tahara, E., Saito, M., Ito, K., Nakamura, H., Nakanishi, T., Ide, T. & Ishikawa, F. (1998). Telomerase activation by hTRT in human normal fibroblasts and hepatocellular carcinomas. Nat. Genet. 18, 65-68.

Nishi, R. (1994). Neurotrophic factors: two are better than one. Science 265, 1052-1053.

Olovnikov, A.M. (1999). The telomere shortening signal may be explained by a fountain mechanism modulating the expression of eukaryotic genes. J. Anti-Aging Med. 2, 59-74.

Olovnikov, A.M., (1971). Principle of marginotomy in template synthesis of polynucleotides [in Russian].

Doklayd Akademii Nauk (SSR) 201, 1496-1499.

Ornish, D., Scherwitz, L.W. Billings, J.H., Brown, S.E., Gould, K.L., Merritt, T.A., Sparler, S., Armstrong, W.T., Ports, T.A., Kirkeeide, R.L., Hogeboom, C. & Brand, R.J. (1998). Intensive lifestyle changes for reversal of coronary heart disease. JAMA 280, 2001-2007.

Prowse, K.R. & Greider, C.W. (1995). Developmental and tissue-specific regulation of mouse telomerase and telomere length. Proc. Natl. Acad. Sci. USA 92, 4818-4822.

Ramirez, R.D., Wright, W.E., Shay, J.W. & Taylor, R.S. (1997). Telomerase activity concentrates in the mitotically active segments of human hair follicles. J. Invest. Dermatol. 108, 113-117.

Rao, R.N. (1996). Targets for cancer therapy in the cell cycle pathway. Current Opinions Oncology 8, 516-524.

Rose, M. & Charlesworth, B. (1980). A test of evolutionary theories of senescence. Nature 287, 141-42.

Rose, M. (1998). Darwinian Anti-Aging Medicine. J. Anti-Aging Med. 1, 105-108.

Rudolph, K.L., Chang, S., Lee, H.W., Blasco, M., Gottlieb, G.J., Greider, C. & DePinho, R.A. (1999). Longevity, Stress Response, and Cancer in Aging Telomerase-Deficient Mice. Cell, 96, 701-712.

Shay J.W. & Bacchetti, S. (1997). A survey of telomerase in human cancer. Eur. J. Cancer 33, 787-791.

Shay, J. (1998). Accelerated telomere shortening in bone-marrow receipients. Lancet 351, 153-154.

Shay, J.W. & Gazdar, A.F. (1997). Telomeres in the early detection of cancer. J. Clin. Pathol., 50, 106-109.

Shelton, D.S., Chang, E., Whittier, P.S., Choi, D. & Funk, W.D. (1999). Microarray analysis of replicative senescence. Current Biology, in press.

Sherr, C.J. (1996). Cancer Cell Cycles. Science 274, 1672-1677.

Singer, M.S. & Gottschling, D.E. (1994). TLC1: template RNA component of Saccharomyces cerevisiae telomerase. Science 266, 404-409.

Starr, P. (1982). The Social Transformation of American Medicine. Basic Books, New York.

Sturrock, R.R. (1976). Changes in neurologia and myelination in the white matter of aging mice. J. Gerontol. 31, 513-522.

van Steensel, B., Smogorzewska, A. & de Lange, T. (1998). TRF2 protects human telomeres from end-to-end fusions. Cell 92, 401-413.

Vaziri, H. (1998). Extension of Life Span in Normal Human Cells by Telomerase Activation: A Revolution in Cultural Senescence. Journal of Anti-Aging Medicine 1, No. 2, 125-130.

Vaziri, H., Schachter F., Uchida, I., Wei, L., Zhu, X., Effros, R., Cohen, D. & Harley, C.B. Loss of telomeric DNA during aging of normal and trisomy 21 human lymphocytes. Am. J. Hum. Genet. 52, 661-667.

Wang, J., Xie, L., Allan, S., Beach, D. & Hannon, G. (1998). Myc activates telomerase. Genes Dev. 12, 1769.

Watanabe, Y., Lee, S.W., Detmar, M., Ajioka, I. & Dvorak, H.F. (1997). Vascular permeability factor/vascular endothelial growth factor (VPF/VEGF) delays and induces escape from senescence in human dermal microvascular endothelial cells. Oncogene 14, 2025-2032.

Watson, J.D. (1968). The Double Helix. Mentor, New York.

Watson, J.D. (1972). Origin of Concatameric T7 DNA. Nature: New Biology 239, 197-201.

Wolfe, J. (1998). Growth hormone: a physiological fountain of youth? J. Anti-Aging Med. 1, 9-26.

Wolters, K.C., Bea, G., Wisman, A., Otto, S.A., de Roda Husman, A.M., Schaft, N., de Wolf, F., Goudsmit, J., Coutinho, R.A., van der Zee, A.G., Meyaard, L., Miedema, F. (1996). T cell telomere length in HIV-1 infection: no evidence for increased CD4+ cell turnover. Science 274, 1543-1547.

Wynn, R.F., Cross, M.A., Hatton, C. & et al. (1998). Accelerated telomere shortening in young recipients of allogenic bone-marrow transplants. Lancet 351:178-181.

Yegorov, Y.E., Akimov, S.S., Hass, R., Zelenin, A.V. & Prudovsky, I.A. (1998). Endogenous Beta-Galactosidase Activity in Continuously Nonproliferating Cells. Exp. Cell Res. 243, 207-211.

Yu, G.L., Bradley, J.D., Attardi, L.D. & Blackburn, E.H. (1990). In vivo alteration of telomere sequences and senescence caused by mutated Tetrahymena telomerase RNAs. Nature 344, 126-132.

Zeichner, S.L., Palumbo, P., Feng, Y., Xiao, X., Gee, D., Sleasman, J., Goodenow, M., Biggar, R. & Dimitrov, D. (1999). Rapid telomere shortening in children. Blood 93(9):2824-2830.

MITOCHONDRIAL OXIDATIVE STRESS AND METABOLIC ALTERATIONS IN NEURODEGENERATIVE DISORDERS

JEFFREY N. KELLER & GORDON W. GLAZNER

Sanders-Brown Center on Aging, University of Kentucky, Lexington KY 40536-0230

Table of Contents

Introduction

Neuronal apoptosis occurs during the development of the nervous system and in numerous neurodegenerative conditions including Alzheimer's disease, Parkinson's disease, and ischemia. Increasing evidence suggests a role for elevated levels of reactive oxygen species (ROS), due to environmental and genetic stressors, in mediating neuronal apoptosis. Once formed, ROS are capable of interacting with cellular proteins, lipids, DNA, and signal transduction pathways influencing neuronal viability. Mitochondria are the primary sites of ROS production and mediate neuronal apoptosis in a variety of neurodegenerative conditions. The focus of this review is to describe the mechanism by which mitochondria generate ROS, discuss genetic and environmental influences that may increase mitochondrial ROS production, and adduce the probable role of mitochondrial ROS in mediating neuronal apoptosis in neurodegenerative conditions.

Interorganellar Signaling in Age-Related Disease, Edited by Mark P. Mattson, 205—237

Neuronal Mitochondria

Mitochondria are present in all eukaryotic cells and have been studied primarily for their role in supplying cellular ATP (Tzagoloff, 1982). The number of individual mitochondrion within a cell is tissue dependent, and appears to be related to the energy requirements of a cell, with relatively high levels of mitochondria (several hundred) present in neurons. In addition to controlling cellular ATP levels, an increasing appreciation for mitochondrial involvement in regulation of calcium and reactive oxygen species (ROS) is emerging.

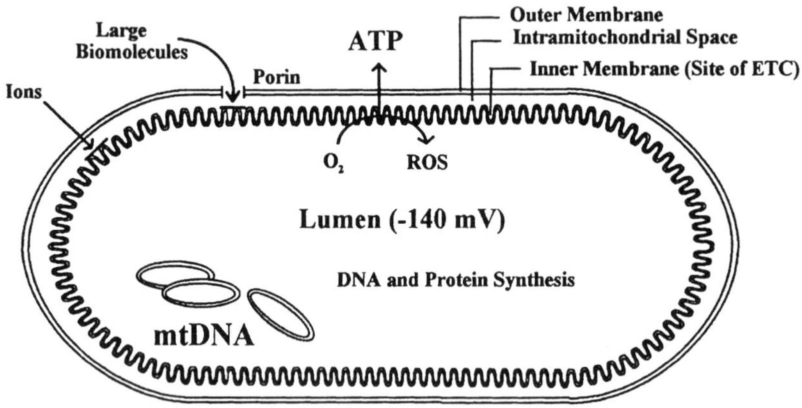

Figure 1. Structure of Mitochondria. Mitochondria contain an inner and outer lipid bilayer membrane. The outer membrane is fairly permeable and contains porins for tranport of large molecules, whereas the inner, which is the site of the ETC, is fairly impermeable even to ions. This allows maintenance of an ionic and electrical gradient which provides the motive force for production of ATP. The inner chamber, called the lumen, is the site of mtDNA, and DNA and protein synthesis. (modified from Glazner, 1999)

Mitochondria are composed of an impermeable inner mitochondrial membrane and a semipermeable outer mitochondrial membrane. The outer mitochondrial membrane is 55-70% protein, while the inner mitochondrial membrane is >75% protein (Tzagoloff, 1982; Lenaz, 1998), leading to an increased importance in membrane lipid composition, requiring tight regulation. The inner mitochondrial matrix houses the components of the electron transport chain (ETC) and between 4-200 copies of mitochondrial DNA (mtDNA). The outer mitochondrial membrane contains monoamine oxidase B and docking proteins, and porin, which allows the outer mitochondrial membrane to be permeable to solutes <5 kDa. The fact that mitochondria and mtDNA are replicated throughout the life of the neuron may have dramatic influences on neuronal physiology and pathology as discussed below.

The most universal and critical function of mitochondria, and the one most widely known, is oxidative phosphorylation (Tzagoloff, 1982). Essentially, this system derives energy by means of electron transfer along the electron transport chain. This is accomplished through discrete metabolic reactions that produce reduced pyridine nucleotide (NADH, NADPH) and /or flavoproteins that transfer electrons from hydrogen to the electron transport chain, ultimately reducing oxygen to water. This reaction in turn produces a proton gradient across the mitochondrial membrane, providing the necessary potential for synthesis of ATP from ADP and phosphate, by the membrane complex ATP synthase.

The oxidative phosphorylation system includes five large multienzyme assemblages, called electron transport chain (ETC) complexes I through IV, and complex V (ATP synthase). These complexes are inserted into the inner mitochondrial membrane, and consist of 77 separate proteins (Kadenbach et al., 1991). Complex I (NADH-ubiquinone oxidoreductase) reduces ubiquinone, and is the largest of the ETC complexes, being composed of 44 proteins (Tzagoloff, 1982). Ubiquinone can also be reduced by either of two flavoproteins, complex II (succinate-ubiquinone oxidoreductase), or electron transfer flavoprotein, which is integral to fatty and amino acid oxidation. Complex III (cytochrome c oxidoreductase) transfers electrons from ubiquinone to reduce cytochrome c. The final step in the ETC is complex IV (cytochrome c oxidase) which reduces oxygen to form H_2O.

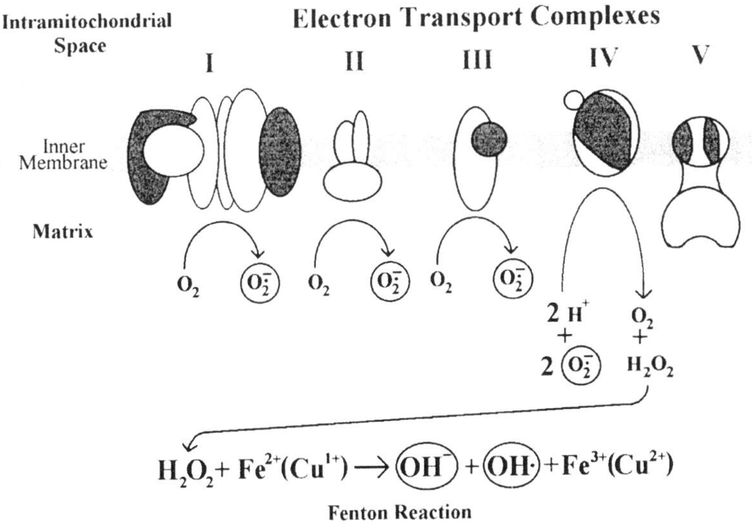

Figure 2. Subunits of Oxidative Phosphorylation. The ETC is composed of 5 complexes (I-V) inserted into the inner mitochondrial membrane. Each complex is composed of multiple polypeptide subunits encoded by either nuclear (white) or mitochondrial (gray) genes. From this figure, one can see the greater risk to complex I and IV function from mtDNA mutations. The electron transfers which occur in complexes I, II and III produce ROS in the form of superoxide ($O^{\cdot-}$). During the last transfer at complex IV, superoxide can react with H^+ to produce hydrogen peroxide (H_2O_2). H_2O_2 in the presence of transition metals such as Cu^+ or Fe^{2+} can form the highly reactive hydroxyl radical ($OH^\cdot$) via the Fenton reaction. (modified from Glazner, 1999)

Calcium

Calcium serves as a second messenger in a variety of signal transduction pathways which culminate in both physiological and pathological responses (Rasmussen, 1983, Woods et al., 1986; Richter and Cass, 1991; Nicotera et al., 1992; McConley and Orrenius, 1995). In order to maintain low cytosolic calcium levels (100-200nM) in the face of high extracellular calcium levels (1-3mM), cells utilize a network of calcium ATPases, calcium transporters, and calcium binding proteins which expel or compartmentalize cytosolic calcium (Carafoli, 1987; Richter and Cass, 1991). Low "resting" calcium levels are observed to exist in a dynamic state, constantly changing in response to genetic, environmental, and intrinsic factors. By utilizing calcium clearance mechanisms, cells are able to maintain extremely tight regulation of calcium levels, and calcium mediated signaling. The reason for this focus on regulation of Ca^{2+} homeostasis is due to the compelling evidence that increased $[Ca^{2+}]i$ precedes cell death in neurons (reviewed by Mattson and Mark, 1996). When elevated $[Ca^{2+}]i$ occurs due to exposure to excitatory neurotransmitters, it is referred to as excitotoxicity. Excitotoxicity has been shown to contribute to most forms of pathological neuronal death, including acute insults such as hypoxia, hypoglycemia and seizures (Olney and de Gubareff, 1978; Rothman et al., 1986; Mattson and Scheff, 1994), and chronic neurodegenerative disorders such as Parkinson's, Alzheimer's and Huntington's diseases (Choi, 1988; Mattson et al., 1993; McDonald and Johnston, 1990; Siesjo et al., 1989).

Mitochondria and Calcium Regulation

Mitochondria have long been known to sequester cytosolic calcium, however the role of such uptake has remained unclear (Gunter and Pfeiffer, 1990; Gunter et al., 1994). Mitochondria have been demonstrated to be concentrated at sites of calcium release and calcium influx (Satoh, 1991; Pralang et al., 1992, 1994; Takei et al., 1992; Babcock et al., 1997), and following neuronal stimulation, mitochondria display marked increases in calcium (Miyata et al., 1991; Gunter et al., 1994). Such sequestration by mitochondria has been proposed to serve as a calcium sink for rapid removal of calcium from cytosol, however recent data suggest a physiological and pathological role for mitochondrial calcium accumulation (Hoek et al., 1995, 1997; Nicholls, 1985; Gunter et al., 1994; Simpson and Russell, 1998).

Increased intra-mitochondrial calcium is dependent on extracellular calcium influx as well as release from intracellular stores, however, it is not dependent upon increases in cytosolic $[Ca^{2+}]$, as shown by recent experiments in our laboratory (Figure 3). Mitochondria are known to respond to, and possibly contribute to calcium oscillations within cells (Miyata et al., 1991; Pralang et al., 1992, 1994; Ichas et al., 1994; Hajnoczky et al., 1995). Because the rate of mitochondrial calcium sequestration exceeds mitochondrial calcium release, mitochondria are observed to exhibit a sustained calcium elevation in response to low frequency stimulation (Ichas et al., 1994; Hajnoczky et al., 1995). Such a mechanism allows cells to rapidly respond to a stimulus by increasing mitochondrial calcium levels without the need of increasing cytosolic calcium levels globally. Such a system likely contributes to the elevation of mitochondrial calcium levels which stimulate ATP production following β-adrenergic stimulation and

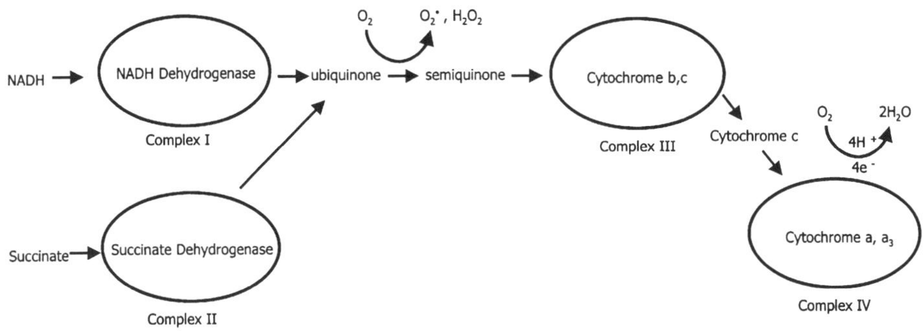

Figure 3. Electron leakage from the electron transport chain. As electrons flow down the electron gradient of the electron transport 1-2% of electrons are lost and cause the formation of oxyradicals in the mitochondria. Intermediates such as NADH or succinate donate electrons to the electron transport chain by the action of dehydrogenases within complex I (NADH dehydrogenase) or complex II (succinate dehydrogenase). Electrons are then transferred to ubiquinone and semiquinone on route to cytochrome b,c segment of complex III. The largest loss of electrons occurs during this transfer and allows for the formation of superoxide and hydrogen peroxide, which is likely derived by the superoxide dismutase. From complex III electrons are transferred to complex IV via cytochrome c, which unlike all other members of the electron transport chain is located on the cytosolic face of the inner mitochondrial membrane.

hormonally induced increases in respiration (McCormack et al., 1990; Miyata et al., 1991; Ichas et al., 1994; Hajnoczky et al., 1995; Hoek et al., 1995). It is likely that mitochondrial calcium accumulation plays a pathological role, whereby in response to environmental stresses neural cells increase mitochondrial calcium uptake, generate increased ROS, and undergo apoptosis.

Mitochondria contain calcium transporters which allow them to rapidly sequester cytosolic calcium, and possess several mechanisms by which mitochondrial calcium may be rapidly released into the cytosol (Gunter and Pfeiffer, 1990; Gunter et al., 1994). Basal levels of inner matrix calcium are maintained at 150-300 nM, as compared to the cytosol 80-120 nM, by the coordinated activity of calcium exchangers, uniporters, and pore complex. Sodium dependent and sodium independent calcium transporters are maintained by the proton gradient across the inner and outer mitochondrial membrane (Gunter and Pfeiffer, 1990; Gunter et al., 1994). The calcium transporters release calcium from the inner mitochondrial matrix, however, loss of the proton gradient can reverse the direction of calcium flow, causing mitochondrial calcium influx. The contribution of these two processes to mitochondrial calcium homeostasis appears to be minimal based on pharmacological studies (Gunter et al., 1994). The primary means of mitochondrial calcium sequestration is mediated via the ruthenium red-sensitive calcium uniporter (Gunter and Pfieffer, 1990). Studies using ruthenium red have identified the importance of uniporter mitochondrial sequestration in mitochondrial dysfunction, mitochondrial ROS accumulation, and cell death (Kruman et al., 1998; Keller et al., 1998b). Pharmacological inhibitors have identified the principle mechanism for mitochondrial calcium release as being the mitochondrial permeability transition (MPT) or megapore (Gunter et al., 1994; Zoratti and Szabo, 1995; Bernardi et al., 1998; Hirsch et al., 1998). Although the identification of the individual components of the MPT complex are unclear, recent studies have identified porin, kinases, and the adenine nucleotide translocase as constituents of the MPT (Zoratti and Szabo, 1995; Bernardi et al., 1998; Hirsch et al., 1998; Ruck et al., 1998). Once induced, the MPT forms a proteinaceous pore allowing the diffusion of solutes <1.5 kD, though prolonged induction of the MPT may result in the release of larger components of the mitochondrial matrix (Gunter and Pfieffer, 1990; Zoratti and Szabo, 1995; Bernardi et al., 1998; Hirsch et al., 1998). Mitochondrial calcium accumulation occurs, and is necessary for neuron death following exposure to metabolic poisons and glutamate (Dugan et al., 1995; Choi, 1996; Jimenez-Jimenez et al., 1996; Keller et al., 1998b). Therefore, it is likely that mitochondrial calcium accumulation plays a central role in mitochondrial dysfunction.

ROS

Mitochondria are the primary sites of free radical production with the primary product being superoxide (Halliwell and Gutteridge, 1986; Beal, 1994; Hockenberry et al., 1994). Constitutive mitochondrial production of superoxide occurs at two main sites within the mitochondria: the flavoprotein (NADH dehydrogenase) and the ubiquinone-cytochrome b segment, which under normal conditions account for 1/3 and 2/3 of O_2 production respectively (Boveris et al., 1976; Turrens and Boveris, 1980; Boveris and Chance, 1984; Cross and Jones, 1991). Superoxide is produced in these reactions by a single electron transfer from the electron transport chain to molecular oxygen. Superoxide is not lipid soluble and is poorly reactive, but is known to give rise to more reactive free radicals including hydrogen peroxide and hydroxyl radical, which can be generated by either the Haber-Weis or the Fenton reaction respectively (Sies, 1986; Coyle and Puttfarken, 1993). The outer mitochondrial membrane contains monoamine oxidase B, which is capable of generating hydrogen peroxide following the deamination of amines (Halliwell and Gutteridge, 1986). Although the importance of monoamine oxidase in neurodegeneration or overall radical production is unclear, recent studies employing monoamine oxidase inhibitors indicate the possible involvement of monoamine oxidase in oxidative stress (Wadia et al., 1998). Recent data indicate the existence of nitric oxide synthase in mitochondria (Bates et al., 1996). While the characterization of mitochondrial nitric oxide synthase remains to be elucidated, it is especially important to note that NO is capable of impairing the ETC and decreasing ATP production (Beckman and Crow, 1993; Zoratti and Szabo, 1995; Nicotera et al., 1997; Bolanos et al., 1997). Furthermore, NO may promote mitochondrial calcium accumulation and mitochondrial ROS production (Beckman and Crow, 1993; Gunter et al., 1994; Bolanos et al., 1997).

Studies of neuronal excitotoxicity have demonstrated the early occurrence and necessity of mitochondrial ROS production to mediate neuron death. Pharmacological agents such as ruthenium red (inhibitor of mitochondrial calcium uptake) or cyclosporin A (an inhibitor of the MPT) prevent calcium cycling and aberrant mitochondrial

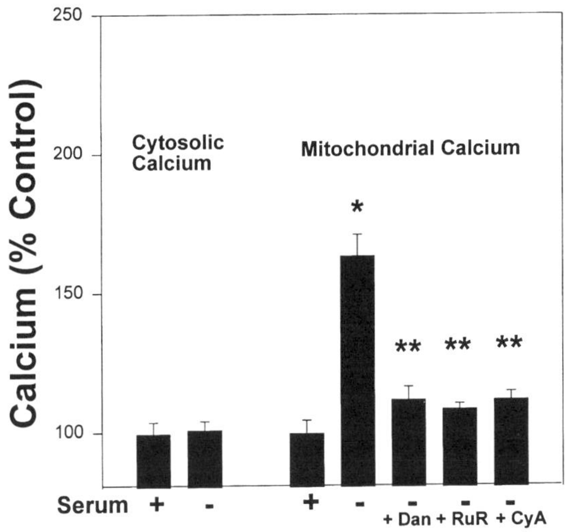

Figure 4. Serum deprivation increases mitochondrial calcium levels in the absence of increased cytosolic calcium. Neural PC6 cells were deprived of serum for 3 h and analyzed for alterations in cytosolic and mitochondrial calcium using fura-2 or Rhod-2, respectively. No detectable increase in cytosolic calcium was detected although mitochondrial calcium increased ~1.75 fold. Pretreatment with the IP3 receptor inhibitor dantrolene (Dan, 1 μM), mitochondrial uniporter inhibitor ruthenium red (RuR, 0.5 μM), or mitochondrial permeability transition inhibitor cyclosporin A (CyA,0.5 μM) reduced increases in mitochondrial calcium accumulation. Data are the mean and SEM of at least 6 cultures. *$P<0.05$ as compared to serum containing cultures, **$P<0.05$ as compared to cultures deprived of serum.

ROS production (Thomas and Reed, 1988; Belmar et al., 1995; Bernardi, 1996; Lin et al., 1996; Waring and Beaver, 1996). In such a scenario, calcium influx causes aberrant ROS generation via uncoupling of the ETC, causing swelling or direct interaction with ETC components. Swelling of mitochondrial matrix increases the rate of electron leakage of the ETC, allowing for increased formation of superoxide. Several components of the ETC are sensitive to calcium fluxes, becoming uncoupled in response to elevated calcium levels (Malis and Bonventre, 1986; Murphy et al., 1990; Gunter et al., 1994). It is important to point out the relationship between the mitochondrial membrane potential and mitochondrial ROS production. By maintaining a high proton gradient on the inner mitochondrial matrix, the transfer of electrons along the ETC is favored over their escape to free oxygen. Therefore, by decreasing mitochondrial membrane potential, the loss of electrons from the ETC is increased.

Similar to calcium, reactive oxygen species (ROS) are observed to mediate an increasing number of physiological and pathological responses (Ames et al., 1993; Slater et al., 1995; McConkey and Orrenius, 1996; Lander, 1997). Once formed, ROS interact with cellular components to effect a growing number of signaling pathways including kinases, phosphatases, and transcription factors (McConkey and Orrenius, 1996; Lander, 1997). Cells are equipped with a number of antioxidant enzyme systems with which to regulate ROS levels (Halliwell and Gutteridge, 1986; Ames et al., 1993;

Jacob and Burri, 1996). These enzymes are highly conserved throughout evolution, consistent with an important role for ROS in cellular signaling (Lander, 1997). Antioxidant enzyme systems are compartmentalized within cells, and can be specific for a given ROS. It is likely that cellular processes regulated by ROS are activated by specific oxyradical exposure. For example, nitric oxide is observed to stimulate the activation of soluble guanylate cyclase, resulting in increased levels of cyclic GMP, while related nitrogen species are unable to activate guanylate cyclase or increase cyclic GMP (Beckman and Koppenol, 1996). Mitochondrial derived ROS appear to be necessary for neuropathological increases in lipid peroxidation. Increased levels of the mitochondrial antioxidant enzyme MnSOD decrease formation of thiobarbituric reactive substances and the lipid peroxidation product 4-hydroxynonenal (HNE) following oxidative stress (Keller et al., 1998a). In addition, cells overexpressing MnSOD are protected from direct HNE toxicity. The neuroprotective properties of elevated MnSOD activity are likely mediated by prevention of mitochondrial calcium and mitochondrial ROS mediated damage. Increased levels of MnSOD, localized to the mitochondria, results in attenuation of mitochondrial ROS formation (Li and Oberly, 1997; Keller et al., 1998a,b). Previous studies have demonstrated a central role for mitochondrial derived ROS in cell loss following TNFα and ceramide (Wong et al., 1989; Shoji et al., 1995; Zamzami et al., 1995b; France-Lanord et al., 1997; Li and Oberley, 1997; Qillet-Mary et al., 1997). In addition, HNE can have inhibitory effects upon the ETC (Esterbauer et al., 1991; Esterbauer, 1993). Because lipid peroxidation and lipid peroxidation products are increased in conditions associated with neural apoptosis they are likely involved in ROS mediated toxicity, however, recent data indicates that lipid peroxidation and HNE occur after initial alterations in mitochondrial ROS (Lenaz, 1998; Wei, 1998). Additionally, lipid peroxidation is not detected in all apoptotic paradigms, and therefore does not appear to be a central means by which all neural apoptosis occurs. For example, at no time following trophic factor withdrawal is there observed to be a detectable increase in lipid peroxidation or HNE (Troy et al., 1997).

Increasing evidence suggest a role for (ROS) in the apoptotic process (Gotz et al., 1994). Although there are multiple means by which ROS may be formed, studies indicate that the mitochondria may serve as the primary source for apoptotic ROS generation. Recent studies support a role for increased mitochondrial ROS production as a primary cause for ROS mediated neuron damage and loss during aging and excitotoxic injury. Alterations in mitochondrial DNA may also contribute towards mitochondrial ROS formation and neuron apoptosis (Esposito et al., 1999). Taken together, these data indicate a pivotal role for mitochondrial ROS production in mediating neuronal apoptosis during development and following genetic and environmental injury.

Influences upon Mitochondrial Reactive Oxygen Species Formation

Because of the interplay between mitochondrial membrane potential (MPT), mitochondrial calcium, and mitochondrial ROS it has been difficult to elucidate the order and relationship of these factors to the apoptotic process. Studies in thymocytes have demonstrated that apoptosis is initiated by a decrease in mitochondrial membrane potential which is sustained by activation of the MPT, leading to accumulation of mitochondrial calcium and mitochondrial ROS (Marchetti et al., 1996a,b). In neurons it appears that small and reversible decreases in mitochondrial membrane potential are the primary events in mitochondrial dysfunction (Gunter and Pfeiffer, 1990). This decrease can be sustained by the activation of the MPT. The ability of calcium and ROS to activate the MPT, and the observation that increases in mitochondrial calcium and mitochondrial ROS occur after initial decreases in membrane potential (Gunter and Pfeiffer, 1990; Gunter et al., 1994; Kruman et al., 1998), suggest a dynamic interplay between these three factors in mitochondrial homeostasis. The ability of cyclosporin A to attenuate prolonged loss of the MMP and elevations in mitochondrial calcium and ROS indicate the involvement of the MPT in these processes (Bernardi et al., 1998; Hirsch et al., 1998; Lemasters et al., 1998). Mitochondrial calcium uptake by the ruthenium red uniporter is also necessary for these events (Keller et al., 1998b; Kruman et al., 1998; Pivovarova et al., 1999. However it is important to point out that accumulations in mitochondrial calcium and ROS are not deleterious *per se*. This is highlighted by recent data demonstrating that increases in mitochondrial calcium and mitochondrial ROS occur in nonpathological conditions (Gunter and Pfeiffer, 1990; Ichas et al., 1994; Brand et al., 1994; Hajnoczky et al., 1995; Hoek et al., 1995, 1997).

Calcium, Reactive Oxygen Species, and Lipid Peroxidation in Neuron Death

Increasing evidence suggests a possible convergence between calcium and ROS signaling. For example, increases in cellular calcium are associated with increases in ROS, antioxidants attenuate calcium mediated toxicity, while calcium chelators and calcium antagonist inhibit ROS toxicity (Richter and Cass, 1991; Bagchi et al., 1997; Goldman et al., 1998; Jimenez-Jimenez, 1996; McConkey and Orrenius, 1996; Kruman et al., 1997, 1998). It is therefore reasonable to hypothesize that calcium mediated events, such as neuronal apoptosis, are carried out in coordination with ROS mediated events. Furthermore, these data suggest that mitochondria may serve as a convergence point between calcium and ROS signaling.

Mitochondrial calcium accumulation occurs concomitantly with increased ROS formation (Malis and Bonventre, 1986; Murphy et al., 1990; Kruman and Mattson, 1999). Agents which prevent increases in mitochondrial calcium prevent increased mitochondrial ROS formation (Belmar, 1995; Bernardi, 1996; Keller et al., 1998a). Under a variety of apoptotic conditions mitochondrial calcium is increased to levels that give rise to ROS, ultimately culminating in apoptotic cell death. Such robust elevations in calcium can result in uncoupling of the ETC and result in increased formation of free radicals, formed by the loss of electrons from ETC (Malis and Bonventre, 1986; Murphy et al., 1990; Keller et al., 1998a,b; Kruman and Mattson, 1999). Additionally, mitochondrial calcium uptake is necessary for ROS formation and ROS toxicity following trophic factor withdrawal (Keller et al., 1998a). Furthermore exogenous application of oxyradicals such as nitric oxide, or lipid peroxidation products such as 4-hydroxynonenal, result in robust increases in mitochondrial ROS (Kruman et al., 1997, 1998; Keller et al., 1998a). Agents which attenuate mitochondrial calcium uptake prevent ROS formation, consistent with mitochondrial calcium mediating ROS formation. In particular, activation of the MPT appears to be central to increased levels of mitochondrial calcium, ROS, and apoptosis following apoptotic stimuli (Zoratti and

Szabo, 1995; Bernardi et al., 1998; Hirsch et al., 1998; Keller et al., 1998a). Increased mitochondrial calcium and mitochondrial ROS precede activation of cell death pathways such as caspases, nitric oxide production, and calcineurin (Zamzami et al., 1995a,b; Ankarcrona et al., 1996; Kroemer et al., 1997; Kruman et al., 1998). It is important to note that lipid peroxidation products accumulate within mitochondria following oxidative injury (Richter et al., 1988; Richter and Cass, 1991; Takeyama et al., 1993). Because of the ability of exogenous lipid peroxidation products to alter mitochondrial calcium and ROS homeostasis it is likely that lipid peroxidation accumulation within the mitochondria is particularly relevant to neuronal degeneration. In particular, the lipid peroxidation product 4-hydroxynonenal (HNE) causes activation of the MPT and exacerbates mitochondrial dysfunction and mitochondrial ROS accumulation following oxidative injury (Kristal et al., 1996; Humphries et al., 1998; Picklo et al., 1999). Future studies will be needed to determine the involvement of mitochondrial lipid peroxidation in neurodegenerative diseases. Taken together, these data indicate a role for mitochondrial calcium and mitochondrial ROS in mediating neural apoptosis.

Apoptosis and MPT

Throughout the course of neuronal development, large numbers of neurons die in an orderly systematic or programmed form of cell death referred to as apoptosis (Oppenheim, 1991; Henderson, 1996; Pettman and Henderson, 1998). Apoptosis occurs in most neuronal populations and accounts for nearly all developmental neuron death. Apoptosis is characterized by the requirement of new protein and DNA synthesis, activation of the caspase family of proteases, and nuclear fragmentation in the absence of organelle swelling and cell lysis (Bredesen, 1995; D'Mello, 1998). Genetic and biochemical studies have identified a limited set of apoptosis inducing factors that compromise a common apoptotic pathway and are ubiquitously expressed in all species of vertebrates and many invertebrates. The most intensively studied of these proteins include members of the BCL-2 family of proto-oncogenes (Merry and Korsmeyer, 1997). The BCL-2 family form homo- and heterodimers with various homologues and cellular transcription factors, to induce or inhibit apoptosis (Reed, 1997). A central event to the pro- and anti-apoptotic actions of BCL-2 member interactions, as well as other apoptotic factors, appears to be regulation of mitochondrial function. For example, overexpression of BCL-2 appears to inhibit mitochondrial ROS production and release of mitochondrial apoptosis inducing factor(s) (Newmeyer et al., 1994; Kluck et al., 1997). Conversely, the pro-apoptotic BCL-2 homologue BAX appears to induce apoptosis in a manner that is dependent upon mitochondrial alterations (Oltvai et al., 1993; Reed, 1997).

Because apoptosis removes specific neurons during neuronal development, without affecting neighboring cells, it is an ideal method for regulating neuronal number. However, in addition to its established role in neuronal development, neuronal apoptosis likely mediates neuron death in numerous diseases. Several neuropathological conditions are associated with immunohistochemical evidence for apoptosis, including Alzheimer's disease, Parkinson's disease, Huntington's disease, and ischemia (Bredesen, 1995; D'Mello, 1998; Pettman and Henderson, 1998). Therefore, the developmental apoptotic machinery does not appear to be removed from developed neurons and may represent a means by which attempts are made to remove mature neurons in response to environmental or genetic injury.

As mentioned previously, activation of the MPT appears to be necessary for mitochondrial dysfunction, mitochondrial calcium accumulation, and mitochondrial ROS accumulation (Zamzami et al., 1995a, 1995b; Skulachev, 1996). Several factors can induce the MPT including increased ADP/ATP ratio, increased intramitochondrial calcium, and increased intramitochondrial ROS accumulation (Pastorino et al., 1993; Gunter et al., 1994; Petronelli et al., 1994; Chernyak and Bernardi, 1996). The ability of each of these factors to induce the MPT independently of the other stimuli suggests that these three factors may act through a common pathway to induce the MPT. Decrease of the mitochondrial membrane potential (MMP) can enhance the MPT activation and decrease the stimuli necessary to induce MPT (Zoratti and Szabo, 1995; Bernardi et al., 1998; Hirsch et al., 1998). Similar results have been demonstrated with the lipid peroxidation product 4-hydroxynonenal HNE (Kristal et al., 1996). These data raise the possibility that MPT activation can occur, but is prolonged in response to loss of the MMP or exposure to lipid peroxidation products. These data could offer an explanation for why under certain conditions MPT activation may result in physiological responses and in others apoptosis. In particular, these data support the possibility that MPT activation is not deleterious *per se*, but may depend upon the duration of MPT activation. In such a scenario, brief activation is beneficial and necessary, whereas chronic activation induces morbidity.

The involvement of the MPT in neurodegenerative processes is highlighted by studies demonstrating a neuroprotective effect of cyclosporin A (Keller et al., 1998a). However, these data must be carefully scrutinized given the ability of other immunosuppressants, which do not inhibit MPT, to decrease neuron death (Schrieber, 1991, 1992). In addition to inhibiting the MPT, cyclosporin A inhibits calcineurin, and causes increased expression of grp 78 and glucocorticoid receptor (Schrieber, 1991, 1992).

A number of cell death pathways are known to be activated during apoptosis, with a large focus of research designed at identifying the contribution of each member to the overall apoptotic process. Activation of members of the caspase family of proteases is known to mediate apoptotic cell loss (Schwartz and Milligan, 1996; Armstrong et al., 1997; D'Mello, 1998). At least ten caspases are known to exist, each synthesized as an inactive cytosolic protein, which undergoes cleavage to form an active heterodimer composed of two chains of approximately 10 and 20 kD (Brancolini and Schneider, 1997; Hofmann, 1999). Increasing evidence suggest a role for a mitochondrial derived apoptosis inducing factor which causes the cleavage of specific caspase members and results in apoptosis (Susin et al., 1996; Kroemer et al., 1997). Mitochondrial cytochrome c, when released, results in caspase activation and apoptosis and as such has been proposed to likely be the mitochondrial derived apoptosis inducing factor (Liu et al., 1996; Kluck et al., 1997). Previous studies have demonstrated a possible dependence upon the MPT mediating cytochrome c release (Bossy-Wetzel et al., 1998). Activation of the MPT results in release of cytochrome c in isolated mitochondria (Kantrow and Piantadosi, 1997). Together these data indicate the possibility that mitochondrial calcium and mitochondrial ROS may serve as triggers of cytochrome c release and subsequent caspase activation and apoptosis. It is interesting to note that while apoptosis is inhibited by pretreatment with antioxidants, antioxidants have been demonstrated to have no effect upon released cytochrome c mediated apoptosis, suggesting that ROS act upstream of cytochrome c release (Bossy-Wetzel et al., 1998).

The involvement of caspase activation in neuron apoptosis is clear (Bredesen, 1995; D'Mello, 1998); however, the timing and mechanism by which caspases are activated and participate in apoptotic process is unclear. For example, activation of caspases is necessary for ROS accumulation following exposure to glutamate (Tan et al., 1998; Tenneti et al., 1998). Alternative studies have demonstrated the requirement of mitochondrial dysfunction prior to caspase activation (Newmeyer et al., 1994; France-Lanord et al. 1997; Kluck et al., 1997). Furthermore, cytochrome c release may be necessary for caspase activation in some apoptotic paradigms (Bredesen, 1995; D'Mello, 1998; Hofmann, 1999). Inhibitors of MPT can attenuate cytochrome c release, however it is important to point out that MPT does not directly release cytochrome c (Bossy-Wetzel et al., 1998). Unlike other components of the ETC, cytochrome c is held electrostatically on the cytoplasmic face of the inner mitochondrial membrane. Halocytochrome c, containing Fe^{2+}, is required for caspase activation (Liu et al., 1996). It is likely that MPT inhibitors prevent cytochrome c release by maintaining mitochondrial membrane potential. This is particularly interesting given the recent studies demonstrating the neuroprotective effect of the loss of membrane potential against subsequent apoptotic injury (Castilho et al., 1998).

Mitochondria as a Convergence Point in Neurodegeneration

Mitochondria do not respond uniformly to stimuli, and exhibit calcium and ROS elevations in distinct microdomains. Previous studies have demonstrated that cytosolic calcium elevations may move in a wave-like pattern through the mitochondrial populations (Pralang et al., 1994; Hajnoczky et al., 1995). This is particularly interesting given that mitochondria are concentrated at the sites of calcium influx or release. These data serve to heighten the importance of calcium elevations and the significance of the location of the cell in which mitochondrial alterations occur. Mitochondria are not stationary and are transported along the cytoskeleton by small ATP driven chaperone proteins (Tzagoloff, 1982). Recent data has demonstrated the existence of mitochondrial associated factors which may transport the mitochondria to the nucleus (Pralang et al., 1994; Kroemer et al., 1997).

The possible role of mitochondria in promoting or initiating cell death in post-mitotic cells, such as neurons, requires special consideration due to the fact that mitochondria continue to be made on a continual basis throughout the life of the neuron. Because mitochondrial genesis requires the catabolism of damaged or defective mitochondria (Tzagoloff, 1982), a selection may take place during the life of a neuron. Mitochondria that possess lower oxidative phosphorylation rates, and hence undergo less oxidative damage, are selected for over the life of a neuron. In such a scenario less ATP is made from an equivalent number of mitochondria over time, and is thought to contribute to the loss of ATP that occurs during aging (Beal, 1994; Wei, 1998). Mitochondrial genesis allows for dramatic differences in mitochondria populations between tissue types and even cell specific alterations in the same tissue (Tzagoloff, 1982).

Alterations in the mitochondrial population are dose-dependent, requiring over 85% permeation before emergence of phenotype. The importance of heteroplasmy and dose dependence in mitochondrial and cellular homeostasis are highlighted below.

Genetic Influences on Mitichondrial Dysfunction

Presenilin Mutations

Mutations in either presenilin-1 (PS-1) located on chromosome 14 or presenilin-2 (PS-2) on chromosome 1 are associate with some cases of early onset autosomal dominant Alzheimer's disease (Hardy, 1997; Mattson et al., 1998). Both PS-1 and PS-2 encode for integral transmembrane proteins whose function has not been identified. However, recent studies have demonstrated that mutations in either PS-1 or PS-2 increase neuronal susceptibility to apoptotic stress (Guo et al., 1997; Mattson et al., 1998; Keller et al., 1998b). Two main theories have emerged as to the mechanism by which PS mutations increase neuron death. The first is based on studies demonstrating that cells expressing PS mutations have altered β amyloid production (Hardy, 1997; Mattson et al., 1998), which causes neuron death. Alternatively, PS mutations may increase neuronal susceptibility by altering calcium and oxyradical homeostasis (Guo et al., 1997). It is important to note that the two theories are not exclusionary and likely involve the mitochondria. For example, mitochondrial dysfunction occurs early in β amyloid toxicity and is associated with accumulation of mitochondrial calcium and mitochondrial ROS (Behl et al., 1994; Mattson et al., 1995, 1996; Keller et al., 1998a). Neural cells transfected with a PS-1 mutation are more susceptible to mitochondrial toxin-induced apoptosis (Keller et al., 1998b), further strengthening the role of mitochondria in PS conferred susceptibility.

Apoplipoprotein Alleles

Recent studies have demonstrated that apolipoprotein A alleles may contribute to neuron death (Cotman and Su, 1996; Markesbery, 1997; Montine et al., 1997; Wisniewski et al., 1997). ApoE is the predominant HDL-like particle in the CNS. There are three alleles for ApoE, termed E2, E3, and E4 which appear to regulate susceptibility to neuron death, with the higher dosage of E4 increasing the chances of late onset AD, and increasing neuron death following injury (Markesbery, 1997; Montine et al., 1997; Wisniewski et al., 1997). By increasing neuron vulnerability to oxidative injury, ApoE4 may increase alterations in mitochondrial stress. In addition the ApoE complex appears to participate in the accumulation of amyloid β protein, as discussed below.

β amyloid Mutations

Increased accumulation of β amyloid (Aβ) protein occurs in AD (Mattson, 1997; Wisniewski et al., 1997). The Aβ fragment is a 40-42 amino acid cleavage product of a larger non-toxic β-amyloid precursor protein (APP). The APP is encoded by at least

10 exons on chromosome 21 and has at least 10 known isoforms. Mutations in the APP gene are thought to increase the amount of overall Aβ and the amount of Aβ aggregation. Increasing these two variables would be expected to increase the levels of oxidative stress in neurons and contribute to mitochondrial dysfunction. The first APP mutation occurs in a small subset of familial AD and occurs among families with hereditary cerebral hemorrhage with amyloidosis, Dutch type (HCHWA-D) (Levy et al., 1990). The increased levels of Aβ-containing plaques, which are noncongophillic, is the result of a substitution of glutamine for glutamic acid on residue 22 of Aβ. Cells expressing this mutation demonstrate increased fibril formation (Wisniewski et al., 1991; Clements et al., 1993), consistent with the ability of HCHWA-D, and other APP mutations, to increase oxidative burden by altering the levels of neurotoxic Aβ.

Mitochondrial DNA Mutations

Mitochondria contain several copies (2-50) of DNA (mtDNA), which are responsible for the transcription of 13 genes for proteins of the respiratory chain (seven subunits of complex I, one subunit of complex III, three subunits of complex IV, and two subunits of complex V) as well as two ribosomal RNA (rRNA) and 22 transfer RNA (tRNA) genes. The mtDNA is small (~16.5 kb), circular, and located in the inner mitochondrial matrix. The first complete 16,569 base pair (bp) sequencing of human mtDNA was reported by Anderson et al. in 1981, and is perhaps the most efficiently packed genome found in nature. In contrast to nuclear DNA, mtDNA is inherited maternally and is continually duplicated throughout the life of the cell. Variations in mtDNA can occur within individual mitochondrion resulting in a diverse population of mitochondria and mtDNA (Wallace, 1994). Such heteroplasmy is a primary influence on both the degree and temporal course of mtDNA phenotype manifestation, with a minimal dosage of a given mtDNA alteration required before phenotype display. Cellular differences in a given tissue often occur, with neighboring cells displaying normal mtDNA adjacent to cells harboring multiple mtDNA mutations. This observation has been proposed to demonstrate that the primary influence of mtDNA alterations is to predispose mitochondria to subsequent environmental factors. It is interesting to note that mitochondria are continually recycled throughout the life of a neuron, and oxidatively damaged mitochondria are primary targets of such recycling. Because the mitochondrion that produces the largest amount of ATP would be expected to contain the highest levels of oxidative damage, it has been proposed that a selection for the least efficient ATP-producing mitochondria occur over time. There are several deleterious factors associated with mtDNA that may contribute to the accumulation of mtDNA. The mtDNA is contained in the matrix which is an oxidative environment, exposing the mtDNA to high levels of oxidative stress. Unlike nuclear DNA, mtDNA lacks histones, increasing the opportunity of oxidative damage. The mitochondria possess diminished DNA repair enzymes, as compared to nuclear DNA repair enzymes (Kunkel and Loeb, 1979, 1981), allowing for oxidative damage to accumulate or alter transcription. Lastly, mtDNA replication is carried out at a lower fidelity than nuclear DNA replication (Richter et al., 1988). Together with genetic mutations described previously, these alterations in mtDNA can exert adverse effects upon mitochondrial and cellular homeostasis.

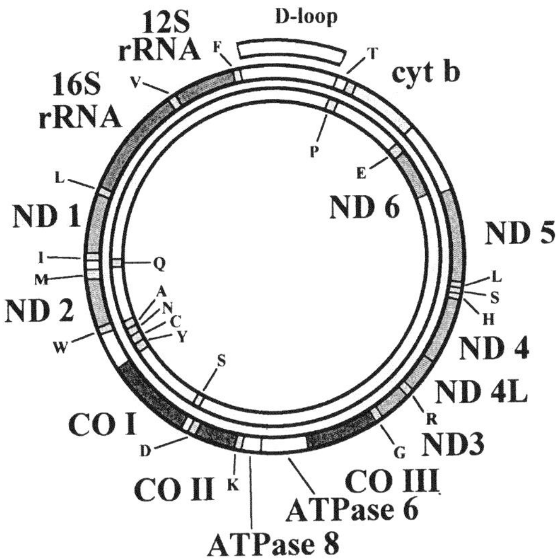

Figure 5. Structure of mtDNA. mtDNA is a closed, circular molecule containing no introns. This molecule codes for 7 subunits of ETC complex I (NADH ubiquinone oxidoreductase, ND 1, 2, 3, 4, 4L, 5, and 6), 3 subunits of complex IV (cytochrome c oxidase, CO I, II, and III), 2 subunits of complex V (ATPsynthase, ATPase 6 and 8) and one subunit of complex III (cytochrome c oxidoreductase, cytochrome b). The 2 mitochondrial rRNAs (16S and 12S) and 22 tRNAs (small letters) allow transcription and translation of these genes to occur in the mitochondria. In addition to encoding m, t, and rRNAs, there is a triple-stranded "D" loop, site of the control region and replication origin. (modified from Glazner, 1999).

For example, mtDNA alterations could cause a transcriptional alteration of a ETC component causing decreased ATP production or enhanced ROS production. Decreased ATP production or increased ROS production may impair the action of ion motive ATPases or oxidative burden, increasing neuronal susceptibility to subsequent stress.

A number of diseases have identified mutations in mtDNA that are capable of eliciting neuropathological manifestations. In 1958, Kearns and Sayre identified a syndrome which included cerebellar dysfunction and elevations in protein level within the CSF (reviewed by Brown and Squire, 1996). Both mtDNA deletions and mtDNA duplication mutations have been ascribed to KS. The deletion mutations identified thus far consist of regions encoding tRNA and oxidative phosphorylation peptide genes. The smallest deletion reported capable of eliciting KSS is a single nucleotide deletion in the tRNA leucine (UUR) at position 3271 (King et al., 1992). The duplication mutation produces two tandemly arranged mtDNA molecules of 16.6 kb. Mitochondrial encephalopathy, lactic acidosis and stroke-like episodes (MELAS) first reported in 1975 (Gardner-Medwin, 1975), is characterized by cerebellar ataxia, migraines, and subsequent strokes. The cause of MELAS is a mtDNA point mutation concentrated in the tRNA leucine (UUR) (Goto et al., 1981), with and A-G mutation at position 3243 found in approximately 80% of MELAS cases (Otabe et al., 1994). This mutation is also associated with diabetes mellitus, occurring in approximately 1% of adult-onset diabetics. Other point mutations associated with MELAS include the tRNA Ser (UCN) (Nakamura et al., 1995) gene and a mutation in the tRNA valine gene (Taylor et al., 1996). The MELAS mutations interfere with the majority of mitochondrial events including protein synthesis (Chomyn et al., 1992; King et al., 1992), mitochondrial membrane potential (James et al., 1996; Moudy et al., 1995), and calcium sequestration (Moudy et al., 1995). Epilepsy, cerebellar ataxia, and dementia characterize myoclonus epilepsy with ragged red fibers (MERRF) (Shoffner et al., 1990, 1991),. The mutation most associated with MERRF is an A-G transition in the tRNA LSY gene at position 8344. Leber's hereditary optic neuropathy (LHON), first described by Leber in 1871, accounts for nearly 3% of blindness in young adult males (reviewed by Newman, 1993; Wallace, 1994). Although 19 different mutations have been reported to be associated with LHON, five appear to be primary sources (G14459A), G11778A), G3640A, T14484C, and G15257A (Wallace, 1994; Brown et al., 1992). Perhaps the most characterized neurodegeneration mtDNA mutation is that of Leigh syndrome, which is frequently lethal and characterized by the progressive degeneration of the basal ganglia (Tatuch et al., 1992; Ortiz et al., 1993). The L8993G (Holt et al., 1990) and L899C (de Vries et al., 1993) mutations causing this syndrome have been demonstrated to cause defects in ATP synthase, complex I, II, and IV.

A number of neurological diseases associated with aging may have mitochondrial dysfunction as a contributing factor to the age of onset and progression of the disease (Wallace, 1994). The most common neuropathies associated with aging are Alzheimer's and Parkinson's diseases, and the role of mitochondrial dysfunction, as well as possible contributions of, and correlations with, mitochondrial dysfunction and mtDNA mutations are currently being investigated.

Alzheimer's Disease

Alzheimer's disease (AD) is a progressive and fatal neurodegenerative disorder characterized by the death of neurons in brain regions involved in learning and memory processes (Markesbery, 1997). Considerable data implicate accumulations of insoluble fibrillar aggregates of a protein called amyloid β-peptide (Aβ) in the pathogenesis of AD (reviewed by Mattson, 1997). Aβ can damage and kill cultured neurons by a mechanism that is dependent upon increases in cytosolic calcium and ROS. Mitochondrial dysfunction results in increased ROS production which may play a role in Aβ toxicity, as it has been observed that amyloidogenic amyloid precursor protein fragments aggregate in the presence of oxidation systems, a phenomenon which is inhibited by antioxidants (Dyrks et al., 1992, 1993). Amyloid β itself may generate free radicals (Hensley et al., 1994), and inhibit mitochondrial function (Shearman et al., 1994; Keller et al., 1997, 1998a), thus leading to another destructive feedback cycle.

Among the factors found to be associated with AD are deficiencies in mitochondrial cytochrome c oxidase (complex IV) activity (Chandrasekaran et al., 1992; Davis et al., 1997; Kish et al., 1992; Mutisya et al., 1994; Parker, 1991; Parker et al., 1994; Sheehan et al., 1997). The experimental and clinical correlations of complex IV activity and the symptomology and progression of AD are many, and exemplify the critical role of mitochondria in neuronal survival and function. Disruption of cytochrome c oxidase activity leads to neurodegenerative events discussed earlier, including accumulation of ROS, decreased ATP production and disruption of Ca^{2+} homeostasis. Inhibition of cytochrome c oxidase with azide leads to defects in learning and memory, as well as altering hippocampal long-term potentiation, a component of memory and learning (Bennett et al., 1992, 1996). Recent reports have shown that impairment of complex IV leads to increased ROS production. Experimental inhibition of complex IV also has been shown to increase production of superoxide (Partridge et al., 1994).

Therefore an evidentiary chain can be constructed, from complex IV defects in AD leading to increased ROS production, further mitochondrial damage, loss of $[Ca^{2+}]i$ homeostasis and neuronal death.

Parkinsonism

Parkinson's disease (PD) is a neurological disorder of movement characterized by the loss of dopaminergic neurons in the substantia nigra. Pathologically, PD is characterized by degeneration of the substantia nigra and the locus coeruleus, and appearance of Lewy bodies in the neuronal cytoplasm. The substantia nigra sends dopaminergic fibers to the striatum, therefore both dopamine content and tyrosine hydroxylase activity are greatly reduced in this target. In 1983, it was observed that the fungal neurotoxin 1-methyl-4-phenyl-1, 2, 3, 6-tetrahydropyridine (MPTP) caused a pathological syndrome remarkably similar to PD (Langston et al., 1983). Later studies showed the target of this toxin to be complex I of the mitochondrial ETC (McNaught et al., 1996). A great deal of interest has been generated of late by these findings, as it has been reported by several groups that there is a regional and disease-specific reduction in mitochondrial complex I activity in the substantia nigra of victims of PD (Schapira et al., 1989; Janetzky et al., 1994). However, this observation alone still leaves the root cause of the disease a mystery.

As is the case for many neurodegenerative diseases, there are likely a number of different factors which eventually terminate at the same end-point. The information indicating a complex I deficit has focused some attention on possible mitochondrial DNA mutations, both inherited and spontaneous (Swerdlow et al., 1996; Graeber and Muller, 1998). Shoffner (1995) has reported increased incidence of an A-to-G mutation in the mitochondrial gene for tRNA (5.3% in PD vs. 0.7% in control). Because so much of the mtDNA is devoted to coding for complex I products, this part of the ETC is the most likely to suffer damage from oxidative stress, mitochondrial tRNA mutations, and other insults, and is the most likely to be inherited.

The possibility remains that mtDNA mutations may be responsible for a subset of PD, and for the severity and progression of those PD cases, which do not have mtDNA mutations as the proximal cause of the disease. Direct evidence for this comes from studies in vitro in which the relative contributions of environmental toxins, complex I nuclear DNA mutations, and mitochondrial mutations were analyzed (Swerdlow et al., 1996). From these studies it was concluded that a portion of the complex I defect found in PD arises from mtDNA mutations. More supporting evidence comes from studies of a rare form of familial dystonia, in which a single point mutation of mtDNA complex I causes a basal ganglia dysfunction with characteristics of PD (Jun et al., 1996; Shoffner, 1995; Miller et al., 1996)

There is significant evidence that overproduction of the enzyme monoamine oxidase B (MAO B) is a primary cause of PD, and that the damage done to the cell is mediated both by neurotransmitter depletion and by production of ROS. Excessive oxidation of dopamine, the principal target of MAO B, leads to a depletion of this neurotransmitter, and an increase in levels of H_2O_2 and other ROS (Riederer et al., 1989). The weight of evidence points to some role of ROS in PD, and it is clear that both complex I and mtDNA is particularly vulnerable to damage by these oxyradicals. (Jesberger and Richardson, 1991). It is hypothesized that the basal ganglion dopaminergic neurons are more at risk from oxidative damage because the deamination of dopamine by MAO B increases with aging (Benedetti and Dostert, 1994), and results in increased formation of H_2O_2 and other toxic byproducts such as free radicals, 6-hydroxydopamine and quinones.

Again, a feedback cycle can be envisioned in which some primary etiology, such as overproduction of MAO B, leads to radical production, inhibiting complex I either directly or by damage to mtDNA. In addition, mtDNA mutations during aging play a key role in the activity of complex I, which decreases over time. The "common mutation" of mtDNA occurs during aging (Wallace, 1994), and as this is a 5 kb deletion between the 13 bp direct repeat sequence encompassing genes for ATPase 6/8, CO 3, ND3, ND4, ND4L, this mutation would cause a loss of complex I. Regardless of the possible increased occurrence of the common mutation in PD patients, it is known to accumulate normally in aging, and may account at least in part to the age of onset of PD.

Though it seems evident that mitochondrial mutations are not the proximal cause of PD (reviewed by Singer et al., 1995), the relative load of mitochondrial mutations carried by an individual may predispose to the disease. Perhaps more importantly, the disease itself induces damaging mtDNA mutations, which may further exacerbate the progression.

Bcl-2 Family Members and Antioxidant Enzymes

Several homologues of the BCL-2 family have been identified in the brain, and may be relevant to the neurodegenerative process. Neuroprotective homologues including BCL-2 and BCL-XL have been demonstrated to exert their neuroprotective effects via maintaining mitochondrial homeostasis (Mah et al., 1993; Martinou et al., 1994). For example, mitochondria expressing increased levels of BCL-2 have an increased capacity for mitochondrial calcium sequestration, inhibited lipid peroxidation, and decreased oxyradical formation (Bredesen, 1995, Kruman et al., 1997). It is interesting to note that BCL-2 has been reported to act in an antioxidant manner to mediate protection from apoptotic stress. Alternatively, the BCL-2 family members BAD and BID may induce neuronal apoptosis. The ability of BAD and BID to induce apoptosis is dependent upon their binding to the outer mitochondrial membrane, presumably altering mitochondrial homeostasis (Knudson et al., 1995). The mitochondrial antioxidant enzyme MnSOD is necessary for attenuation of oxidative injury (Chan et al., 1995; Gonzalezuleta et al., 1998; Keller et al., 1998a; Murakami et al., 1998). Loss of MnSOD due to genetic mutations in its promoter or encoding region would be expected to result in deleterious neuronal alterations.

Non Genetic Influences on Mitochondrial Homeostasis

As mentioned previously, alterations in mitochondrial homeostasis occur early in a variety of neurodegenerative conditions including traumatic injury, ischemia, epilepsy, metabolic poisoning, and HIV. As with genetic influences upon mitochondrial homeostasis, these conditions are associated with increased levels of cytosolic calcium and ROS. Increasing the levels of either cytosolic calcium or ROS would be expected to alter mitochondrial homeostasis. These conditions are associated with inflammation, which can increase the level of neurotoxic cytokines or increase the level of ROS in a neuron, which may also contribute to mitochondrial alterations. The synapse is a site of repeated elevations in calcium and ROS, and as such mitochondria within the synapse may be most vulnerable to dysfunction. Studies from our laboratory have demonstrated that apoptotic signals originating in the synapse are sufficient to induce neuronal apoptosis. Many neurodegenerative conditions are associated with the loss of glucose utilization and trophic factor support (Coyle and Puttfarken, 1993; Blass, 1993, Markesbery, 1997), which may in turn increase mitochondrial dysfunction.

Conclusion

Mitochondria are primary sites of ROS production, generating superoxide, hydrogen peroxide, and nitric oxide by the electron transport chain (ETC), monoamine oxidase, and mitochondrial nitric oxide synthase respectively (Halliwell and Gutteridge, 1986). Elevations in calcium and ROS have been implicated as being key events in the neurodegenerative process (Mattson et al., 1996). Furthermore, it is likely that calcium and ROS cooperatively mediate neurodegeneration. Because mitochondria are involved in both calcium and ROS homeostasis it is likely that they play a pivotal role in mediating neuron death. The density and localization of mitochondria within a neuron likely influence the involvement of mitochondria in neuronal calcium and ROS homeostasis (Pralang et al., 1994).

Although mitochondria are known to be essential to apoptotic cell death, the identification of their specific role(s) in mediating or participating in apoptotic process has remained elusive. As with physiological mitochondrial ROS formation, pathological ROS formation can occur via uncoupling of the electron transport chain (ETC) (Halliwell and Gutteridge, 1986; McCormack et al., 1990; Hajnoczky et al., 1995; Jimenez-Jimenez et al., 1996). Such uncoupling would enhance the normal loss of electrons along the ETC, and allow for a marked increase in free radical formation. Aberrant mitochondrial ROS production has been demonstrated to account for nearly all ROS observed in pathological conditions (Halliwell and Gutteridge, 1986; Richter, 1993; Benzi and Moretti, 1995). A number of factors have been demonstrated to uncouple ETC, however two agents appear to underlie or induce ETC uncoupling in most paradigms studied to date, calcium and ROS. Three calcium sensitive mitochondrial dehydrogenases are known to exist suggesting a direct role for calcium mediated mitochondrial alterations, while indirectly elevated mitochondrial calcium levels have been demonstrated to regulate the ETC by influencing mitochondrial volume and mitochondrial pH (Gunter et al., 1994; Hajnoczky et al., 1995). Application of ROS have been demonstrated to increase uncoupling of the ETC in a reversible manner (Richter and Cass, 1991; Beckman and Koppenol, 1996; Bolanos et al., 1997). It has been suggested that such inhibition is the result of direct oxidation by ROS of redox sensitive sites located upon members of the ETC. In addition to ETC-derived oxyradicals, increased mitochondrial ROS can occur via activation of ROS generating enzymes within the mitochondria, such as phospholipase A_2 (Malis and Bonventre, 1986; Levrat and Louisot, 1996; Ghagfourifar and Richter, 1997; Madesh and Balasubramanian, 1997).

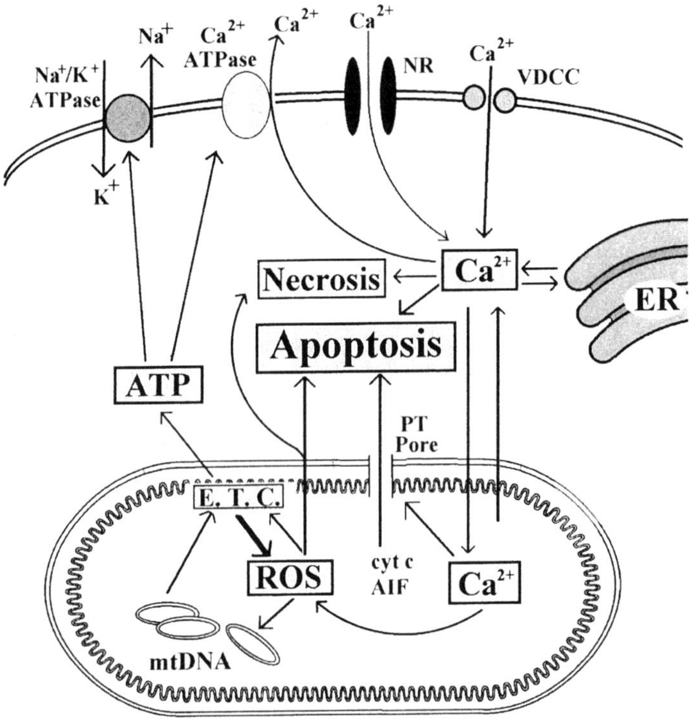

Figure 6. Pathways of Neuronal Death from Mitochondrial Dysfunction. Mutation in mtDNA lead to production of aberrant subunits of the electron transport chain (ETC), thus decreasing efficiency and increasing production of reactive oxygen species (ROS). These ROS can then damage cellular biomolecules, including ETC subunits and mtDNA, further exacerbating ROS production and decreasing ATP production. Decreased availability of ATP to membrane Na+/K+ ATPase and Ca^{2+} ATPase causes increased levels of intracellular Ca^{2+} and membrane depolarization. This depolarization in turn activates voltage-dependent Ca^{2+} channels (VDCC) and allows activation of NMDA receptors, further increasing $[Ca^{2+}]_i$, which can induce cellular necrosis or apoptosis. Mitochondria containing damaged mtDNA and inefficient ETC have a decreased Ca^{2+}-buffering ability, leading to a drop in mitochondrial membrane potential, which results in decoupling of the ETC, membrane permeability transition, and opening of pores in the inner membrane. Proteins which may induce apoptosis, such as cytochrome c and AIF, are then released through these openings. (modified from Glazner, 1999).

Mitochondrial calcium accumulation and ROS formation are observed to be dependent upon the mitochondrial permeability transition (MPT). Such dependence is evident by making use of the immunosuppressant cyclosporin A, an extremely potent inhibitor of the MPT (Schrieber, 1991, 1992; Bernardi et al., 1998; Hirsch et al., 1998). Pretreatment with cyclosporin A inhibits both mitochondrial calcium accumulation and ROS formation suggesting a central role for the MPT in such processes. Prevention of mitochondrial calcium uptake results in attenuation of mitochondrial calcium and mitochondrial ROS accumulation. Cells within the central nervous system are able to exhibit the activation of MPT suggesting a role for such a system within the brain (Kristal and Dubinsky, 1997). The MPT can exist as a persistent pore opening associated with loss of mitochondrial membrane potential or may be transitory (Bernardi et al., 1994; Gunter et al., 1994; Ichas et al., 1994; Marchetti et al., 1996a, b). This information forms the creddenda of a model in which mitochondrial calcium and ROS alterations are due to both calcium release from the MPT and mitochondrial calcium uptake. Such a system allows for a convergence of calcium and ROS mediated signaling. For example, oxidants are known to increase mitochondrial calcium release in a manner dependent upon the MPT (Broekenmeier et al., 1992; Richter, 1993; Richter and Schlegel, 1993; Packer and Murphy, 1995). Likewise, calcium mediated ROS generation has been observed to be dependent upon MPT (Bagchi et al., 1997; Takeyama et al., 1993; Van de Water et al., 1994). Activation of MPT may also be the indirect result of calcium and ROS mediated inhibition of the ETC. For example, inhibition of ETC inhibits fatty acid oxidation and results in accumulation of long chain acyl-CoA, in particular palmitoyl-CoA, which is a potent inducer of the MPT (Petronilli et al., 1993, 1994). Oxidant induced alterations of mitochondrial function have been observed to work synergistically with mitochondrial calcium (Richter and Cass, 1991; Slater et al., 1995; McConkey and Orrenius, 1996; Bagchi et al., 1997). Therefore, calcium and ROS mediate a number of physiological and pathological effects, which may be the result of cooperative interactions through the MPT (Figure 4).

In neuropathological conditions, the levels of calcium and ROS are dramatically elevated and often remain elevated for prolonged periods of time. Studies utilizing pharmacological or transgenic methods, which inhibit calcium or ROS accumulation, have demonstrated the necessity for such elevations in mediating neuron death. Increased calcium and ROS accumulation in neuropathological conditions is associated with increased levels of lipid peroxidation. Lipid peroxidation results from the hydrogen abstraction of a polyunsaturated fatty acid causing the formation of a conjugated diene. Under most instances the ROS generated diene combines with singlet oxygen to generate a peroxyl radical (ROO·) which is capable of abstracting hydrogen from adjacent polyunsaturated fatty acids (Halliwell and Gutteridge, 1986). Once formed, peroxy radicals may undergo further modifications including the addition of hydrogen, or secondary hydrogen abstraction resulting in formation of a variety of lipid peroxidation products. Lipid peroxidation products have been observed to be necessary in ROS-mediated events suggesting a role for lipid peroxidation and lipid peroxidation products (Esterbauer et al., 1991; Esterbauer, 1993).

There is overwhelming evidence for the involvement of calcium and ROS in mediating neural apoptosis. Increased expression of antioxidant enzymes or calcium binding proteins are associated with increased resistance to neural loss while decreased levels of such proteins confers susceptibility (Beal et al., 1995; Chan et al., 1996; Greenlund et al., 1995; Jimenez-Jimenez et al., 1996; Keller et al., 1998a; Mattson et al., 1996). Pretreatment with calcium chelators and antioxidants attenuates neural loss and activation of cell death pathways (Keller et al., 1998a,b; Kruman et al., 1998; Mattson et al., 1995, 1996). Increased levels of calcium and ROS are observed in neural apoptosis and occur early in the process suggesting that such alterations are involved in mediating observed cell loss (Mattson et al., 1995, 1996; Greenlund et al., 1996; Zamzami et al., 1995b; Keller et al., 1998a,b; Kruman et al., 1998). Additionally, previous studies have demonstrated evidence for calcium and ROS working synergistically in mediating their given effects (Slater et al., 1995; Mattson et al., 1996; McConley and Orrenius, 1995). Calcium elevations are associated with increased ROS formation. Pretreatment with antioxidants attenuates calcium-mediated responses while application of calcium chelators attenuates ROS effects. However, a mechanism by which these two signaling pathways converge to mediate such effects has remained elusive. Data from our laboratory indicate that mitochondrial calcium and ROS accumulation as a primary means by which such convergence occurs. Additionally, recent studies identify mitochondrial calcium and an ROS elevation as the primary means by which mitochondria mediate neural apoptosis. Central to both processes is the MPT that is intimately linked as a contributor to neural apoptosis. Environmental and genetic factors may predispose or exacerbate neuronal mitochondria to accumulations in calcium or ROS. By identifying mechanisms with which to prevent sustained elevations in mitochondrial calcium and mitochondrial ROS useful interventions to neurodegenerative conditions can be realized.

References

Ames, B. N., Shingenaga, M. K. & Hagan, T. M. (1993). Oxidants, antioxidants, and the degenerative diseases of aging. Proc. Nat. Acad. Sci. USA 90, 7915-7922.

Anderson, S., Bankier, A.T., Barrell, B.G., de Bruijn, M.H., Coulson, A.R., Drouin, J., Eperon, I.C., Nierlich, D.P., Roe, BA., Sanger, F., Schreier, PH., Smith, A.J., Staden R. & Young, I.G. (1981). Sequence and organization of the human mitochondrial genome. Nature 290, 457-465.

Ankarcrona, M., Dypbukt, J.M., Orrenius, S. & Nicotera, P. (1996). Calcineurin and mitochondrial function in glutamate-induced neuronal cell death. FEBS Lett. 39, 321-324.

Armstrong, R.C., Aja, T.J., Hoang, K.D., Gaur, S., Bai, X., Alnemri, E. S., Litwack, G., Karanewsky, D.S. & Tomaselli, K. J. (1997). Activation of the CED3/ICE-related protease CPP32 in cerebellar granule neurons undergoing apoptosis but not necrosis. J. Neurosci. 17, 553-562.

Babcock, D.F., Herrington, J. & Goodwin, P.C., Park, Y.B., Hille, B. (1997). Mitochondrial participation in the intracellular Ca^{2+} network. J. Cell Biol. 136, 833-844.

Bagchi, D., Wetscher, G.J., Bagchi, M., Hinder, P.R., Perdikis, G., Stohs, S.J., Hinder, R.A. & Das, D.K. (1997). Interrelationship between cellular calcium homeostasis and free radical generation in myocardial reperfusion injury. Chem. Biol. Interact. 104, 65-85.

Bates, T.E., Loesch, A., Burnstock, G. & Clark J.B. (1996). Mitochondrial nitric oxide synthase: a ubiquitous regulator of oxidative phosphorylation? Biochem. Biophys. Res. Commun. 218, 40-44.

Beal, M.F. (1994). Aging, energy, and oxidative stress in neurodegenerative diseases. Ann. Neurol. 19, 357-366.

Beal, M.F., Ferrante, R.J., Henshaw, R., Matthews, R.T., Chan, P.H., Kowall, N.W. & Schulz, J.B. (1995). 3-nitropropionic acid neurotoxicity is attenuated in copper/zinc superoxide dismutase transgenic mice. J. Neurochem. 65, 919-922.

Beckman, J.S. & Crow, J.P. (1993). Pathological implications of nitric oxide, superoxide and peroxynitrite formation. Biochem. Soc. Trans. 21, 330-334.

Beckman, J.S. & Koppenol, W. H. (1996). Nitric oxide, superoxide, and peroxynitrite: the good, the bad, and the ugly. Am. J. Physiol. 271, C1424-C1437.

Behl, C., Davis, J., Lesley, R. & Schubert, D. (1994). Hydrogen peroxide mediates amyloid β protein toxicity. Cell 77, 817-827.

Belmar, E., Garcia-Uglalde, G. & Tapia, R. (1995). Motor alterations and neuronal damage induced by intracerebral administration of Ruthenium red: effect of NMDA receptor antagonist and other anticonvulsant drugs. Mol. Chem. Neuropath. 26, 285-299.

Benedetti, M.S. & Dostert, P. (1994). Contribution of amine oxidases to the metabolism of xenobiotics. Drug Metab. Rev. 26, 507-355.

Bennett, M.C., Diamond, D.M., Stryker, S.L., Parks, J.K., Parker, W.D Jr., Drouin, J., Eperon, I.C., Nierlich, D.P., Roe, B.A., Sanger, Schreier, P.H., Smith, A.J., Staden, R. & Young, I.G. (1992). Cytochrome oxidase inhibition, a novel model of Alzheimer's disease. J. Geriatr. Psychiatry Neurol. 5, 93-101.

Bennett, M.C., Mlady, G.W., Fleshner, M. & Rose, G.M. (1996). Synergy between chronic corticosterone and sodium azide treatments in producing a spatial learning deficit and inhibiting cytochrome oxidase activity. Proc. Natl. Acad. Sci. 93, 1330-1334.

Benzi, G. & Moretti, A. (1995). Are reactive oxygen species involved in Alzheimer's disease? Neurobiol. 16, 661-674.

Bernardi, P. (1996). The permeability transition pore. Control points of a cyclosporin A- sensitive mitochondrial channel involved in cell death. Biochim. Biophys. Acta 1275, 5-9.

Bernardi, P., Broekemeir, K.M. & Pfeiffer, D.R. (1994). Recent progress on regulation of the mitochondrial permeability transition pore, a cyclosporin-sensitive pore in the inner mitochondrial membrane. J. Bioenerg. Biomembr. 26, 509-517.

Bernardi, P., Colonna, R., Constantini, P., Eriksson, O., Fontaine, E., Ichas, F., Massari, S., Nicolli, A., Petronilli, V. & Scorrano, L. (1998). The mitochondrial permeability transition. Biofactors 8, 273-281.

Blass, J.P. (1993). Metabolic alterations common to neural and non-neural cells in Alzheimer's disease. Hippocampus 3, 45-53.

Bolanos, V.P., Almedia, A., Stewart, V., Peuchen, S., Land, J.M., Clark, J.B. & Heales, S.J.R. (1997). Nitric oxide-mediated mitochondrial damage in the brain: mechanisms and implications for neurodegenerative diseases. J. Neurochem. 68, 227-240.

Bossy-Wetzel, E., Newmeyer, D.D. & Green, D.R. (1998). Mitochondrial cytochrome c release in apoptosis occurs upstream of DEVD-specific caspase activation and independently of mitochondrial transmembrane depolarization. EMBO J. 17, 37-49.

Boveris, A., Cadenas, E. & Stoppani, A. O. M. (1976). Role of Ubiquinone in the mitochondrial generation of hydrogen peroxide. Biochem. J. 156, 435-444.

Boveris, A. & Chance, B. (1984). The mitochondrial generation of hydrogen peroxide. Biochem. J. 134, 707-716.

Bredesen, D.E. (1995). Neural apoptosis. Ann. Neurol. 38, 839-851.

Brancolini, C. & Schneider, C. (1997). Cut and die: proteolytic cascades regulating apoptosis. Adv. Clin. Path. 1, 177-189.

Brand, M.D., Chien, L.F., Ainscow, L.F., Rolfe, D.F. & Porter, R.K. (1994). The causes and functions of mitochondrial proton leak. Biochim. Biophys. Acta. 1187, 132-139.

Broekenmeier, K.M., Carpenter-Deyo, L., Reed, D.J. & Pfeiffer, D.R. (1992). Cyclosporin A protects hepatocytes subjected to high Ca^{2+} and oxidative stress. FEBS Lett. 304, 192-194.

Brown, G.K. & Squier, M.V. (1996). Neuropathology and pathogenesis of mitochondrial diseases. J. Inherit Metab. Dis. 19, 553-572.

Brown, M.D., Voljavec, A.S., Lott, M.T., MacDonald, I. & Wallace, D.C. (1992). Leber's hereditary optic neuropathy: a model for mitochondrial neurodegenerative diseases. FASEB J. 6, 2791-2799.

Carafoli, E. (1987). Intracellular calcium homeostasis. Ann. Rev. Biochem. 56, 395-433.

Castilho, R.F., Kowaltowski, A.J. & Vercesi, A.E. (1998). 3,5,3'-triiodothyronine induces mitochondrial permeability transition mediated by reactive oxygen species and membrane protein thiol oxidation. Arch. Biochem. Biophys. 354, 151-157.

Chan, P.H., Epstein, C.J., Li, Y., Huang, T.T., Carlson, E., Kinouchi, H., Yang, G., Kamii, H., Mikawa, S. & Kondo, T. (1995). Transgenic mice and knockout mutants in the study of oxidative stress in brain injury. J. Neurotrauma 12, 815-824.

Chan, P.H., Epstein, C.J., Li, Y., Huang, T.T., Carlson, E., Mikawa, S., Kondo, T., Chen, S.F. & Reola, L. (1996). Studies of neuronal injury mechanism in focal stroke using mitochondrial manganese superoxide dismutase-deficient mice, In (Kriegelstein, J., ed) Pharmacolgy of Cerebral Ischemia. Medpharm, Stuttgart, pp. 573-579.

Chandrasekaran, K., Stoll, J., Giordano, T., Atack, J.R., Matocha, M.F., Brady, D.R. & Rapoport, S.I. (1992). Differential expression of cytochrome oxidase. (COX). genes in different regions of monkey brain. J. Neurosci. Res. 32, 415-423.

Chernyak, B.V. & Bernardi, P. (1996). The mitochondrial permeability transition pore is modulated by oxidative agents through both pyridine nucleotides and glutathione at two separate sites. Eur. J. Biochem. 238, 623-630.

Choi, D.W. (1988). Glutamate neurotoxicity and diseases of the nervous system. Neuron 1:623-34.

Choi, D.W. (1996). Ischemia-induced neuronal apoptosis. Curr. Opin. Neurobiol. 6, 667-672.

Chomyn, A., Martinuzzi, A., Yoneda, M., Daga, A., Hurko, O., Johns, D., Lai, S.T., Nonaka, I., Angelini, C. & Attardi, G. (1992). MELAS mutation in mtDNA binding site for transcription termination factor causes defects in protein synthesis and in respiration but no change in levels of upstream and downstream mature transcripts. Proc. Natl. Acad. Sci. USA 89, 4221-4225.

Clements, A., Walsh, D.M., Williams, C.H. & Allsop, D. (1993). Effects of the mutation Glu22 to Gln and Ala21 to Gly on the aggregation of a synthetic fragment of the Alzheimer's amyloid β/A4 peptide. Neurosci. Lett. 161, 17-20.

Cotman, C.W. & Su, J.H. (1996). Mechanisms of neuronal death in Alzheimer's disease. Brain Pathol. 6, 493-506.

Coyle, J.T. & Puttfarken, P. (1993). Oxidative stress, glutamate, and neurodegenerative disorders. Science 262, 689-695.

Cross, A.R. & Jones, O.T.G. (1991). Enzymatic mechanisms of superoxide generation. Biochem. Biophys. Acta. 1057, 281-298.

Davis, R.E., Miller, S., Herrnstadt, C., Ghosh, S.S., Fahy, E., Shinobu, L.A., Galasko, D., Thal, L.J., Beal, M.F., Howell, N. & Parker, W.D Jr. (1997). Mutations in mitochondrial cytochrome c oxidase genes segregate with late-onset Alzheimer disease. Proc. Natl. Acad. Sci. USA 94, 4526-4531.

D'Mello, S.R. (1998). Molecular regulation of neuronal apoptosis. Curr. Top. Dev. Biol. 39, 187- 213.

de Vries, D.D., van Engelen, B.G., Gabreelsm, F.J., Ruitenbeek, W. & van Oost, B. (1993). A second missense mutation in the mitochondrial ATPase 6 gene in Leigh's syndrome. Ann. Neurol. 34, 410-422.

Dugan, L.L., Sensi, S.L., Canzoniero, L.M., Handran, S.D., Rothman, S.M., Goldberg, M.P. & Choi, D.W. (1995). Mitochondrial production of reactive oxygen species in cortical neurons following exposure to N-methyl-D-aspartate. J. Neurosci. 15, 6377-6388.

Dyrks, T., Dyrks, E., Hartmann, T., Masters, C. & Beyreuther, K. (1992). Amyloidogenicity of β A4 and β A4-bearing amyloid protein precursor fragments by metal-catalyzed oxidation. J. Biol. Chem. 267, 18210-18217

Dyrks, T., Dyrks, E., Masters, C.L. & Beyreuther K. (1993). Amyloidogenicity of rodent and human β A4 sequences. FEBS Lett. 324, 231-236.

Esposito, L.A., Melov, S., Panov, A., Cottrell, B.A. & Wllace, D.C. (1999). Mitochonrial disease in mouse results in increased oxidative stress Proc. Natl. Acad. Sci. USA 96, 4820-4825.

Esterbauer, H. (1993). Cytotoxicity and genotoxicity of lipid peroxidation products. Am. J. Clin. Nutr. 57, 779S-786S.

Esterbauer, H., Schauer, R.J. & Zollner, H. (1991). Chemistry and biochemistry of 4-hydroxynonenal, malonaldehyde and related aldehydes. Free Rad. Biol. Med. 11, 81-128.

France-Lanord, V., Brugg, B., Michel, P.P. & Ruberg, M. (1997). Mitochondrial free radical signal in ceramide dependent apoptosis: a putative mechanism for neuronal death in Parkinson's disease. J. Neurochem. 69, 1612-1621.

Gardner-Medwin, D., Dale, G. & Parkin, J.M. (1975). Lactic acidosis with
mitochondiral myopathy and recurrent coma. (Abstr). First International Congress of Child Neurology No 47.

Ghagfourifar, P. & Richter, C. (1997). Nitric oxide synthase activity in mitochondria. FEBS Lett. 418, 291-296.

Glazner, G.W. (1999). Mitochondrial mutations and neurodegenerative disease. In Genetic Aberrancies and Neurodegenerative Disorders (Mattson, M.P., ed). Advances in Gerontology, JAI Press, Stamford CT. In press.

Goldman, R., Moshonov, S. & Zor, U. (1998). Generation of reactive oxygen species in a human keratinocyte cell line: role of calcium. Arch. Biochem. Biophys. 350, 10-18.

Gonzalezuleta, M., Enez, L.M., Mukhina, G., Lebovitz, R.M., Zwacka Z.M., Engelhardt, J.F., Oberley, L.W., Dawson, V.L. & Dawson, T. (1998). Manganese superoxide dismutase protects nNOS neurons from NMDA and nitric oxide-mediated neurotoxicity. J. Neurosci. 18, 2040-2055.

Goto, Y., Nonaka, I. & Horai, S. (1981). A mutation in the tRNALeucine (UUR) gene associated with the MELAS subgroup of mitochondrial encephalomyopathies. Nature 348, 651-653.

Gotz, M.E., Kunig, G., Riederer, P. & Youdim, M. B. H. (1994). Oxidative stress: Free radical production in neural degeneration. Pharmac. Ther. 63, 37-122.

Graeber, M.B. & Muller, U. (1998). Recent developments in the molecular genetics of mitochondrial disorders. J. Neurol. Sci. 153, 251-263.

Greenlund, L.J., Deckworth, T.L. & Johnson, E.M. (1995). Superoxide dismutase delays neuronal apoptosis: a role for reactive oxygen species in programmed cell death. Neuron 14, 303-315.

Gunter, T.E. & Pfeiffer, D.R. (1990). Mechanisms by which mitochondria transport calcium. Am. J. Physiol 258, C755-C786.

Gunter, T.E., Gunter, K. K., Sheu, S.S. & Gavin, C.E. (1994). Mitochondrial calcium transport: physiological and pathological relevence. Am. J. Physiol. 267, C313-C339.

Gunter, T.E., Buntinas, L., Sparagna, G.C. & Gunter, K.K. (1998). The Ca^{2+} transport mechanisms of mitochondria and Ca^{2+} uptake from physiological-type Ca^{2+} transients. Biochim. Biophys. Acta 1366, 5-15.

Guo, Q., Sopher, B.L., Furukawa, K., Pham, D.G., Robinson, N., Martin, G.M. & Mattson, M.P. (1997). Alzheimer's presenilin mutation sensitizes neural cells to apoptosis induced by trophic factor withdrawal and amyloid beta-peptide: involvement of calcium and oxyradicals. J. Neurosci. 17, 4212-22.

Hajnoczky, G., Robb-Gaspers, L. D., Seitz, M. B. & Thomas, A.P. (1995). Decoding of cytosolic calcium oscillations in the mitochondria. Cell 82, 415-424.

Halliwell, B. & Gutteridge, J.M. C. (1986). Oxygen free radicals and iron in relation to biology and medicine: some problems and concepts. Arch. Biochem. Biophys. 246, 501-514.

Hardy, J. (1997). Amyloid, the presenilins and Alzheimer's disease Trends Neurosci. 20,154-159.

Henderson, C.E. (1996). Programmed cell death in the developing nervous system. Neuron 17, 579-585.

Hensley, K., Carney, J.M., Mattson, M.P., Aksenova, M., Harris, M., Wu, J.F., Floyd, R.A. & Butterfield, D.A. (1994). A model for β-amyloid aggregation and neurotoxicity based on free radical generation by the peptide, relevance to Alzheimer disease. Proc. Natl. Acad. Sci. 91, 3270-3274.

Hirsch, T., Susin, S.A., Marzo, I., Marchetti, P., Zamzami, N. & Kroemer, G. (1998). Mitochondrial permeability transition in apoptosis and necrosis Cell. Biol. Toxicol. 14, 141-145.

Hockenberry, D.M., Oltvai, Z.N., Xiao-Ming, Y., Milliman, C.L. & Korsemeyer, S.J. (1994). Bcl-2 functions in an antioxidant pathway to prevent apoptosis. Cell 75, 241-251.

Hoek, J.B., Farber, J.L., Thomas, A.P. & Wang, X. (1995). Calcium ion-dependent signaling and mitochondrial dysfunction: mitochondrial calcium uptake during hormonal stimulation in intact liver cells and its implication for the mitochondrial permeability transition. Biochem. Biophys. Acta. 1271, 93-102.

Hoek, J.B., Rode, E. W. & Wang, X. (1997). Hormonal stimulation, mitochondrial calcium accumulation, and the control of the mitochondrial permeability transition in intact hepatocytes. Mol. Cell. BioL. Chem. 174, 173-179.

Holt, I.J., Harding, A.E., Petty, R.K. & Morgan-Hughes. (1990). A new mitochondrial disease associated with mitochondrial DNA heteroplasmy. Am. J. Hum. Genet. 46, 428-433.

Hofmann, K. (1999). The modular nature of apoptotic signaling proteins. Cell. Mol. Life Sci. 55, 1113-1128.

Humphries, K.M., Yoo, Y. & Szweda, L.I. (1998). Inhibition of NADH-linked mitochondrial respiration by 4-hydroxy-2- nonenal. Biochemistry 37, 552-557.

Ichas, F., Jouaville, L.S., Sidash, S.S., Mazat, J.P. & Holmuhamedov, E.L. (1994). Mitochondrial calcium spiking: a transduction mechanism based on calcium-induced permeability transition involved in cell calcium signalling. FEBS Lett. 348, 211-215.

James, A.M., Wei, Y.H., Pang, C.Y. & Murphy, M.P. (1996). Altered mitochondrial function in fibroblasts containing MELAS or MERRF mitochondrial DNA mutations. Biochem. J. 318, 401-407.

Jacob, R.A. & Burri, B.J. (1996). Oxidative damage and defense. Am. J. Clin. Nutr. 63, 985S- 990S.

Janetzky, B., Hauck, S., Youdim, M.B., Riederer, P., Jellinger, K., Pantucek, F., Zochling, R., Boissl K.W. & Reichmann, H. (1994). Unaltered aconitase activity, but decreased complex I activity in substantia nigra pars compacta of patients with Parkinson's disease. Neurosci. Lett. 169, 126-128.

Jesberger, J.A. & Richardson, J.S. (1991). Oxygen free radicals and brain dysfunction. Int. J. Neurosci. 57, 1-17.

Jimenez-Jimenez, F. J., Garcia-Ruiz, P.J. & Bustos, F. (1996). Calcium, neuronal death, and neurological disease. Rev. Neurol. 24, 1199-1209.

Jun, A.S., Trounce, I.A., Brown, M.D., Shoffner, J.M. & Wallace, D.C. (1996). Use of transmitochondrial cybrids to assign a complex I defect to the mitochondrial DNA-encoded NADH dehydrogenase subunit 6 gene mutation at nucleotide pair 14459 that causes Leber hereditary optic neuropathy and dystonia. Mol. Cell. Biol. 16, 771-777.

Kadenbach, B., Schneyder, B., Mell, O., Stroh, S., and Reimann, A. (1991). Respiratory chain proteins. Rev. Neurol. 147, 436-442.

Kantrow, S.P. & Piantadosi, C.A. (1997). Release of cytochrome c from liver mitochondria during permeability transition. Biochem. Biophys. Res. Comm. 232, 669-671.

Kearns, N.P. & Sayre, G.P. (1958). Retinitis pigmentosa, external ophthalmoplegia, and complete heart block. Arch. Ophthalmol. 60, 280-289.

Keller, J.N., Pang, Z., Geddes, J.W., Begley, J.G., Germeyer, A., Waeg, G. & Mattson, M.P. (1997). Impairment of glucose and glutamate transport and induction of mitochondrial oxidative stress and dysfunction in synaptosomes by amyloid b-peptide: role of the lipid peroxidation product 4-hydroxynonenal. J. Neurochem. 68, 273-284.

Keller, J.N., Kindy, M.S., Holtsberg, F.W., St. Clair, D.K., Yen, H.C., Germeyer, A., Steiner, S.M., Hutchins, J.B. & Mattson, M.P. (1998a). Mn-SOD prevents neural apoptosis by suppression of peroxynitrite production and consequent lipid peroxidation and mitochondrial dysfunction, and reduces ischemic brain injury in vivo. J. Neurosci. 18, 687-697.

Keller, J.N., Guo, Q., Holtsberg, F.W. & Mattson, M.P. (1998b). Expression of the L286V presenilin-1 mutation confers increased susceptibility to mitochondrial toxin-induced apoptosis. J. Neurosci. 18, 6997-7003.

King, M., Koga, Y., Davidson, M. & Schon, E. (1992). Defects in mitochondrial protein synthesis and respiratory chain activity segregate with the tRNALeucine (UUR) mutation associated with mitochondrial myopathy, encephalopathy, lactic acidosis, and strokelike episodes. Mol. Cell. Biol. 12, 180-190.

Kish, S.J., Bergeron, C., Rajput, A., Dozic, S., Mastrogiacomo, F., Chang, L.J., Wilson, J.M., DiStefano L.M. & Nobrega J.N. (1992). Brain cytochrome oxidase in Alzheimer's disease. J. Neurochem. 59, 776-779.

Kluck, R.M., Bossy-Wetzel, E., Green, D.R. & Newmeyer, D.D. (1997). The release of cytochrome c from mitochondria: a primary site for Bcl-2 regulation of apoptosis. Science 275, 1132-1136.

Knudson, C.M., Tung, K.S., Tourtelle, W.G., Brown, G.A. & Korsmeyer, S.J. (1995). Bax-deficient mice with lymphoid hyperplasia and male germ cell death. Science 270, 96-99.

Kristal, B.S. & Dubinsky, J.M. (1997). Mitochondrial permeability transition in the central nervous system: induction by calcium cycling dependent and independent pathways. J. Neurochem. 69, 524-538.

Kristal, B.S., Park, B.K. & Yu, B.P. (1996). 4-hydroxyhexenal is a potent inducer of the mitochondrial permeability transition. J. Biol. Chem. 271, 6033-6038.

Kroemer, G., Zamzami, N. & Susin, S.A. (1997). Mitochondrial control of apoptosis. Immunol. Today 18, 44-51.

Kruman, I.I. & Mattson, M.P. (1999). Pivotal role of mitochondrial calcium uptake in neuronal cell apoptosis and necrosis. J. Neurochem. 72, 529-540.

Kruman, I., Bruce, A.J., Bredesen, D.E., Waeg, G. & Mattson, M.P. (1997). Evidence that 4 hydroxynonenal mediates oxidative stress-induced neuronal apoptosis. J. Neurosci. 17, 5089-5100.

Kruman, I., Guo, Q. & Mattson, M.P. (1998). Calcium and reactive oxygen species mediate staurosporine induced mitochondrial dysfunction and apoptosis in PC12 cells. J. Neurosci. Res. 51, 293-308.

Kunkel, T.A. & Loeb, L.A. (1979). On the fidelity of DNA replication. Effect of divalent metal ion activators and deoxyribonucleoside triphosphate pools on in vitro mutagenesis. J. Biol. Chem. 254, 5718-5725.

Kunkel, T.A. & Loeb, L.A. (1981). Fidelity of mammalian DNA polymerases. Science 213, 765-767.

Langston, J.W., Ballard, P., Tetrud, J.W. & Irwin I. (1983). Chronic Parkinsonism in humans due to a product of meperidine-analog synthesis. Science 219, 979-980.

Lander, H.M. (1997). An essential role for free radicals and derived species in signal transduction. FASEB J. 11, 118-124.

Leber, T. (1871). Ueber hereditare und congenital angelegte Sehnopsinlieden. Albrecht v. Graefes Arch. Ophthalmol. 17, 249-291.

Lemasters, J.J., Nieminen, A.L., Qian, T., Trost, L.C., Elmore, S.P., Nishimura, Y., Crowe, R.A., Cascio, W.E.,

Bradham, C.A., Brenner, D.A. & Herman, B. (1998). The mitochondrial permeability transition in cell death: a common mechanism in necrosis, apoptosis and autophagy. Biochim. Biophys. Acta 1366, 177-96.

Lenaz, G. (1998). Role of mitochondria in oxidative stress and aging. Biochim. Biophys. Acta 11366, 53-67.

Levrat, C. & Louisot, P (1996). Increase of mitochondrial PLA_2-released fatty acids is an early Event in tumour necrosis factor alpha-treated WEHI-164 cells. Biochem. Biophys. Res. Comm. 221, 531-538.

Levy, E., Carman, M.D., Fernandez-Madrid, I., Lieberburg, I., Power, M.D., vanDuinen, S.G., Bots, G.T., Luyendijk, W. & Frangione, B. (1990). Mutation of the Alzheimer's disease amyloid gene in hereditary cerebral hemorrhage, Dutch type. Science 248, 1124-1126.

Li, J.J. & Oberley, L.W. (1997). Overexpression of manganese containing superoxide dismutase confers resistance to the cytotoxicity of tumour necrosis factor α and/or hypothermia. Cancer Res. 57, 1001-1008.

Lin, M. & Lin-Shiau, M.R. (1996). Ruthenium red, a novel enhancer of K^+ currents at mouse motor nerve terminals. Neuropharmacology 35, 615-626.

Liu, X., Kim, N.C., Yang, J., Jemmerson, R. & Wanf, X. (1996). Induction of apoptotic program in cell free extract: requirement for dATP and cytochrome C. Cell 86, 147-157.

Madesh, M. & Balasubramanian, K.A. (1997). Activation of mitochondrial phospholipase A_2 by superoxide. Arch. Biochem. Biophys. 15, 187-192.

Mah, S.P., Zhong, L.T., Liu, Y., Roghani, A., Edwards, R.H. & Bredesen, D.E. (1993). The protooncogene bcl-2 inhibits apoptosis in PC12 cells. J. Neurochem. 60, 1183-1186.

Malis, C.D. & Bonventre, J.V. (1986). Mechanism of calcium potentiation of oxygen free radical injury to renal mitochondria. J. Biol. Chem. 261, 14201-14208.

Marchetti, P., Susin, S.A., Decaudin, D., Gamen, S., Castedo, M., Hirsch, T., Zamzami, N., Naval, J., Senik, A. & Kroemer, G. (1996a). Mitochondrial permeability transition is a central coordinating event of apoptosis. J. Exp. Med. 184, 115-125.

Marchetti, P., Hirsch, T., Zamzami, N., Castedo, M., Decaudin, D., Susin, S.A., Masse, B. & Kroemer, G. (1996b). Mitochondrial permeability transition triggers Lymphocyte apoptosis. J. Immunol. 157, 4830-4836.

Markesbery, W.R. Oxidative stress hypothesis in Alzheimer's disease. Free Rad. Biol. Med. 23, 134-147, 1997.

Martinou, J.C., Dubois-Dauphin, M., Staple, J.K., Rodriguez, I., Frankowsky, H., Missotten, M., Albertini, P., Talabot, D., Catsicas, S., Pietra, C. & Huarte, J. (1994). Overexpression of Bcl-2 in transgenic mice protects neurons from naturally occurring cell death and experimental ischemia. Neuron 13, 1017-1030.

Mattson, M.P. (1997). Cellular actions of β-amyloid precursor protein and its soluble and Fibrillogenic derivatives. Physiol. Rev. 77, 1081-1132.

Mattson, M.P. & Scheff, S.W. (1994). Endogenous neuroprotection factors and traumatic brain injury: mechanisms of action and implications for therapy. J. Neurotrauma 11, 3-33.

Mattson, M.P. & Mark, R.J. (1996). Excitotoxicity and excitoprotection *in vitro*. Adv. Neurol. 71, 1-30.

Mattson, M.P., Barger, S.W., Cheng, B., Lieberburg, I., Smith-Swintosky, V.L. & Rydel R.E. (1993). β-Amyloid precursor protein metabolites and loss of neuronal Ca^{2+} homeostasis in Alzheimer's disease. Trends Neurosci. 16, 409-414.

Mattson, M.P., Barger, S.W., Begley, J.G. & Mark, R.J. (1995). Calcium, free radicals and excitotoxic neuronal death in primary cell culture. Meth. Cell Biol. 46, 187-216.

Mattson, M.P., Furukawa, K., Bruce, A.J., Mark, R.J. & Blanc, E.M. (1996). Calcium Homeostasis and free radical metabolism as convergence points in the pathophysiology of dementia. Molecular Mechanisms of Dementia, Humana Press, pp. 103-143.

Mattson, M.P., Guo, Q., Furukawa, K. & Pedersen W.A. (1998). Presenilins, the Endoplasmic reticulum, and neuronal apoptosis in Alzheimer's disease. J. Neurochem. 70, 1-14.

McConley, D.J. & Orrenius, S. (1995). The role of calcium in the regulation of apoptosis. J. Leukoc. Biol. 59, 775-83.

McConkey, D.J. & Orrenius, S. (1996). Signal transduction pathways in apoptosis. Stem Cells 14, 619-631.

McCormack, J.G., Halestrap, A.P. & Denton, R.M. (1990). Role of calcium ions in regulation of mammalian intramitochondrial metabolism. Physiol. Rev. 70, 391-425.

McDonald, J.W. & Johnston, M.V. (1990). Physiological and pathophysiological roles of excitatory amino acids during central nervous system development. Brain Res. Rev. 15, 41-70.

McNaught, K.S., Thull, U., Carrupt, P.A., Altomare, C., Cellamare, S., Carotti, A., Testa, B., Jenner, P. & Marsden, C.D. (1996). Inhibition of [H^3]dopamine uptake into striatal synaptosomes by isoquinoline derivatives structurally related to 1-methyl-4-phenyl-1, 2, 3, 6-tetrahydropyridine. Biochem. Pharmacol. 52, 29-34.

Merry, D.E. & Korsmeyer, S.J. (1997). Bcl-2 gene family in the nervous system. Annu. Rev. Neurosci. 20, 245-267.

Miller, S.W., Trimmer, P.A., Parker, W.D, Jr. & Davis, R.E. (1996). Creation and characterization of mitochondrial DNA-depleted cell lines with "neuronal-like" properties. J. Neurochem. 67, 1897-1907.

Miyata, H., Silverman, H.S., Sollott, S.J., Lakatta, E.G., Stern, M.D. & Hansford, R.G. (1991). Measurement of mitochondrial free Ca^{2+} concentration in living single rat cardiac myocytes. Am. J. Physiol. 261, H1123-H1134.

Montine, K.S., Olson, S.J., Amarnath, V., Whetsell, W.O., Graham, D.G. & Montine, T.J. (1997). Immunohistochemical detection of 4-hydroxy-2-nonenal adducts in Alzheimer's diseases associated with inheritance of ApoE4. Am. J. Pathol. 150, 437-443.

Moudy, A.M., Handran, S.D., Goldberg, M.P., Ruffin, N., Karl I. & Kranz-Eble, P. (1995). Abnormal calcium homeostasis and mitochondrial polarization in a human encephalomyopathy. Proc. Natl. Acad. Sci. USA 92, 729-733.

Murakami, K., Kondo, T., Kawase, M., Li, Y., Sato, S., Chen, S.F. & Chan, P.H. (1998). Mitochondrial susceptibility to oxidative stress exacerbates cerebral infarction that follows permanent focal brain ischemia in mutant mice with manganese superoxide dismutase deficiency. J. Neurosci. 18, 205-213.

Murphy, A.N., Kelleher, J.K. & Fiskum, G. (1990). Submicromolar Ca^{2+} regulates phosphorylating respiration by normal rat liver and AS-30D hepatoma mitochondria by different mechanisms. J. Biol. Chem. 265, 10527-10534.

Mutisya, E.M., Bowling, A.C. & Beal, M.F. (1994). Cortical cytochrome oxidase activity is reduced in Alzheimer's disease. J. Neurochem. 63, 2179-2184.

Nakamura, M., Nakano, S., Goto, Y., Ozawa, M., Nagahama, Y., Fukuyama, H., Akiguchi, I., Kaji, R. & Kimura, J. (1995). A novel point mutation in the mitochondrial tRNAser(UCN) gene detected in a family with MERRF/MELAS overlap syndrome. Biochem. Biophys. Res. Commun. 214, 86-93.

Newman, N.J. (1993). Neuro-ophthalmology, the afferent visual system. Curr. Opin Neurol. 738-746.

Newmeyer, D.D., Farschon, D.M. & Reed, J.C. (1994). Cell free apoptosis in Xeonopus extracts: inhibition by Bcl-2 and requirement for an organelle fraction enriched in mitochondria. Cell 79, 353-364.

Nicholls, D.G. (1985). A role for mitochondria in the protection of cells against calcium overload? Prog. Brain Res. 63, 97-106.

Nicotera, P., Bellomo, G. & Orrenius, S. (1992). Calcium-mediated mechanisms in chemically induced cell death. Annu. Rev. Pharmacol. Toxicol. 32, 449-470.

Nicotera, P., Ankarcrona, M., Bonfoco, E., Orrenius, S. & Lipton, S.A. (1997). Neuronal necrosis and apoptosis: two distinct events induced by exposure to glutamate or oxidative stress. Adv. Neuro. 72, 95-101.

Olney, J.W. & de Gubareff, T. (1978). Glutamate neurotoxicity and Huntington's chorea. Nature 271, 557-559.

Oltvai, Z.N., Milliman, C.L. & Korsmeyer, S.J. (1993). Bcl-2 heterodimerizes in vivo with a conserved homolog, Bax, that accelerates programmed cell death. Cell 74, 609-619.

Otabe, S., Sakura, H., Shimokawa, K., Mori Y., Kadowaki, H., Yasuda, K., Nonaka, K., Hagura, R., Akanuma, Y. & Yazaki Y. (1994). The high prevalence of the diabetic patients with a mutation in the mitochondrial gene in Japan. J. Clin. Endocrinol. Metab. 79, 768-771.

Oppenheim, R.W. (1991). Cell death during development of the nervous system. Annu. Rev. Neurosci. 14, 453-501.

Ortiz, R.G., Newman, N.J., Shoffner, J.M., Kaufman, A.E., Koontz, D.A. & Wallace, D.C. (1993). Variable retinal and neurologic manifestations in patients harboring the mitochondrial DNA 8993 mutation. Arch. Ophthalmol. 111, 1525-1530.

Packer, M.A. & Murphy, M.P. (1995). Peroxynitrite formed. by simultaneuos nitric oxide and superoxide generation causes cyclosporin-A sensitive mitochondrial calcium efflux and depolarization. Eur. J. Biochem. 234, 231-239.

Parker, W.D Jr. (1991). Cytochrome oxidase deficiency in Alzheimer's disease. Ann. N.Y. Acad. Sci. 640, 59-64.

Parker, W.D. Jr., Mahr, N.J., Filley, C.M., Parks, J.K., Hughes, D., Young, D.A. & Cullum C.M. (1994). Reduced platelet cytochrome c oxidase activity in Alzheimer's disease. Neurology 44(6), 1086-90

Partridge, R.S., Monroe, S.M., Parks, J.K., Johnson, K., Parker, W.D. Jr., Eaton, G.R. & Eaton, S.S. (1994). Spin trapping of azidyl and hydroxyl radicals in azide-inhibited rat brain submitochondrial particles. Arch. Biochem. Biophys. 310, 210-217.

Pastorino, J.G., Snyder, J.W., Serroni, A., Hoek, J.B. & Farber, J.L. (1993). Cyclosporin A and carnitine prevent anoxic death of cultured hepatocytes by inhibiting the mitochondrial permeability transition. J. Biol. Chem. 268, 13791-13798.

Petronilli, V., Cola, C. & Bernardi, P. (1993). Modulation of the mitochondrial cyclosporin A-sensitive permeability transition pore: the minimal requirements for pore induction underscore a key role for transmembrane electrical potential, matrix pH, and matrix Ca^{2+}. J. Biol. Chem. 268, 1011-1016.

Petronilli, V., Nicolli, A., Costantini, P., Colonna, R. & Bernardi, P. (1994). Regulation of the permeability transition pore, a voltage-dependent mitochondrial channel inhibited by cyclosporin A. Biochim. Biophys. Acta 1187, 255-259.

Pettman, B. & Henderson, C.E. (1998). Neuronal cell death. Neuron 20, 633-647.

Picklo, M.J., Amarnath, V., McIntyre, J.O., Graham, D.G. & Montine, T.J. (1999). 4-Hydroxy-2(E)-nonenal inhibits CNS mitochondrial respiration at multiple sites. J. Neurochem. 72, 1617-24.

Pralang, W.F., Hunyady, L., Varnai, P., Wollhiem, C.B. & Spat, A. (1992). Pyridine nucleotide redox state parallels production of aldosterone in potassium-stimulated adrenal glomerulosa cells. Proc. Nat. Acad. Sci. USA 89, 132-136.

Pivovarova, N.B., Hongpaisan, J., Andrews, S.B. & Friel, D.D. (1999). Depolarization-induced mitochondrial Ca^{2+} accumulation in sympathetic neurons: spatial and temporal characteristics J. Neurosci. 19, 6372-6384.

Pralang, W.F., Spat, A. & Wollhiem, C.B. (1994). Dynamic spacing of cell metabolism by intracellular Ca^{2+} transients. J. Biol. Chem. 269, 27310-27314.

Quillet-Mary, A., Jaffrezout, J.P., Mansat, V., Bordier, C., Naval, J. & Lauren, G. (1997). Implications of mitochondrial hydrogen peroxide generation in ceramide-induced apoptosis. J. Biol. Chem. 272, 21388-21395.

Rasmussen, H. (1983). Modulation of cell function in the calcium messenger system. Rev. Physiol. Biochem.

Pharmacol. 95, 111-148.

Reed, J.C. (1997). Double identity for proteins of the Bcl-2 family. Nature 387, 773-776.

Richter, C. (1993). Prooxidants and mitochondrial calcium, their relationship to apoptosis and oncogenesis. FEBS Lett. 325, 104-107.

Richter, C. & Cass, G.E.N. (1991). Oxidative stress in mitochondria: its relationship to cellular calcium homeostasis, cell death, proliferation, and differentiation. Chem. Biol. Interact. 77, 1-23.

Richter, C. & Schlegel, J.Z. (1993). Mitochondrial calcium release induced by prooxidants. Toxicol. Lett. 67, 119-127.

Richter, C., Park, J.W. & Ames, B.N. (1988). Normal oxidative damage to mitochondrial and nuclear DNA is extensive. Proc. Natl. Acad. Sci. USA 85, 6465-6467.

Riederer, P., Konradi, C., Hebenstreit, G. & Youdim, M.B. (1989). Neurochemical perspectives to the function of monoamine oxidase. Acta. Neurol. Scand. Suppl. 126, 41-45.

Rothman, S.M. (1986). Glutamate and anoxic neuronal death *in vitro*. Adv. Exp. Med. Biol. 203, 687-695.

Ruck, A., Dolder, M., Wallimann, T. & Brdiczka, D. (1998). Reconstituted adenine nucleotide translocase forms a channel for small molecules comparable to the mitochondrial permeability transition pore. FEBS Lett. 426, 97-101.

Satoh, T., Ross, C.A. Villa, A., Supattapone, S., Pozzan, T., Snyder, S.H. & Meldolesi, J. (1991). The inositol 1, 4, 5-triphosphate receptor in cerebellar Purkinje cells: quantitative immunogold labeling reveals concentration in an ER subcompartment. J. Cell Biol. 111, 615-624.

Schapira, A.H., Cooper, J.M., Dexter, D., Jenner, P., Clark, J.B. & Marsden, C.D. (1989). Mitochondrial complex I deficiency in Parkinson's. Lancet 1, 1269.

Shearman, M.S., Ragan, C.I. & Iversen, L.L. (1994). Inhibition of PC12 cell redox activity is a specific, early indicator of the mechanism of b-amyloid-mediated cell death. Proc. Nat.l Acad. Sci. USA 91, 1470-1474.

Schrieber, S.L. (1991). Chemistry and biology of the immunophilins and their immunosuppressive ligands. Science 253, 283-287.

Schrieber, S.L. (1992). Immunophilin sensitive protein phosphatase action in cell signaling pathways. Cell 70, 365-368.

Schwartz, L.M. & Milligan, C.E. (1996). Cold thoughts of death: the role of ICE proteases in neuronal cell death. Trends Neurosci. 10, 555-563.

Sheehan, J.P., Swerdlow, R.H., Miller, S.W., Davis, R.E., Parks, J.K., Parker, W.D. & Tuttle, J.B. (1997). Calcium homeostasis and reactive oxygen species production in cells transformed by mitochondria from individuals with sporadic Alzheimer's disease. J. Neurosci. 17, 4612-4622.

Shoffner, J.M. (1995). Mitochondrial defects in basal ganglia diseases. Curr. Opin. Neurol. 8, 474-479.

Shoffner, J.M., Lott, M.T., Lezza, A.M.S., Seibel, P., Ballinger, S.W. & Wallace, D.C. (1990). Myoclonic epilepsy and tagged-red fiber disease (MERRF) is associated with a mitochondrial DNA tRNA (Lys) mutation. Cell 61, 931-937.

Shoffner, J.M., Lott, M.T. & Wallace, D.C. (1991). MERRF: a model disease for understanding the principles of mitochondrial genetics. Rev. Neurol. 147, 431-435.

Shoji, Y., Uedono, Y., Ishikura, H., Takeyama, N. & Tanaka, T. (1995). DNA damage induced by tumor necrosis factor alpha in L929 cells is mediated by mitochondrial oxygen radical formation. Immunology 84, 543-548.

Sies, H. (1986). Biochemistry of oxidative stress. 25, 1058-1071. Agnew Chem. Int. England.

Siesjo, B.K., Agardh, C.D. & Bengtsson, F. (1989). Free radicals and brain damage. Cerebrovasc. Brain Metab. Rev. 1, 165-211.

Simpson, P.B. & Russell, J.T. (1998). Role of mitochondrial Ca^{2+} regulation in neuronal and glial cell signaling. Brain Res. Rev. 26, 72-81.

Singer, T.P., Ramsay, R.R. & Ackrell, B.A. (1995). Deficiencies of NADH and succinate dehydrogenases in degenerative diseases and myopathies. Biochim. Biophys. Acta. 1271, 211-219.

Skulachev, V.P. (1996). Why are mitochondria involved in apoptosis. FEBS Lett. 397, 7-10.

Slater, A. F., Stefan, C., Nobel, I., van deb Dobbelsteen, D.J. & Orrenius, S. (1995). Signaling mechanisms and oxidative stress in apoptosis. Toxicol. Lett. 82-83, 149-153.

Susin, S.A., Zamzami, N., Castedo, M., Hirsch, T., Marcheti, P., Macho, A., Daugas, E., Geukens, M. & Kroemer, G. (1996). Bcl-2 inhibits the release of an apoptogenic protease. J. Exp. Med. 184, 1331-1341.

Swerdlow, R.H., Parks, J.K., Miller, S.W., Tuttle, J.B., Trimmer, P.A., Sheehan, J.P., Bennett, J.P., Davis, R.E. & Parker, W.D. Jr. (1996). Origin and functional consequences of the complex I defect in Parkinson's disease. Ann. Neurol. 40, 663-671.

Takei, K., Stukenbrok, H., Metcalf, A., Mignery, G.A., Sudhof, T. C., Volpe, P. & De Camilli, P. (1992). Ca^{2+} stores in Purkinje neurons: endoplasmic reticulum subcompartments demonstrated by the heterogeneous distribution of the $InsP_3$ receptor, Ca^{2+} ATP-ase, and calsequestrin. J. Neurosci. 12, 489-505.

Takeyama, N., Matsuo, N. & Tanaka, T. (1993). Oxidative damage to mitochondria is mediated by the Ca ($^{2+}$) dependent inner membrane permeability transition. Biochem. J. 294, 719-725.

Tan, S., Sagara, Y., Liu,Y., Maher, P. & Schubert, D.T. (1998). The regulation of reactive oxygen species production during programmed cell death. J. Cell. Biol. 141, 1423-32.

Tatuch, Y., Christodoulou, J., Feigenbaum, A., Clarke, J.T., Wherret, J., Smith, C., Rudd, N., Petrova-Benedict, R. & Robinson, B.H. (1992). Heteroplasmic mtDNA mutation. (T---G) at 8993 can cause Leigh disease when the percentage of abnormal mtDNA is high. Am. J. Hum. Genet. 50, 852-858.

Taylor, R.W., Chinnery, P.F., Haldane, F., Morris, A.A., Bindoff, L.A., Wilson, J. & Turnbull, D.M. (1996). MELAS associated with a mutation in the valine transfer RNA gene of mitochondrial DNA. Ann. Neurol. 40, 459-462.

Tenneti, L., D'Emilia, D.M., Troy, C.M. & Lipton, S.A. (1998). Role of caspases in N-methyl-D-aspartate-induced apoptosis in cerebrocortical neurons. J. Neurochem. 71, 946-59.

Thomas, E. & Reed, D.J. (1988). Effects of extracellular calcium omission on isolated hepatocytes: loss of mitochondria membrane potential and protection by inhibitors of uniport calcium transduction. J. Pharmacol. Exp. Ther. 245, 501-507.

Troy, C.M., Stefanis, L., Greene, L.A. & Shelanski, M.L. (1997). Nedd2 is required for apoptosis after trophic factor withdrawal, but not superoxide dismutase (SOD1) downregulation, in sympathetic neurons and PC12 cells. J. Neurosci. 17, 1911-1918.

Turrens, J.F. & Boveris, A. (1980). Generation of superoxide anion by the NADH dehydrogenase of bovine heart mitochondria. Biochem. J. 191, 421-427.

Tzagoloff, A. (1982). Mitochondria. New York, Plenum Press.

Van de Water, B., Zoeteweiji, J.P., Bont, H.J.G.M., Mulder, G.J. & Nagelkerke, J.F. (1994). Role of mitochondrial calcium in the oxidative stress induced dissipation of the mitochondrial membrane potential. J. Biol. Chem. 269, 14546-14552.

Wadia, J.S., Chalmers-Redman, R.M.E., Ju, W.J., Carlile, G.W., Phillips, J.L., Fraser, A.D. & Tatton, W.G. (1998). Mitochondrial membrane potential and nuclear changes in apoptosis caused by serum and nerve growth factor withdrawal: timecourse and modifications by (-) -Deprenyl. J. Neurosci. 18, 932-947.

Wallace, D.C. (1994). Mitochondrial DNA sequence variation in human evolution and disease. Proc. Natl. Acad. Sci. USA 91, 8739-8746.

Waring, P. & Beaver, J. (1996). Cyclosporin A rescues thymocytes from apoptosis induced by very low concentrations of thapsigargin: effects on mitochondrial function. Exp. Cell Res. 227, 264- 276.

Wei, Y.H. (1998). Oxidative stress and mitochondrial DNA mutations in human aging. Proc. Soc. Exp. Biol. Med. 217, 53-63.

Wisniewski, T., Ghiso, J. & Frangione, B. (1991). Peptides homologous to the amyloid protein of Alzheimer's disease containing a glutamine for glutamic acid substitution have accelerated amyloid fibril formation. Biochem. Biophys. Res. Comm. 179, 1247-1254.

Wisniewski, T., Ghiso, J. & Frangione, B. (1997). Biology of Aβ amyloid in Alzheimer's disease. Neurobiol. Dis. 4, 313-328.

Wong, G.H., Elwell, J.H., Oberley, L.W. & Goeddel, D.V. (1989). Manganese superoxide dismutase is essential for cellular resistance to tumor necrosis factor. Cell 58, 923-931.

Woods, N.M., Cuthbertson, K.S.R. & Cobbold, P.H. (1986). Repetitive transient rises in cytoplasmic free calcium in hormone stimulated hepatocytes. Nature 319, 600-602.

Yuan, J. (1997). Transducing signals of life and death. Curr. Opin. Cell. Biol. 9, 247-251.

Zamzami, N., Marchetti, P., Castedo, M., Zanin, C., Vayssiere, J.L., Petit, P.X. & Kroemer, G. (1995a). Reduction of the mitochondrial membrane potential constitutes an early reversible step of programmed. lymphocyte death *in vivo*. J. Exp. Med. 181, 1661-1672.

Zamzami, N., Marchetti, P., Decaudin, D., Macho, A., Hirsch, T., Susin, S.A., Petit, P.X. Zamzami, N., Marchetti, P., Decaudin, D., Macho, A., Hirsch, T., Susin, S.A., Petit, P.X., Mignotte, B. & Kroemer, G. (1995b). Sequential reduction of mitochondrial transmembrane potential and generation of reactive oxygen species in early programmed. cell death. J. Exp. Med. 182, 367-377.

Zorati, M. & Szabo, I. (1995). The mitochondrial permeability transition Biochim. Biophys. Acta. 1241, 139-176.

IMPORTANCE OF PROTEASOME-MEDIATED PROTEOLYSIS AND ALTERATIONS OF THE UBIQUITIN/PROTEASOME PATHWAY IN CANCER CELLS

VITO J. SPATARO & CHRIS J. NORBURY

Servizio Oncologico Ticinese, Ospedale San Giovanni CH-6500 Bellinzona, Switzerland

Imperial Cancer Research Fund Molecular Oncology Laboratory, University of Oxford Institute of Molecular Medicine, Headley Way, The Radcliffe, Oxford OX3 9DS, UK

Table of Contents

Introduction

Two major proteolytic pathways exist in all eukaryotic cells. The first pathway acts on extracellular proteins which have been internalized via the processes of endocytosis or pinocytosis and are degraded in the lysosome. The other pathway accounts for the turn=over of most, if not all the intracellular proteins synthesized by the cell and involves the targeting of a substrate protein by conjugation to a small protein called ubiquitin followed by degradation in a specialized cellular particle called the proteasome. This ubiquitin/proteasome pathway of protein degradation appeared initially as a simple mechanism of destruction for proteins that were abnormal, misfolded or damaged; however, it has emerged in recent years as an important mechanism of regulation in many cellular processes, such as cell cycle progression, regulation of gene transcription

Interorganellar Signaling in Age-Related Disease, Edited by Mark P. Mattson, 239—270

and antigen presentation (Hochstrasser, 1995; King, 1996; Pahl and Baeuerle, 1996). As a mechanism of regulation of gene expression, proteolysis is characterized by the rapidity with which the cellular level of a given protein can be reduced and the irreversibility of the loss of function which results from degradation. In addition, protein degradation can act both as a downregulation and as an upregulation mechanism, depending on whether the degraded substrate is an activator or an inhibitor of a given process.

The main player in the ubiquitin/proteasome pathway is a multiprotein complex called the 26S proteasome which degrades both nuclear and cytosolic proteins in an ATP-dependent manner. All substrate proteins, with few exceptions, need to be targeted for degradation by the addition of multiple molecules of ubiquitin, a 76 amino acid polypeptide, to specific lysine residue in a multistep reaction requiring three classes of enzymes called E1, E2 and E3. E1 enzymes, also called ubiquitin-activating enzymes, activate ubiquitin at its C-terminal glycine residue by creating a thiol ester intermediate. Enzymes of the E2 class are able to transfer ubiquitin from E1 enzymes to the substrate and are therefore called ubiquitin-conjugating enzymes. For most substrates transfer of ubiquitin to the substrate requires the presence of an E3 enzyme, also known as ubiquitin-ligase, which interacts physically with the substrate. The first ubiquitin molecule is bound to the substrate by an isopeptide bond between the C-terminal glycine of ubiquitin and an ε-NH_2 group of a lysine residue of the substrate. For substrates, with many lysine residues, ubiquitination can occur at several positions along the sequence. However, addition of single ubiquitin monomers is usually not sufficient to allow substrate degradation and a polybiquitin chain is required, which is formed in multiple cycles of the E1, E2, E3 multistep reaction by addition of another ubiquitin molecule to the lysine at position 48 of the previously conjugated ubiquitin. Ubiquitination/de-ubiquitination is currently considered to be a highly regulated process, in which de-ubiquitinating enzymes called isopeptidases or ubiquitin C-terminal hydrolases (UCH) are involved. Indeed, the function of UCHs is thought to be important not only in recycling monomers after substrate degradation but also in the recovery of poorly or incorrectly ubiquitinated substrates (Shaeffer and Cohen, 1996). Once a substrate is correctly and sufficiently ubiquitinated, it can be recognized by the 26S proteasome. The mechanisms of substrate recognition are not yet elucidated but seem to involve at least one specialized subunit called S5a which binds ubiquitin conjugates (Deveraux, 1994). The 26S proteasome is a complex of an estimated molecular mass of 2.0 MDa containing more than 50 protein subunits. It consists of three large subcomplexes, namely a 700 kDa core particle called the 20S proteasome and two identical cap structures called 19S proteasome, regulatory complex or PA700 (for proteasome activator of 700 kDa). The structure of the 20S proteasome has been studied in great detail by X-ray crystallography in lower organisms such as the archaebacterium Thermoplasma acidophilum (Löwe et al., 1995) and the budding yeast Saccharomyces cerevisiae (Groll et al., 1997). The 20S particle has a structure which is similar to a hollow cylinder composed of a stack of four rings of seven related subunits and containing a central channel with three cavities. The subunits of the inner rings are called beta subunits and contain proteolytic active sites which are oriented towards the central cavity, whereas the subunits of the outer rings called alpha subunits lack proteolytic activity and are thought to control the interaction with the 19S cap or the access of substrates to the central cavity. Electron microscopy reveals that cells contain both the full 26S proteasome and free 20S and 19S particles and shows that the 26S proteasome is formed by the addition of two 19S particles in opposite orientation at the two ends of the 20S core particle (Peters et al., 1993) (Figure 1). It is not clear whether isolated 20S proteasomes might function in vivo, since this particle has very limited proteolytic activity in vitro. The 19S regulatory complex consists of at least 15 subunits, which can be classified as ATPases and non-ATPases (Dubiel et al., 1995a,b) and is thought to act in recognition, unfolding and translocation of the substrates into the 20S proteasome for proteolysis (Rubin and Finley, 1995). The composition and the function of the regulatory complex is not yet fully characterized. The analysis of 19S regulatory subunits from yeast mutants reveals that the ATPases appear to be stringently organized in the 26S complex, while peripheral non-ATPases might serve as subunits which shuttle substrates to the enzyme (Seeger et al., 1997). The dynamic structure of the 26S proteasome is also characterized by transient associations with components such as isopeptidases, capable of deubiquitinating substrates.

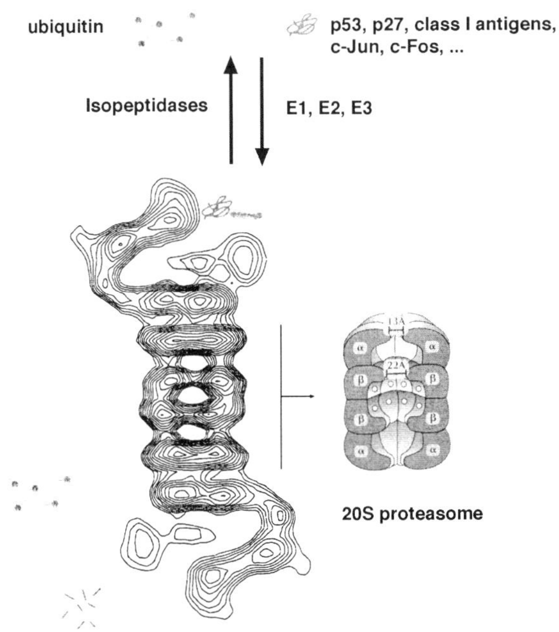

Figure 1. (adapted with permission from Rubin & Finley, Current Biology 1995, Vol. 5, No 8, pages 855-858). The 26S proteasome is a multiprotein complex which acts as a multicatalytic protease degrading proteins that have been targeted by the ubiquitin pathway. Proteins are ubiquitinated in a cascade reaction involving three classes of ubiquitinating enzymes called E1, E2 and E3 and can be deubiquitinated by isopeptidases. The 20S proteasome consists of a stack of 4 rings of 7 subunits. The inner rings made of β-subunits display the catalytic sites on the inner surface. At each end the 20S proteasome can be capped by a regulatory complex called 19S or PA700 which contains ATPases and is probably involved in recognition, unfolding and translocation of the substrate into the 20S proteasome (Rubin and Finley, 1995).

Proteasomes are localized both in the nucleus and in the cytoplasm of eukaryotic cells (Palmer et al., 1994, 1996). They appear to be either free in the cytoplasm or bound to the endoplasmic reticulum (Yang et al., 1995), where they deliver antigenic peptides to specialized carrier proteins called TAP1 and TAP2 (transporter associated with antigen processing). The dynamics of their distribution in the cell have been studied recently using a fusion protein of the proteasome subunit LMP2 and the green fluorescent protein (GFP) (Reits et al., 1997). This work has shown that proteasomes diffuse rapidly within both the nuclear and the cytoplasmic compartments. In addition, during interphase proteasomes are transported slowly unidirectionally from the cytoplasm to the nucleus and they can enter or leave the nucleus by rapid diffusion during mitosis when the nuclear envelope has disintegrated. These data suggest that proteasomes may perform quality control of intracellular proteins by continuous collision and degrade those proteins that are properly tagged or misfolded. Moreover, the transport of the proteasomes across the nuclear membrane could have an important functional relevance given that the degradation of certain substrates and their nucleocytoplasmic transport appear to be coupled. For instance, nuclear import of NF-κB is coupled to its activation by the ubiquitin/proteasome mediated degradation of its inhibitor IκBα (Palombella et al., 1994). In a similar fashion, the ubiquitin/proteasome mediated degradation of p53 has been shown recently to be coupled with the nuclear export of MDM2 (Roth et al., 1998).

Selectivity of the ubiquitin/proteasome pathway is probably achieved by both ubiquitin ligases (E3s) and by regulatory subunits forming the 19S complex (Dubiel et al., 1995b; Seeger et al., 1997). The 19S regulator subunit S5a is responsible for binding ubiquitin conjugates (Deveraux et al., 1994). Other subunits of the 19S regulator might exhibit binding sites for ubiquitin-independent proteolysis, as indicated by direct degradation of IκBα (Krappmann et al., 1996) and c-JUN (Jariel Encontre et al., 1995) by the 26S proteasome. In processes such as cell cycle regulation and signal transduction, other cellular complexes have been identified which co-operate with the ubiquitin/proteasome pathway. During late mitosis and G1, a complex named cyclosome or anaphase promoting complex (King et al., 1995) regulates the degradation of mitotic cyclins by the 26S proteasome. Likewise, a multiprotein complex has been recently identified which appears to be involved in signal transduction and has kinase activity against IκBα and c-JUN, two substrates of the 26S proteasome (Seeger et al., 1998). Interestingly, this complex named signalosome, which consists of at least 8 subunits, contains three subunits which exhibit sequence similarities to regulatory components of the 26S proteasome; immunofluorescence staining reveals that this new complex shows a subcellular distribution similar to that of the 26S proteasome. The composition and subunit identities of the signalosome are almost identical to the COP9 complex, a multiprotein complex involved in signal transduction in the plant Arabidopsis thaliana. Another mammalian multiprotein complex, the translation initiation factor 3 (eIF3), which binds to the 40S ribosome and promotes the binding of methyonil-tRNA and mRNA, consists of ten subunits; two of them, eIF3-p47 and eIF3-p40 are clearly related by sequence respectively to the subunits S12/p40 and S13/POH1 of the 19S proteasome (Asano et al., 1997).

Table 1 Subunits of the human JAB1-containing complex ("signalosome"), their homologues in other species and similarities with subunits of the 26S proteasome

Subunit	Origin	Comments	Approximative molecular mass (kDa)
Sgn1	Human	MAPKK activation loop motif	55
Gps1	Human		
FUS6	*A. thaliana*		
COP11	*A. thaliana*		
Sgn2	Human		50
Trip15	*Drosophila*		
Sgn3	Human	Homology to 26S proteasome S3	45
LC15	*L. chilense*		
Sgn4	Human		42
Sgn5	Human	Homology to yeast and mouse Pad1,	38
JAB1	Human	human POH1 corresponding to 26S proteasome S13	
Sgn6	Human	Homology to 26S proteasome S12, MAPKK activation loop motif	36
Sgn7	Human	MAKK activation loop motif	32
Sgn8	Human		20
COP9	Human		
COP9	*A. thaliana*		

Like many regulatory mechanisms with an involvement in fundamental biochemical processes, the ubiquitin-proteasome pathway is a potential target for cancer related alterations of function that could confer a selective advantage in cell growth and survival. Alterations of proteasome function were first described in events such as cellular transformation by oncogenic viruses (Ciechanover et al., 1994; Scheffner et al., 1993, 1990) and have been found to contribute to tumor escape from the immune system (Restifo et al., 1993; Rotem Yehudar et al., 1996; Seliger et al., 1996; Sibille et al., 1995). More recently, alterations of proteasome activity have been reported to confer a phenotype of clinical aggressiveness associated with poor prognosis in common cancer types such as colon, lung, prostate and breast cancer (Catzavelos et al., 1997; Cote et al., 1998; Esposito et al., 1997; Loda et al., 1997; Porter et al., 1997; Singh et al., 1998; Yang et al., 1998). Furthermore, several lines of evidence suggest that the proteasome is involved in modulating the response to several stresses, including exposure to anticancer agents in both unicellular eukaryotes used as model organisms (Gordon et al., 1993; Gordon et al., 1996; Wilkinson et al., 1997) and in human cells (Beg and Baltimore, 1996; Delic et al., 1998; Spataro et al., 1997; Van Antwerp et al., 1996; Wang et al., 1996). As a result, the ubiquitin-proteasome pathway has become a promising new target for anticancer therapy, which is currently being actively investigated at the preclinical level.

Table 2. Summary of settings in which the ubiquitin/proteasome pathway plays a role in cancer (see text for references)

Substrate	Proteolysis dysregulation	Functional effect	Biological effect	Type of cancer/comment
p53	Increased	p53 inactivation	Transformation by "high-risk" HPV, mediated by E6 oncoprotein	Anogenital cancers
			Tumor initiation or progression mediated by amplification of the MDM2 oncogene	Sarcomas
			Tumor initiation or progression mediated by loss of the p19/ARF protein	Suggested in high-grade neuroendocrine lung cancer
p27	Increased	p27 inactivation	Tumor progression	Unfavorable prognosis in cancers of the breast, lung, prostate, colon, esophagus
Cyclin D1, E and B	Decreased?	Cyclin D1, E or B	Tumor progression	In cell lines and surgical specimens of various tumor types
MHC-I restricted antigens	Decreased	Defective antigen presentation	Escape from immune surveillance	In tumor cells of various types, induced by oncogenic transformation, preliminary *in vivo* evidence
NF-κB inbitor, I-κB	Increased	Increased NF-κB activation	Resistance to TNF-α killing	Overcome by proteasome inhibitors
Transcription factors of the AP-1 family	Decreased	Increased AP-1 activity	Multidrug resistance (MDR)	Evidence in yeast, highly conserved pathway in human, role in tumor MDR?

Regulation of the Tumor Suppressor Protein p53

Tumors frequently inactivate p53 tumor suppressor protein by gene mutation or deletion. The p53 protein transcriptionally regulates numerous genes as part of signal transduction pathways initiated by cellular stresses (Harris, 1996; Levine, 1997). p53 is an unstable nuclear protein with a half-life of about 20-35 minutes in normal cells. Following cellular stress or DNA damage p53 is stabilized, leading to growth arrest or apoptosis. The rise in p53 protein level is detectable almost immediately after DNA damage and the absolute level of p53 protein and duration of the response depend on the nature of the damage (reviewed in Cox & Lane, 1995). This accumulation of p53 is thought to occur mainly via the down-regulation of its degradation by the ubiquitin/proteasome pathway (Harris, 1996; Maki et al., 1996). This was first supported by experiments performed using a cell line containing a thermolabile E1 ubiquitin-activating enzyme, in which p53 accumulates at the non-permissive temperature, whereas this accumulation was prevented by introduction of the wild-type E1 gene (Chowdary et al., 1994). Recent research has provided more insights into the regulation of p53 degradation, suggesting that increased p53 degradation might be a frequent mechanism of p53 inactivation in tumors. It has been known since 1992 that the product of the proto-oncogene MDM2, which is amplified in 30-40% of sarcomas (Oliner et al., 1992) was able to bind and inactivate the transcriptional activity of p53, resulting in abrogation of p53's antiproliferative and apoptotic effects (Oliner et al., 1993). In addition, deletion of the MDM2 gene in mice results in early embryonic lethality, which can be rescued by the simultaneous deletion of p53, indicating that p53 is the major target, if not the only target, of MDM2 in development (Jones et al., 1995; Montes de Oca Luna et al., 1995). In 1997, it was shown that MDM2 is in fact a potent inducer of p53 degradation by the proteasome and that MDM2-induced p53 degradation plays a critical role in regulating the level of the p53 protein (Haupt et al., 1997; Kubbutat et al., 1997). Regulation of p53 activity appears to be primarily at the level of protein abundance and MDM2 appears to be the main partner of an autoregulatory feedback loop, in which p53 induces MDM2 by transactivation of the MDM2 gene and the MDM2 protein inactivates p53 by promoting p53 degradation by the proteasome. This loop allows a rapid rise of p53 levels by arresting p53 degradation upon various cellular stresses and a rapid drop of p53 levels upon removal of the stress.

More recently it has been shown that the MDM2 protein is also a target of proteasome-mediated degradation and that a third protein named ARF can bind MDM2, promote MDM2 degradation, and neutralize MDM2's inhibition of p53 (Pomerantz et al., 1998; Zhang et al., 1998). The ARF protein is the product of the INK4a-ARF locus which encodes the two unrelated proteins p16 and human $p19^{ARF}$. P16 binds and inhibits the activity of the cyclin-dependent kinases CDK4 and CDK6, and human $p19^{ARF}$ arrests the cycle in G1-S and G2-M in a p53-dependent manner. Inactivation of $p19^{ARF}$ through deletion of the INK4a-ARF locus is a frequent event in tumors. Other mechanisms of inactivation of $p19^{ARF}$ are suggested by recent work, which has shown that lung tumors with inactivated p16 may have expression of the $p19^{ARF}$ protein and vice versa (Gazzeri et al., 1998). It has been found that sarcomas with MDM2 amplification retain the wild-type p53 gene, suggesting that MDM2 amplification is

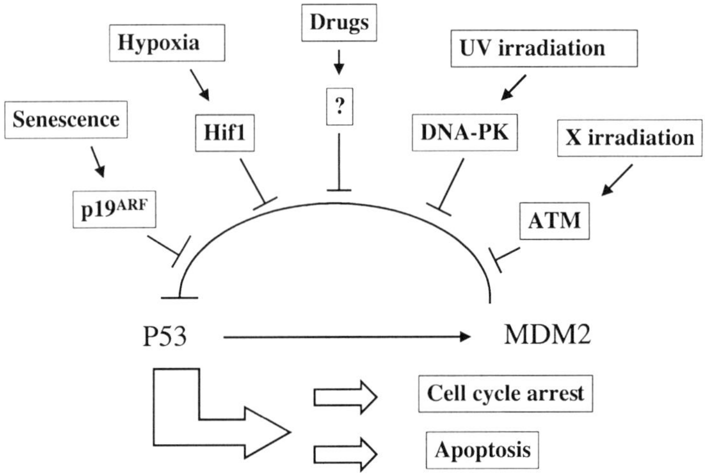

Figure 2. The tumour suppressor protein induces MDM2 synthesis by transactivation of the MDM2 gene and the MDM2 protein inactivates p53 by promoting its rapid proteasome-mediated degradation, establishing an autoregulatory loop. The model represented in the figure suggests that various cellular stresses result in a rapid rise of the p53 protein level by blocking the negative feed-back loop through distinct signaling mediators (Hif1 = Hypoxia inducible factor 1; DNA-PK= DNA-dependent protein kinase; ATM = protein mutated in Ataxia Telangiectasia).

biologically equivalent to p53 inactivation by mutation. Likewise, it is likely that $p19^{ARF}$ inactivation by mutation or deletion in tumors serves to inactivate p53 function in tumors that retain wild-type p53 and do not have MDM2 amplification (Zhang, et al., 1998). As discussed in more detail in the section on drug resistance below, the pathway by which MDM2 promotes the degradation of p53 seems to require nucleo-cytoplasmic transport of the MDM2 protein via the pathway used by the human immunodeficiency virus rev protein (Roth, et al., 1998). The product of the MDM2 gene localizes predominantly in the nucleus but shuttles constantly between the nucleus and the cytoplasm. The shuttling is abolished by mutation of a nuclear export signal (NES) at positions 184-192 and by competitive inhibitors of rev export. Very interestingly, inhibition of MDM2 export stabilizes p53 by decreasing its proteasome-mediated degradation. However, the biochemical mechanism underlying MDM2-promoted p53 degradation, as well as ARF-promoted MDM2 degradation, remains to be fully elucidated.

Proteasome-mediated degradation of p53 is responsible for p53 inactivation not only in tumors with amplification of the MDM2 oncogene or inactivation of the INK4a-ARF locus, but also in cancers due to infection by the oncogenic human papilloma virus (HPV). HPV is involved in the etiology of most cancers of the uterine cervix and most likely of many human carcinomas of the anogenital area. Its oncogenicity is mediated by upregulation of p53 degradation by the ubiquitin/proteasome pathway. The E6 oncoprotein encoded by high-risk HPV (e.g. HPV-16, -18, -5 and -8) binds to p53 and promotes its degradation by the proteasome (Scheffner, et al., 1990), a property which is critical for immortalization of human cells by HPV. In contrast, low-risk HPV (e.g. HPV-6 and -11) encode an E6 protein which does not bind to p53 and does not promote its degradation.

At the molecular level, cancer research is providing an increasingly complex picture of how the degradation of a given target protein by the ubiquitin/proteasome pathway is regulated. The formation of the E6-p53 complex requires a cellular E6-binding protein called E6-AP (E6-associated protein) (Scheffner, et al., 1993), which forms thiol ester complexes with ubiquitin in the presence of enzymes of the E2 category such as UBC4 (Rolfe et al., 1995) or E2-F1 (Ciechanover, et al., 1994). E6-AP acts as an E3 enzyme which ubiquitinates p53, leading to its rapid degradation by the 26S proteasome. It could be hypothesized that E6-AP acts as an E3-like ligase for p53 under physiological conditions. However, in the absence of E6, E6-AP does not bind to p53 (Huibregtse et al., 1991; Scheffner, et al., 1993). In addition, the inhibitory activity of MDM2 on p53 is maintained in vitro in the absence of E6 and E6-AP (Haupt, et al., 1997; Kubbutat, et al., 1997), suggesting that MDM2 utilizes a different mechanism to target p53 degradation by the proteasome. In conclusion, available data suggest that p53 degradation can be regulated independently by at least two cellular proteins, MDM2 and E6-AP, which can both act as E3 ubiquitin ligases. The current view is that the MDM2-p53 interaction seems to be the final level of regulation of p53 stability by multiple signals mediated by various pathways (see Figure 2), whereas the role of E6-AP in the absence of a HPV infection is not known.

Alteration of p27 Degradation as a Prognostic Factor

Progression through the cell cycle is promoted by oscillations in the activity of cyclin-dependent kinases (CDK) and proteolysis by the ubiquitin-proteasome pathway regulates CDK activity by degrading CDK activators and inhibitors. In addition, proteolysis by the proteasome triggers the transition from metaphase to anaphase during mitosis (reviewed in King, 1996). One of the best-characterized proteasome substrates among all CDK activators and inhibitors is p27. The activity of the p27 CDK inhibitor is upregulated in response to extracellular signals such as cell-cell contact and TGF-beta (Polyak, 1994; Slingerland, 1994). Loss of contact inhibition is a hallmark of cellular transformation, which may imply an alteration of function of p27 during oncogenesis, even though p27 mutations in human tumors are extremely rare (Ferrando, 1996; Hunter and Pines, 1994; Morosetti et al., 1995). Recent research provided an elegant explanation to this paradox, when it was shown that unlike p21, which is also a member of the family of cip/kip CDK inhibitors acting in G1 and appears to be regulated principally at the transcriptional level, p27 is critically regulated posttranslationally by proteolysis by the ubiquitin-proteasome pathway (Hengst and Reed, 1996; Pagano et al., 1995). Alteration of p27 levels have been observed subsequently in breast, lung, colon, esophageal and prostate cancer, and decreased levels of p27 were associated with worse prognosis (Catzavelos, et al., 1997; Cote, et al., 1998; Esposito, et al., 1997; Loda, et al., 1997; Porter, et al., 1997; Singh, et al., 1998; Yang, et al., 1998). A comparison of immunohistochemical analysis and in situ hybridization was performed in breast and colon cancer (Catzavelos, et al., 1997; Loda, et al., 1997) and showed a discordance between p27 mRNA and protein levels, suggesting that in tumors p27 levels is also regulated posttranslationally. Moreover, in two of the above studies on colon and lung cancer it was clearly shown that increased proteasome-dependent degradation was responsible for low p27 levels in tumor samples (Esposito, et al., 1997; Loda, et al., 1997). Total cellular extracts from frozen tumor samples were tested for p27 proteasome-mediated degradation using recombinant p27 as a substrate, and a very good correlation was found between low levels of p27 and increased proteasome activity. Degradation was abolished by proteasome depletion and resumed following proteasome readdition. Downregulation of p27 by the proteasome was found in tumors regardless of clinical stage. In breast cancer Catzavelos et al. showed that increased p27 proteolysis can be an early event in tumorigenesis as suggested by analysis of high-grade ductal carcinoma in situ (DCIS) or alternatively can occur upon progression as shown by reduced p27 levels in axillary lymph node metastasis as compared to primary tumors assessed simultaneously. Interestingly, one study in Barrett's associated adenocarcinoma of the oesophagus has found that the expression of p27 messenger RNA and protein parallel one another in premalignant lesions (metaplastic and dysplastic epithelium), whereas p27 protein is often inactivated at a post-transcriptional level in invasive adenocarcinomas derived from these lesions, which suggests again that p27 inactivation can be an early event in tumorigenesis (Singh, et al., 1998). In breast cancer, low p27 immunohistochemical staining correlated with lack of expression of estrogen receptors (Catzavelos, et al., 1997; Porter, et al., 1997) and in prostate cancer, decreased p27 staining correlated with higher tumor grade and with seminal vesicle involvement (Tsihlias et al., 1998). However, p27 staining maintained its prognostic value on multivariate analysis in many of the above cited studies, where such statistical analysis was reported (Catzavelos, et al., 1997; Cote, et al., 1998; Loda, et al., 1997; Porter, et al., 1997; Tsihlias, et al., 1998). Even taking into account the caveats associated with the methodology of retrospective studies on prognostic factors in cancer, these recent studies conclude that p27 protein level (and its proteasome-dependent degradation, which was shown to be inversely related) are powerful independent prognostic factors of survival in most common solid tumor types and show clearly that deregulation of gene products involved in clinical tumor progression can occur via alterations of ubiquitin/proteasome proteolysis. Furthermore, they show that this is not a rare event, given that the unfavorable phenotype of decreased p27 levels (defined as immunostaining in $< 50\%$ of the cells) involves the majority of the studied population for tumors of the colon, breast and esophagus (Catzavelos, et al., 1997; Loda, et al., 1997; Porter, et al., 1997; Singh, et al., 1998) and about half of the patients in the case of prostate adenocarcinoma (Cote, et al., 1998). Thus, the frequency of the phenotype of decreased p27 and its distribution, which seems to be independent of most other prognostic factors, make p27 levels a very promising new prognostic factor to be evaluated further in confirmatory studies. In tumor types such as colon and breast cancer, assessment of p27 levels by immunohistochemistry could potentially prove useful in the selection for adjuvant treatments of patients without other unfavorable prognostic factors.

At the molecular level, p27 down-regulation does not seem to be due to a general activation of the ubiquitin/proteasome pathway towards several substrates, but rather to a specific activation of the degradation of the p27 protein. Indeed, Loda et al. have shown that p27 degradation activity is not correlated with degradation by the proteasome of other substrates such as p21 and cyclin A, which underscores that the substrate specificity of the ubiquitin/proteasome pathway is highly regulated (Hochstrasser, 1995). Identification of the element(s) responsible for targetting p27 to the ubiquitin-proteasome pathway would of course be extremely important for unravelling this novel pathway associated with tumor progression and identifying new important therapeutic targets in anticancer treatment. Interestingly, some data suggest that inactivation of the p27 protein can occur in adenocarcinoma of the esophagus by both loss and altered subcellular localization of p27. In one study, 83% (45/54) of invasive tumors had decreased p27 nuclear staining and a significant subset of those tumors (27%, 12/45) also showed cytoplasmic staining (Singh, et al., 1998). This observation, together with other data discussed elsewhere in this chapter, support a model where degradation of a given substrate by the ubiquitin/proteasome pathway is coupled to its nucleo-cytoplasmic export.

Apart from p27, many other proteins involved in cell cycle regulation are proteasome substrates and are regulated mainly, if not exclusively, at a post-translational level. Like p27, other elements of the cell cycle machinery, which are substrates of ubiquitin/proteasome degradation, are potential targets for deregulation in tumors. Rapid proteasome-mediated degradation of cyclin B is required for exit from mitosis in the normal cell cycle (Glotzer et al., 1991), and recent evidence shows that continuing proteolysis maintains low levels of cyclin B until the onset of S phase (Amon et al., 1994; Brandeis, 1996). Because cyclin B has been found to be overexpressed in breast cancer cell lines (Keyomarsi and Pardee, 1993), it is possible, but it has not yet been tested, that decreased proteolysis by the proteasome is involved in its overexpression. The same could apply for the cyclins of the D and E type. Cyclin E is also overexpressed in breast cancer cell lines and in surgical specimens of breast tumors (Keyomarsi et al., 1995; Keyomarsi et al., 1994) and cyclin D1 is frequently overexpressed in many common tumor types (Betticher, 1996). Recent evidence suggests that cyclins D1 and E are substrates of the ubiquitin/proteasome pathway (Clurman et al., 1996; Diehl et al., 1997) and decreases in their degradation could contribute to the overexpression of these cyclins in tumors.

Antigen Presentation and Tumor Escape from the Immune System

The 26S proteasome is responsible for the processing of MHC-restricted class I antigens. Peptides derived from endogenously expressed cytoplasmic proteins are carried by MHC class I molecules from the endoplasmic reticulum to the surface for recognition by cytotoxic T lymphocytes. The hypothesis that the proteasome could be the proteolytic system that degrades cytosolic proteins was put forward when it was found that the genes encoding subunits LMP-2 and LMP-7 of the proteasome complex were included in the MHC gene cluster (see for example Beck et al., 1992). Subsequently, experiments performed in a mutant cell with a thermolabile E1 ubiquitinating enzyme (Michalek et al., 1993) and with proteasome inhibitors (Rock et al., 1994) demonstrated that the proteasome activity is required for class I-restricted antigen presentation. This was finally confirmed by the analysis of mice lacking LMP-7 which have decreased surface expression of MHC class I molecules and present antigens inefficiently (Fehling et al., 1994). In recent years, it has been shown that induction by interferon-γ results in the synthesis of the so-called immuno-proteasome, where three of the 28 subunits composing the 20S catalytic core, namely subunits X, Y and Z are changed respectively with the alternative subunits LMP-2, LMP-7 and LMP-10 (Belich et al., 1994; Fruh et al., 1994; Groettrup et al., 1996; Hisamatsu et al., 1996; Nandi et al., 1996). These substitutions result in an increase of the quantity and alteration in the quality of generated peptides due to an enhancement of peptidase activity (Gaczynska et al., 1996; Kuckelkorn et al., 1995), and eventually in a more efficient antigen presentation. Interferon-γ also induces the binding to the 20S catalytic core of the proteasome of a complex called 11S regulator or PA28, which may increase further the spectrum of peptides generated (Groettrup et al., 1995). The 11S/PA28 is present only in higher eukaryotes and its role seems to be restricted to antigen presentation. There is strong evidence that MHC class I-restricted peptide presentation is modified in tumors and may contribute to escape from immune surveillance. Alterations of ubiquitin/proteasome degradation have been reported among other alterations in this pathway. Three different small cell lung carcinoma lines with low to undetectable levels of mRNA for LMP-2 and LMP-7 and functional deficiencies in antigen presentation have been described (Restifo, et al., 1993). More recently, cell lines of other tumor types, such as Wilms' tumors, neuroblastomas and prostate carcinoma, have been found to show down-regulation of multiple components of the class I MHC antigen-processing pathway including LMP-2, LMP-7, LMP-10 and the peptide transporters TAP-1, TAP-2 (Johnsen et al., 1998). Interestingly, the expression of the subunits of the 11S proteasome activator PA28α and PA28β was not altered in any of these lines, in contrast with the subunits of the 20S proteasome. In addition, simultaneous down-regulation of the LMP-10 gene located on human chromosome 16 and of the LMP-2 and LMP-7 genes located in the MHC cluster on chromosome 6 suggests that this combined deficiency is not due to a deletion but is probably mediated by a common regulatory mechanism. Treatment with interferon-γ can reverse these deficiencies in expression detected in these tumor cell lines. This corresponds to a functional activation of the antigen-processing pathway, as it has been shown in the mouse T-cell lymphoma line SP-3 (Sibille et al., 1995). This murine line displays underexpression of LMP-2 and is defective for antigen presentation, whereas LMP-2 expression and antigen presentation to cytotoxic T-lymphocytes are restored upon expression of interferon-γ by transfection. Similar studies on tumor samples are few in numbers. An analysis of expression of both LMP-2 and LMP-7 proteasome subunits together with other elements of the antigen presentation machinery has been carried out on a primary renal cancer and lymph node metastasis of the same patient and compared with normal kidney. Deficiencies, at all levels, including the expression of LMP-2 and LMP-7 proteasome subunits, were associated with transformation and progression. Interferon-α and especially interferon-γ could partially suppress these defects (Seliger et al., 1996). The potential importance of subunits LMP-2 and LMP-7 for MHC class I-restricted antigen presentation is also underscored by the fact that they are specifically downregulated following viral transformation in vitro by oncogenic viruses (Rotem Yehudar et al., 1996). Given the frequency of activation of proto-oncogenes in human tumors, it is likely that down-regulation of the pathway of MHC class I-restricted antigen-presentation is altered in many tumors. Indeed, down-regulation of the peptide transporters TAP-1 and TAP-2 and the proteasome subunits LMP-2 and LMP-7 has been recently observed in RAS-transformed murine fibroblasts in comparison to parental NIH3T3 cells (Seliger et al., 1998). Activating mutations of the proto-oncogene RAS are detected in at least 30% of all human tumors and are particularly frequent in cancer types, which are highly resistant to anticancer drugs such as adenocarcinoma of the pancreas (90%), colon carcinoma (50%) and bronchogenic carcinoma (30%).

Regulation of Transcription Factors by Proteolysis and Its Relationship with the Response to Cytotoxic Agents

An important body of evidence exists which shows that the proteasome participates in events that control gene transcription. Several transcription factors, including nuclear factor-kappa B (NF-κB), p53 (see above), c-JUN, c-FOS and others have been shown to be regulated by proteolysis, resulting in either activation or inactivation of gene expression (for a review see Pahl and Baeuerle, 1996). In response to various signals, NF-κB is involved in the activation of genes encoding products such as cytokines, chemokines, growth factors, cell adhesion molecules and surface receptors and therefore has a central role in mediating the immune/inflammatory responses. NF-κB has also been reported to be activated by cytotoxic agents such as TNF-α, doxorubicin, etoposide and ionizing radiation and by oxidative stress (Wang et al., 1996). The activation of NF-κB requires two steps of proteasome-dependent proteolysis. Active NF-κB is a nuclear heterodimer consisting of two subunits called p50 and p65. The biogenesis of the subunit p50 requires the degradation by the ubiquitin-proteasome system of the precursor p105, and the translocation of the active dimer into the nucleus (Palombella et al., 1994). This requires the degradation of the inhibitory factor IkB in the cytoplasm. It has been recently shown that NF-κB has an antiapoptotic role in response to some cytotoxic agents (Barger et al., 1995; Beg and Baltimore, 1996; Mattson et al., 1997; Van Antwerp et al., 1996; Wang, et al., 1996). In some experiments, TNF-α was more toxic for immortalized embryonic cells of NF-κB knockout mice than for controls (Beg and Baltimore, 1996) and in other experiments expression of the super-repressor IkB-α (inhibiting NF-κB activation) increased moderately the sensitivity to TNF-α, daunorubicin and ionizing radiation (Wang et al., 1996). Consistent with this, the proteasome inhibitor MG132 (preventing NF-κB activation) strongly enhanced in a dose dependent fashion the killing of HT1080V cells by TNF-α. The strategy of inhibiting NF-κB activation with proteasome inhibitors in order to sensitize tumor cells to TNF-α induced apoptosis has also been applied recently with success to chemo- and radioresistant human chronic lymphocytic leukemia cells (Delic et al., 1998). It is hoped that this strategy will be transposable in the clinic as it could constitute an important means of overcoming tumor drug resistance.

The AP-1 (activating protein-1) family of transcription factors consists of homodimers and heterodimers of proteins of the JUN, FOS or activating transcription factor (ATF) subfamilies (Karin et al., 1997). The best characterized members of these families, c-JUN and c-FOS, have been shown to be substrates of the ubiquitin/proteasome system (Hermida-Matsumoto et al., 1996; Jariel-Encontre et al., 1995; Musti et al., 1997; Treier et al., 1994; Tsurumi et al., 1995b). The regulation of proteasome degradation has been particularly well studied in the case of c-JUN. The degradation of c-JUN requires a segment of 27 amino acids called the delta domain, which is necessary for both ubiquitination and degradation. The delta region, and hence this mechanism of down-regulation, is lacking in v-JUN, the transforming retroviral counterpart of c-JUN. It is likely that this increased stability contributes to the oncogenicity of v-JUN (Treier et al., 1994). More recently, it has been shown that the ubiquitin/proteasome pathway is involved in the regulation of c-JUN activity upon signaling through MAP kinases (Musti et al., 1997). Other data suggest that c-JUN activity is, at least in part, regulated at the level of degradation, with the presense of c-FOS and the dimerization state influencing the ubiquitination state and the susceptibility to degradation (Hermida-Matsumoto et al., 1996; Tsurumi et al., 1995b). AP-1 transcription factors appear to be important in the response of mammalian cells to oxidative stress (Pinkus et al., 1996; Schreiber et al., 1995) and in the induction of several genes encoding detoxification enzymes such as the isozymes α, π and Ya of glutathione-S-transferase and γ-glutamyl-cysteine synthetase. Upregulation of AP-1 activity has been associated with drug resistance in several instances such as; in a multidrug resistant derivative of MCF7 cells obtained after vincristine selection (Moffat et al., 1994), in etoposide resistant human leukemia cell lines (Ritke et al., 1994), and in cisplatin-resistant ovarian cancer lines (Yao et al., 1995). However, the regulatory mechanisms of AP-1 activity are not known. Because several lines of evidence currently suggest that the proteasome could be involved in the regulation of AP-1 activity, this provides a working model for the data discussed in the following section that link the proteasome and multidrug resistance (Spataro et al., 1997).

Drug Resistance

We recently found that the product of a human gene identified as a novel potential multidrug resistance gene was in fact an as yet unidentified subunit of the regulatory complex (19S) of the 26S proteasome, thereby indicating a link between ubiquitin-dependent proteolysis and resistance to anticancer drugs. The gene was originally identified in the fission yeast Schizosaccharomyces pombe and named pad1 (for pap1 dependent 1), and encodes a highly conserved protein. The human homologue was named POH1 (for Pad One Homologue 1) and corresponds to subunit 13 (S13) in the proposed nomenclature of human proteasome subunits (Dubiel, 1995); the recently characterized homologue in budding yeast has been named Rpn11/Mpr1 (Glickman et al., 1998; Rinaldi et al., 1995). Very interestingly, POH1/Pad1/Rpn11 all contain a putative cysteine active site which is typical of isopeptidases and are in the same range of molecular weight of an intrinsic isopeptidase found in the proteasome (Lam et al., 1997). Overexpression of the fission yeast Pad1 protein confers resistance to compounds unrelated by structure or mechanism of action such as staurosporine, caffeine, thiabendazole and leptomycin B, but only in the presence of the pap1 gene encoding a transcription factor which is related to mammalian AP-1 transcription factor (Shimanuki et al., 1995). The human Pad1 homologue, POH1, is 68% identical in the amino acid sequence and fully substitutes for Pad1 in yeast. When overexpressed in transfection experiments, POH1 also confers multidrug resistance in mammalian cells to antiproliferative agents (7-hydroxy-staurosporine and paclitaxel) and DNA-damaging agents (UV radiation) (Spataro et al., 1997). The hypothesis that POH1 could be a proteasome subunit was based on its sequence similarity to the subunit S12/p40 of the 26S proteasome (Tsurumi et al., 1995a) and on the pattern of tissue expression, which was very similar to that previouly described for other subunits of the 26S proteasome (Tsurumi, et al., 1995a). Co-purification of POH1 in immunoprecipitation experiments with antibodies raised against a subunit of the proteasome and with full 26S proteasomes obtained by biochemical fractionation confirmed that POH1 was located within the regulatory complex of the proteasome. POH1 is also related by sequence, though to a much lower degree, to JAB1 (Jun-activation-domain-binding protein 1), which has been characterized as a coactivator of AP-1 transcription factor (Claret et al., 1996) (Figure 3). Like POH1, JAB1 displays sequence similarity, though to a much lower degree, with fission yeast Pad1 (57% amino acid identity in the N-terminal region), and confers a drug-resistance phenotype when highly overexpressed in fission yeast, from the powerful nmt1 promoter. In addition, in the presence of fission yeast pap1$^+$, JAB1 induces pap1-dependent transcription, like pad1$^+$ and POH1. However, unlike POH1, JAB1 can not complement a pad1-deficient strain, suggesting that POH1 and not JAB1 is the genuine pad1 homologue (Claret et al., 1996). JAB1 was identified in a two-hybrid screen using the c-JUN activation domain as a bait; it interacts physically with two members of the family of JUN proteins, c-JUN and JUN D, but not with JUN B and v-JUN, and potentiates their transactivation activity. JAB1 has been recently found to be a subunit of the signalosome, a 450kDa multiprotein complex of at least 8 subunits with putative role in signal transduction (Seeger, et al., 1998). The JAB1-containing particle has kinase activity towards several substrates including c-JUN, which is phoshorylated in the activation domain on Ser63 and Ser73. Interestingly, other subunits of this complex have sequence similarities with proteasome subunits and immunofluorescence staining shows a similar subcellular distribution to that of the 26S proteasome.

Because POH1 is associated with the 26S proteasome and with the support of the above data on JAB1, one hypothesis, which is being tested, is that POH1 is involved in the regulated degradation of JUN proteins by the proteasome. Because both POH1 and JAB1 positively regulate pap1$^+$ in yeast, POH1 could also be involved in the activation of AP-1 transcription factors by influencing either the relative abundance of the various JUN proteins or their stability. Interestingly, JAB1-JUN interaction in the two-hybrid system requires the presence of the amino acids 31-57 of c-JUN, the region corresponding to the δ region which is also required for ubiquitin/proteasome-mediated degradation of JUN proteins (Treier, et al., 1994). These data support a model in which POH1 also interacts physically with this region, and binds JUN proteins within the proteasome. The existence of a JUN-binding site within the regulatory complex (19S) of the proteasome could also explain the observation that degradation of c-JUN by

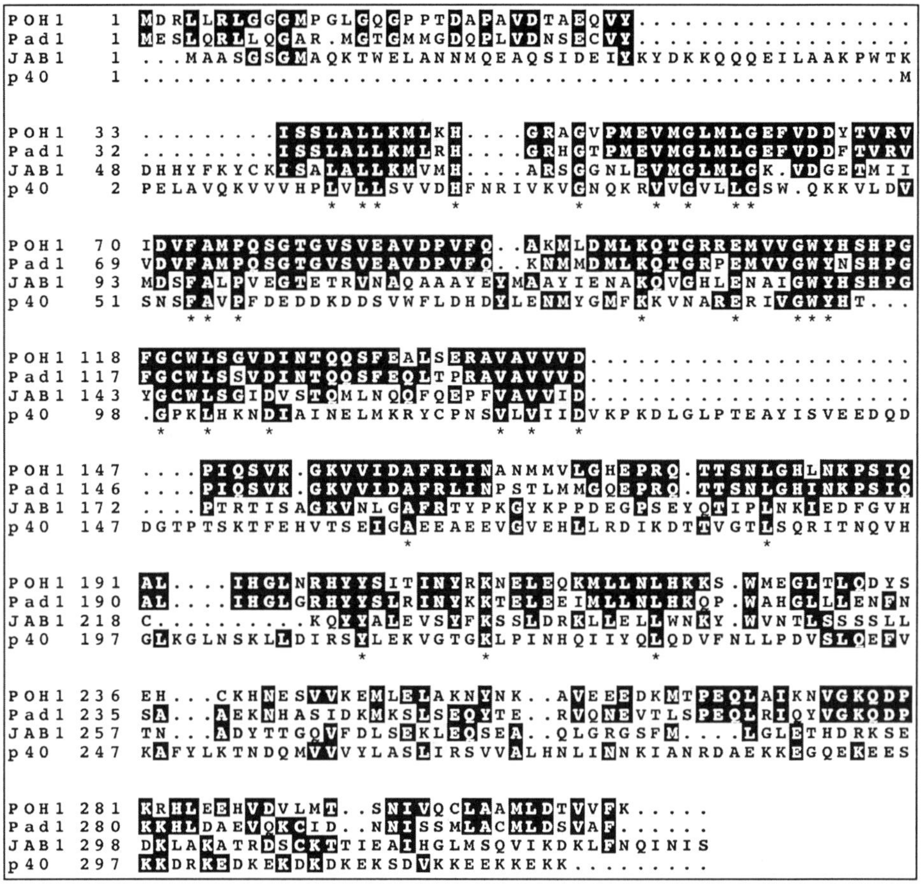

Figure 3. Alignment of the POH1 protein sequence with those of Schizosaccharomyces pombe Pad1, human JAB1 and the human proteasome subunit S12/p40. Residues present at identical positions in all four proteins are highlighted in black, and conservative substitutions are in gray. Residues present at identical positions in all four proteins are marked with*. Residues 117-125 of POH1 display a conserved Cysteine-box motif.

the 26S proteasome does not require ubiquitination (Jariel-Encontre, et al., 1995). However, this model is still being tested and the mechanisms which mediate drug resistance upon POH1 overexpression have not yet been elucidated. Results of transfection experiments in COS cells suggest that transient POH1 overexpression does not activate the expression of the membrane pump P-glycoprotein and does not alter intracellular accumulation of the anticancer drug doxorubicin (Spataro, et al., 1997).

POH1 is widely expressed in human tumor cell lines and work in progress is assessing its contribution to tumor drug resistance in vivo. We have recently assessed in our laboratory the levels of the POH1 protein in 60 tumor cell lines of the U.S. National Cancer Institute (NCI) drug-screening panel. This set of cell lines includes lines from 9 tumor types, namely melanomas, leukemias, and cancers of breast, prostate, lung, colon, ovary, kidney, and central nervous system origin. The vast majority of these lines has never been exposed to chemotherapy and are, therefore, suitable for studies on intrinsic drug resistance. Information derived from the large scale drug screening program (with more than 70,000 new compounds already screened on the 60 lines) is accessible in an activity database, and can be related to the structure database containing information on the molecular structure of the compound and to a target database containing information on possible molecular targets or modulators of activity within the cells (Weinstein et al., 1997). We have found that the POH1 protein is expressed at highly variable levels (over a 100-fold range) by these 60 human tumor cell lines and that POH1 levels are on average higher in cell lines with lower degrees of sensitivity to cytotoxic drugs (Spataro et al., 1998). When looking at the correlation between POH1 levels and GI_{50}s (concentrations giving a 50% growth inhibition in cytotoxicity assays) of 175 clinically used compounds and analogues, there is a statistically significant correlation between high POH1 levels and resistance to 36 compounds. The vast majority of these compounds are alkylating agents or DNA-damaging agents and analysis of variance (ANOVA) suggests that this association is dramatic (unpublished results). The analysis of POH1 protein levels in the lines of the NCI panel suggests that the role of POH1 is independent of other known mechanisms of drug-resistance.

Interestingly, in recent years three other subunits of the 19S regulatory complex of the proteasome called Mts2, Mts3 and Mts4 have been identified in fission yeast through a screen for mutants resistant to the mitotic spindle poison carbendazim (MBC) (Gordon, et al., 1993, 1996; Wilkinson, et al., 1997). Thus, the 26S proteasome plays an important role in determining multidrug resistance in fission yeast. This pathway is highly conserved in mammals, can confer drug resistance to anticancer agents in vitro and could potentially be involved in drug resistance in human tumors.

Another fission yeast gene called Crm1 is involved like Pad1/POH1 in Pap1/AP-1 dependent multidrug resistance (Kumada et al., 1996; Toda et al., 1992) and its mammalian homologue has been recently cloned and characterized by several groups (Fornerod et al., 1997; Fukuda et al., 1997; Kudo et al., 1997; Neville et al., 1997; Ossareh-Nazari et al., 1997; Stade et al., 1997). Human CRM1 encodes a protein of a predicted size of 112 kDa, with a 52% amino acid identity with the fission yeast protein. Human CRM1 has been characterized as a nuclear export factor acting as a receptor for leucine rich sequences called nuclear export signal (NES). Interestingly, Northern blotting of multiple tissues extracts reveals that POH1 and CRM1 have a similar pattern of tissue expression, with expression in all the tissues examined, high mRNA levels in skeletal and heart muscle and low mRNA levels in the kidney (Kudo, et al., 1997; Spataro, et al., 1997), which is typical of several proteasome subunits (Tsurumi, et al., 1995a). It has become increasingly clear in recent years that the control of nuclear import and export is important for regulating the function of many proteins. Several transcription factors involved in the response to chemotherapy in humans are regulated by nuclear translocation or nuclear export, such as NF-κB, YB-1 and p53. As discussed in the previous section, it is well known that the activation of NF-κB requires the ubiquitin/proteasome mediated degradation of its inhibitor IκB in the cytoplasm, followed by nuclear translocation. The Y-box binding protein YB-1 is a transcription factor that binds a specific sequence referred to as the Y-box and which might be a critical regulator element of several genes including the MDR1 gene encoding the P-glycoprotein (Ohga et al., 1996; Sundseth et al., 1997). It has been shown that the YB-1 protein in cancer cells is located predominantly in the cytosol but is rapidly translocated to the nucleus in response to stresses such as UV irradiation (Koike et al., 1997). The nuclear translocation of YB-1 appears to play an important role in vivo, as in 27 untreated primary human breast cancers YB-1 was detected immunohistochemically in the cytoplasm of all specimens but in the nucleus only of 9 specimens that expressed high levels of P-glycoprotein (Bargou et al., 1997). The MDM2 oncogene is overexpressed in tumors and inactivates the tumor suppressor protein p53, by binding of the N-terminal region of MDM2 to the N-terminal transactivation domain of p53 (Oliner et al., 1993). It has been recently shown that the MDM2 protein inactivates p53 by accelerating its proteasome-mediated degradation (Haupt et al., 1997; Kubbutat et al., 1997). Recent evidence (reviewed above) suggests that acceleration of p53 degradation by MDM2 requires NES-mediated MDM2 nucleo-cytoplasmic shuttling via the pathway used by the human immunodeficiency virus rev protein (Roth et al., 1998), which involves Exportin 1 (hCRM1). Recent evidence also suggests that the AP-1 homologue Pap1 is regulated in fission yeast by a mechanism acting primarily on nuclear export (Toone et al., 1998)

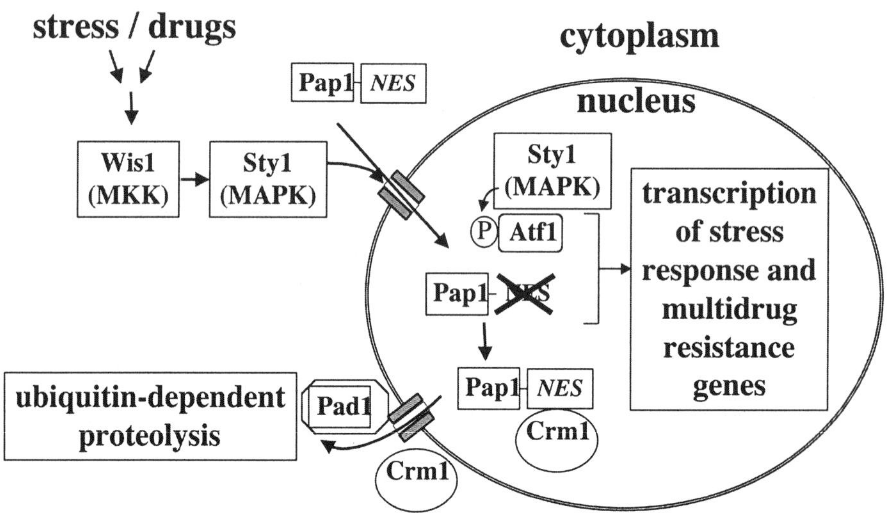

Figure 4. Model showing the involvement of both nucleo-cytoplasmic transport and ubiquin/proteasome-dependent proteolysis in the regulation of the transcription factor Pap1 involved in stress response and in drug resistance in the fission yeast *S. pombe* (see references in the text)

Nuclear accumulation of Pap1 protein was observed upon oxidative stress and in cells carrying crm1 mutations and a similar Crm1-dependent Yap1 activation has been observed in budding yeast. In S. cerevisiae the cytoplasmic localization of Yap1 in unstressed cells requires a C-terminal region named CRD (Cysteine rich domain) which appears to be a specialized NES. In human cells, AP-1 transcription factors are nuclear proteins and data are lacking on their regulation by nucleo-cytoplasmic transport. However, several lines of evidence suggest that hCRM1 could well be involved in modulating sensitivity to anticancer drugs in vitro. It is of interest that the group of Gerard Grosveld found hCRM1 as a protein interacting with the fusion protein DEK-CAN of acute myeloid leukemia with translocation (6; 9) (Fornerod et al., 1996) which is associated with poor prognosis (Lillington et al., 1993), i.e. relative drug resistance.

Other Areas of Cancer Biology

Because of its broad involvement in most cellular processes, the ubiquitin-proteasome pathway can be predicted to play a role in most areas of cancer biology, even in fields where alterations of the regulation of this pathway have not been described so far. One area, which is becoming increasingly important from a therapeutic point of view in which the role of the ubiquitin-proteasome pathway could be highly relevant, is that of growth factors receptors and their signaling pathways. Several cell surface receptors have been shown to be ubiquitinated, suggesting that proteasome-mediated proteolysis is involved in their turnover. One promising family of therapeutic targets are the tyrosine kinase receptors, such as insulin-like growth factor receptors (IGFR) and epidermal growth factor receptors (EGFR). Interestingly, herbimycin A, which targets tyrosine kinase activated signal transduction by inhibiting multiple tyrosine protein kinases and has in vitro and in vivo antitumor activity, acts through an enhancement of receptor degradation by the proteasome (Sepp Lorenzino et al., 1995). This paradigm seems to apply also to other kinase inhibitors, such as the partial agonist protein kinase C inhibitor bryostatin, which promotes the degradation of PKC by the proteasome (Lee et al., 1996).

Not only does the ubiquitin/proteasome pathway seem to be involved in the degradation of signaling molecules such as tyrosine kinase receptors or PKC, but also some evidence suggests a direct role of ubiquitination in the first steps of signaling at the cell surface. For instance, when the T-cell receptor is occupied, one of its subunits becomes ubiquitinated on its cytoplasmic domain (Hou et al., 1994). Also, the platelet -derived-growth-factor (PDGF) receptor-β undergoes polyubiquination as a result of binding of its ligand and the proteasome is involved in the degradation of the ligand-activated receptor (Mori et al., 1995).

Several lines of evidence suggest that DNA repair is another area in which the ubiquitin/proteasome pathway might play a major role. The first evidence was obtained in S. cerevisiae, in which the rad6 DNA repair mutant is defective in the ubiquitin-conjugating enzyme UBC2 and the DNA repair gene RAD-23 encodes a protein containing a ubiquitin-like domain. This domain appeared to be essential for Rad23 function (Watkins et al., 1993) and is conserved in the human homologue HHR23B

(Masutani et al., 1994). Further evidence came from experiments in mammalian cells, in particular using a temperature sensitive mutant from the mouse mammary carcinoma cell line FM3A, which contains a thermosensitive ubiquitin-activating enzyme (E1). E1 mutants grown at the restrictive temperature are more sensitive to DNA damage induced by ultraviolet radiation (Ikehata et al., 1997). These data suggest that ubiquitn conjugation contributes to DNA repair in mammalian cells either directly or indirectly by an influence of ubiquitin-binding proteins such as E1 or E2 enzymes on DNA repair proteins carrying ubiquitin-like domains such as RAD23/HHR23B. In fact recent biochemical data tend to support a physical interaction between the amino-terminal ubiquitin-like domain of Rad23 and the proteasome, given that Rad23 cofractionates with Mss1 and Sug1, two subunits of the 19S regulatory complex (Schauber et al., 1998).

Another area where data await further elucidation is the potential role of isopeptidases or ubiquitin hydrolases (deubiquitinating enzymes) in oncogenic transformation and tumor progression. There have been reports claiming that deubiquitinating enzymes could act as oncogenes as the yeast Doa4 isopeptidase is related to the product of the human Tre-2, which has been found to be tumorigenic when expressed at high levels (Papa and Hochstrasser, 1993), and the human homologue of the murine ubiquitin-releasing enzyme unp (UNPH) had been found to be overexpressed in lung cancer cell lines (Verovski et al., 1996). However, recent work on UNPH has confirmed its role as a deubiquitinating enzyme but has failed to confirm that UNPH is upregulated in lung cancer cell lines and has found that both UNPH mRNA and protein were underexpressed in four small cell lung cancer (Frederick et al., 1998). This finding, together with the fact that the locus 3p21.3 where the gene resides is often deleted in lung cancer, would support a role as a tumor suppressor gene rather than as an oncogene. Interestingly, another gene encoding a ubiquitin hydrolase and that could act as a tumor suppressor has been recently identified in the same chromosomal region at locus 3p21.31. The product of this gene has been named BAP1 and binds the Breast/Ovarian Cancer Susceptibility gene product, BRCA1 (Jensen et al., 1998). BAP1 and BRCA1 are coexpressed during murine breast development and remodeling. Because BAP1 binds to the wild-type BRCA1 protein but not to BRCA1 mutants found in breast cancer kindreds, BAP1 is likely to play a role in the BRCA1 growth control pathway. Finally, a novel isopeptidase identified recently, UBPY, appears to be growth regulated and to play an important role in cell cycle progression, as the cellular abundance of UBPY correlates with the protein ubiquitination and cell proliferation (Naviglio et al., 1998).

It is currently widely recognized that anticancer drugs kill tumor cells by activating apoptosis or programmed cell death under most circumstances and it would therefore be important to ascertain the role of the ubiquitin/proteasome pathway in this context. In fact this issue appears complex because the ubiquitin/proteasome system seems to be involved at several levels, not only within pathways leading from cellular stress to apoptosis. Proteasome-dependent proteolysis has been found to act upstream of apoptosis-specific aspartyl proteases (caspases) in nerve growth factor-deprived neurons, where specific proteasome inhibitors inhibit apoptosis (Sadoul et al., 1996). In contrast, in a cell line with a temperature-sensitive defect in the ubiquitin-activating enzyme E1, there is an induction of apoptosis at the restrictive temperature which can not be prevented by caspase inhibitors, suggesting the existence of short-lived proteins that are substrates of the proteasome and can induce caspase-independent apoptosis (Monney et al., 1998). Because the net effect of proteasome inhibition may vary depending on the cellular context and because of existing alterations of the ubiquitin/proteasome pathway in cancer cells, it would be very important to investigate proteasome involvement in apoptosis induced by anticancer drugs.

Finally, it has been demonstrated that the ubiquitin/proteasome-dependent degradation of several protein substrates of the ubiquitin/proteasome pathway require association with the molecular chaperone Hsc70 (Bercovich et al., 1997). Interestingly, the expression of Hsc70, is induced up to 30-fold by a proteasome inhibitor, unlike other members of the hsp family (Zhou et al., 1996), suggesting that Hsc70 is itself a substrate of the proteasome. This establishes another strong link between the ubiquitin/proteasome pathway and tumor response to treatment, given that Hsc70 is involved in stress response and has a role in tumor resistance to anticancer agents (Ciocca et al., 1992).

Table 3. Summary of proteins involved in the ubiquitin/proteasome pathway with a potential role in cancer

Protein	Biological function	Role in tumors	Reference
UNPH	Ubiquitin-C-terminal Hydrolase	Growth control	(Frederick et al., 1998) (Gray et al., 1995)
BAP1	Ubiquitin-C-terminal Hydrolase	Growth control	(Jensen et al., 1998)
UBPY	Ubiquitin-C-terminal Hydrolase	Growth control	(Naviglio et al., 1998)
POH1	Subunit of the regulatory complex of the proteasome, contains a putative isopeptidase site	Response to anticancer treatment	(Spataro et al., 1997) (Shimanuki et al., 1995)
HSC70	Molecular chaperone, required for ubiquitin/proteasome mediated degradation of several substrates	Response to anticancer treatment	(Bercovich et al., 1997) (Ciocca et al., 1992) (Zhou et al., 1996)
RAD23	DNA repair protein, contains a ubiquitin-like domain interacting with the 26S proteasome	Response to anticancer treatment	(Masutani et al., 1994) (Schauber et al., 1998) (Watkins et al., 1993)

Drugs Acting on the Proteasome

Pharmacological intervention to activate or inhibit one or several proteasome functions could be therapeutically advantageous in anticancer treatment and also in other fields such as immunology, for example, with the intent of targeting activation by the proteasome of NF-κB, which has a key role in mediating the inflammatory and immune responses. Interestingly, the clinically used antitumor drug aclacinomycin A or aclarubicin, known as a DNA-intercalative agent, has been shown to inhibit the degradation of ubiquitinated protein by inhibiting selectively the chymotrypsin-like activity of the proteasome (Figueiredo-Pereira et al., 1996). It is not clear whether this could contribute to the antitumor activity of this drug. The best known proteasome inhibitor is lactacystin, a Streptomyces metabolite discovered on the basis of its ability to induce neurite outgrowth in the Neuro 2A mouse neuroblastoma cell line (Fenteany et al., 1994). This inhibitor was subsequently shown to covalently modify a critical threonine residue of the subunit X/MB1 of the proteasome core (Fenteany, 1995). Lactacystin was found to inhibit cell cycle progression in human osteosarcoma cells (Fenteany et al., 1994) and to induce apoptosis in human monoblast cells (Imajoh-Ohmi et al., 1995). However, we are not aware of any data on the antitumor activity of lactacystin. Aside from lactacystin, most of the proteasome inhibitors developed so far are synthetic protease inhibitors of the family of peptidyl aldehydes (Rock et al., 1994). Some of them, such as n-acetyl-leucinyl-leucinyl-norleucinal (ALLN) and benzyloxycarbonyl (Z)-leucinyl-leucinyl-leucinal (ZLLL) are cell-penetrating, display proteasome specificity and have been reported to induce apoptosis in human tumor cell lines (Fujita et al., 1996; Shinohara et al., 1996). Due to the broad involvement of proteasomes in normal cellular physiology, any attempt to target the proteasome non-specifically might be associated with prohibitive in vivo toxicity. However, the complexity and specificity of proteasome regulation indicate that specific inhibitors of individual proteasome-mediated processes might ultimately become available or that non-specific proteasome inhibitors might be active in particular tumors with altered proteasome function. For example, the novel proteasome inhibitor benzyloxycarbonyl-leucyl-leucyl-phenylalaninal (Z-LLF-CHO), a cell-permeable peptidyl aldehyde inhibitor of the chymotrypsin-like activity of the proteasome has been shown recently to have in vivo antitumor activity with acceptable toxicity on a model of Burkitt lymphoma in mice (Orlowski et al., 1998). Other compounds with proteasome inhibitory activity are being actively screened as anticancer agents to be used alone or in combination with standard chemotherapy. Overall, it is likely that the the rapidly expanding knowledge about the role of proteasomes in normal and tumor cells could provide in the future a rational basis for the use of proteasome-targeting drugs.

Conclusions

Although it has been previously a rather neglected area of research in cancer biology, the ubiquitin/proteasome pathway is emerging as an important area both in terms of the basic knowledge of the mechanisms of oncogenic transformation and cancer progression and in terms of the mechanisms that modulate the response to the pharmacological treatment of cancer. To date, research has found that deregulation of this pathway in cancer can be responsible for crucial phenomena such as oncogenic transformation in HPV-related malignancies, adverse prognosis in colorectal, lung, breast and prostate carcinomas and that it is involved in modulating response to anticancer drugs. New anticancer strategies based on the modulation of proteasome activity are being developed and compounds that inhibit the proteasome are being screened as anticancer agents, with some positive results in terms of antitumor activity in vitro and in animal models. These promising but preliminary results await further confirmation. Understanding the complexity of the ubiquitin/proteasome pathway, and in particular how the specificity for a given substrate is regulated, should allow us in the future to develop more specific therapeutic strategies.

Summary

Regulated proteolysis is involved in a broad range of biological processes such as cell cycle control, antigen presentation and regulation of gene expression. In most cases studied to date, controlled protein degradation is carried out by the 26S proteasome following ubiquitination of the target protein. The ubiquitin/proteasome pathway is the major intracellular non-lysosomal proteolytic mechanism and is involved in a broad range of processes such as cell cycle progression, antigen presentation and control of gene expression. Recent work reviewed here has shown that this pathway is often the target of cancer-related deregulation and can underlie processes such as oncogenic transformation, tumor progression, escape from immune surveillance and drug resistance.

Acknowledgements

The authors' work is funded by the Swiss Cancer League (Krebsforschung Schweiz-KFS 520-8-1997), the Swiss National Science Foundation (grant 31-55642.98) and the Imperial Cancer Research Fund. We thank Dr. D. Finley for his kind permission to adapt Figure 1.

References

Amon, A., Irniger, S. & Nasmyth, K. (1994). Closing the cell cycle circle in yeast: G2 cyclin proteolysis initiated at mitosis persists until the activation of G1 cyclins in the next cycle. Cell 77, 1037-1050.

Asano, K., Vornlocher, H.-P., Richter-Cook, N., Merrick, W., Hinnebusch, A. & Hershey, J. (1997). Structure of cDNAs encoding human eukaryotic initiation factor 3 subunits. Possible roles in RNA binding and macromolecular assembly. J. Biol. Chem. 272, 27042-27052.

Barger, S.W., Horster, D., Furukawa, K., Goodman, Y., Krieglstein, J. & Mattson, M.P. (1995). Tumor necrosis factors α and β protect neurons against amyloid β-peptide toxicity: Evidence for involvement of a κB-binding factor and attenuation of peroxide and Ca^{2+} accumulation. Proc. Natl. Acad. Sci. USA 92, 9328-9332.

Bargou, R.C., Jurchott, K., Wagener, C., Bergmann, S., Metzner, S., Bommert, K., Mapara, M.Y., Winzer, K.J., Dietel, M., Dorken, B. & Royer, H.D. (1997). Nuclear localization and increased levels of transcription factor YB-1 in primary human breast cancers are associated with intrinsic MDR1 gene expression. Nat. Med. 3, 447-450.

Beck, S., Kelly, A., Radley, E., Khurshid, F., Alderton, R. P. & Trowsdale, J. (1992). DNA sequence analysis of 66 kb of the human MHC class II region encoding a cluster of genes for antigen processing. J. Mol. Biol. 228, 433-441.

Beg, A. A. & Baltimore, D. (1996). An essential role for NF-kappaB in preventing TNF-alpha-induced cell death. Science 274, 782-784.

Belich, M.P., Glynne, R.J., Senger, G., Sheer, D. & Trowsdale, J. (1994). Proteasome components with reciprocal expression to that of the MHC- encoded LMP proteins. Curr. Biol. 4, 769-776.

Bercovich, B., Stancovski, I., Mayer, A., Blumenfeld, N., Laszlo, A., Schwartz, A. L. & Ciechanover, A. (1997). Ubiquitin-dependent degradation of certain protein substrates in vitro requires the molecular chaperone Hsc70. J. Biol. Chem. 272, 9002-9010.

Brandeis, M. (1996). The proteolysis of mitotic cyclins in mammalian cells persists from the end of mitosis until the onset of S phase. EMBO J. 15, 5280-5289.

Catzavelos, C., Bhattacharya, N., Ung, Y.C., Wilson, J.A., Roncari, L., Sandhu, C., Shaw, P., Yeger, H., Morava-Protzner, I., Kapusta, L., Franssen, E., Pritchard, K. I. & Slingerland, J.M. (1997). Decreased levels of the cell-cycle inhibitor p27Kip1 protein: prognostic implications in primary breast cancer. Nat. Med. 3, 227-230.

Chowdary, D.R., Dermody, J.J., Jha, K. K. & Ozer, H.L. (1994). Accumulation of p53 in a mutant cell line defective in the ubiquitin pathway. Mol. Cell. Biol. 14, 1997-2003.

Ciechanover, A. (1994). The ubiquitin-proteasome proteolytic pathway. Cell 79, 13-21.

Ciechanover, A., Shkedy, D., Oren, M. & Bercovich, B. (1994). Degradation of the tumor suppressor protein p53 by the ubiquitin- mediated proteolytic system requires a novel species of ubiquitin- carrier protein, E2. J. Biol. Chem. 269, 9582-9589.

Ciocca, D.R., Fuqua, S.A., Lock Lim, S., Toft, D.O., Welch, W.J. & McGuire, W.L. (1992). Response of human breast cancer cells to heat shock and chemotherapeutic drugs. Cancer Res. 52, 3648-3654.

Claret, F.X., Hibi, M., Dhut, S., Toda, T. & Karin, M. (1996). A new group of conserved coactivators that increase the specificity of AP-1 transcription factors. Nature 383, 453-457.

Cote, R.J., Shi, Y., Groshen, S., Feng, A. C., Cordon-Cardo, C., Skinner, D. & Lieskovosky, G. (1998). Association of p27Kip1 levels with recurrence and survival in patients with stage C prostate carcinoma. J. Natl. Cancer Inst. 90, 916-920.

Cox, L.S. & Lane, D.P. (1995). Tumor suppressors, kinases and clamps: how p53 regulates the cell cycle in response to DNA damage. Bioessays 17, 501-508.

Delic, J., Masdehors, P., Omura, S., Cosset, J.M., Dumont, J., Binet, J.L. & Magdelenat, H. (1998). The proteasome inhibitor lactacystin induces apoptosis and sensitizes chemo- and radioresistant human chronic lymphocytic leukaemia lymphocytes to TNF-alpha-initiated apoptosis. Br. J. Cancer 77, 1103-1107.

Deveraux, Q., Ustrell, V., Pickart, C. & Rechsteiner, M. (1994). A 26 S protease subunit that binds ubiquitin conjugates. J. Biol. Chem. 269, 7059-7061.

Dubiel, W., Ferrell, K., Dumdey, R., Standera, S., Prehn, S. & Rechsteiner, M. (1995a). Molecular cloning and expression of subunit 12: a non-MCP and non- ATPase subunit of the 26 S protease. FEBS Lett. 363, 97-100.

Dubiel, W., Ferrell, K. & Rechsteiner, M. (1995b). Subunits of the regulatory complex of the 26S protease. Mol. Biol. Rep. 21, 27-34.

Esposito, V., Baldi, A., De Luca, A., Groger, A.M., Loda, M., Giordano, G.G., Caputi, M., Baldi, F., Pagano, M. & Giordano, A. (1997). Prognostic role of the cyclin-dependent kinase inhibitor p27 in non- small cell lung cancer. Cancer Res. 57, 3381-3385.

Fehling, H.J., Swat, W., Laplace, C., Kühn, R., Rajewsky, K., Müller, U. & von Boehmer, H. (1994). MHC class I expression in mice lacking the proteasome subunit LMP-7. Science 265, 1234-1237.

Fenteany, G. (1995). Inhibition of proteasome activities and subunit-specific amino-terminal threonine modification by lactacystin. Science 268, 726-731.

Fenteany, G., Standaert, R.F., Reichard, G.A., Corey, E.J. & Schreiber, S.L. (1994). A beta-lactone related to lactacystin induces neurite outgrowth in a neuroblastoma cell line and inhibits cell cycle progression in an osteosarcoma cell line. Proc. Natl. Acad. Sci. U. S.A. 91, 3358-3362.

Ferrando, A. (1996). Mutational analysis of the human cyclin-dependent kinase inhibitor p27kip1 in primary breast carcinomas. Hum. Genet. 97, 91-94.

Figueiredo Pereira, M.E., Chen, W.E., Li, J. & Johdo, O. (1996). The antitumor drug aclacinomycin A, which inhibits the degradation of ubiquitinated proteins, shows selectivity for the chymotrypsin-like activity of the bovine pituitary 20 S proteasome. J. Biol. Chem. 271, 16455-16459.

Fornerod, M., Boer, J., van Baal, S., Morreau, H. & Grosveld, G. (1996). Interaction of cellular proteins with the leukemia specific fusion proteins DEK-CAN and SET-CAN and their normal counterpart, the nucleoporin CAN. Oncogene 13, 1801-1808.

Fornerod, M., Ohno, M., Yoshida, M. & Mattaj, I. (1997). CRM1 is an export receptor for leucine-rich nuclear export signals. Cell 90, 1051-1060.

Frederick, A., Rolfe, M. & Chiu, M.I. (1998). The human UNP locus at 3p21.31 encodes two tissue-selective, cytoplasmic isoforms with deubiquitinating activity that have reduced expression in small cell lung carcinoma cell lines. Oncogene 16, 153-165.

Fruh, K., Gossen, M., Wang, K., Bujard, H., Peterson, P.A. & Yang, Y. (1994). Displacement of housekeeping proteasome subunits by MHC-encoded LMPs: a newly discovered mechanism for modulating the multicatalytic proteinase complex. EMBO J. 13, 3236-3244.

Fujita, E., Mukasa, T., Tsukahara, T., Arahata, K., Omura, S. & Momoi, T. (1996). Enhancement of CPP32-like activity in the TNF-treated U937 cells by the proteasome inhibitors. Biochem Biophys. Res. Commun. 224, 74-79.

Fukuda, M., Asano, S., Nakamura, T., Adachi, M., Yoshida, M., Yanagida, M. & Nishida, E. (1997). CRM1 is responsible for intracellular transport mediated by the nuclear export signal. Nature 390, 308-311.

Gaczynska, M., Goldberg, A.L., Tanaka, K., Hendil, K.B. & Rock, K.L. (1996). Proteasome subunits X and Y alter peptidase activities in opposite ways to the interferon-gamma-induced subunits LMP2 and LMP7. J. Biol. Chem. 271, 17275-17280.

Gazzeri, S., Della Valle, V., Chausade, L., Brambilla, C., Larsen, C.-J. & Brambilla, E. (1998). The human p19ARF protein encoded by the beta transcript of the p16INK4a gene is frequently lost in small cell lung cancer. Cancer Res. 58, 3926-3931.

Glickman, M.H., Rubin, D.M., Fried, V. A. & Finley, D. (1998). The regulatory particle of the Saccharomyces cerevisiae proteasome. Mol. Cell. Biol. 18, 3149-3162.

Glotzer, M., Murray, A.W. & Kirschner, M.W. (1991). Cyclin is degraded by the ubiquitin pathway. Nature 349, 132-138.

Gordon, C., McGurk, G., Dillon, P., Rosen, C. & Hastie, N.D. (1993). Defective mitosis due to a mutation in the gene for a fission yeast 26S protease subunit. Nature 366, 355-357.

Gordon, C., McGurk, G., Wallace, M. & Hastie, N.D. (1996). A conditional lethal mutant in the fission yeast 26 S protease subunit mts3+ is defective in metaphase to anaphase transition. J. Biol. Chem. 271, 5704-5711.

Groettrup, M., Kraft, R., Kostka, S., Standera, S., Stohwasser, R. & Kloetzel, P.M. (1996). A third interferon-gamma-induced subunit exchange in the 20S proteasome. Eur. J. Immunol. 26, 863-869.

Groettrup, M., Ruppert, T., Kuehn, L., Seeger, M., Standera, S., Koszinowski, U. & Kloetzel, P. M. (1995).: The interferon-gamma-inducible 11 S regulator (PA28) and the LMP2/LMP7 subunits govern the peptide production by the 20 S proteasome in vitro. J. Biol. Chem. 270, 23808-23815.

Groll, M., Ditzel, L., Löwe, J., Stock, D., Bochtler, M., Bartunik, H.D. & Huber, R. (1997). Structure of 20S proteasome from yeast at 2.4 A resolution. Nature 386, 463-471.

Harris, C.C. (1996). Structure and function of the p53 tumor suppressor gene: clues for rational cancer therapeutic strategies. J. Natl. Cancer Inst. 88, 1442-1455.

Haupt, Y., Maya, R., Kazaz, A. & Oren, M. (1997). Mdm2 promotes the rapid degradation of p53. Nature 387, 296-303.

Hengst, L. & Reed, S. I. (1996). Translational control of p27Kip1 accumulation during the cell cycle. Science 271, 1861-1864.

Hermida Matsumoto, M.L., Chock, P.B., Curran, T. & Yang, D.C. (1996). Ubiquitinylation of transcription factors c-Jun and c-Fos using reconstituted ubiquitinylating enzymes. J. Biol. Chem. 271, 4930-4936.

Hisamatsu, H., Shimbara, N., Saito, Y., Kristensen, P., Hendil, K.B., Fujiwara, T., Takahashi, E., Tanahashi, N., Tamura, T., Ichihara, A. & Tanaka, K. (1996). Newly identified pair of proteasomal subunits regulated reciprocally by interferon gamma. J. Exp. Med. 183, 1807-1816.

Hochstrasser, M. (1995). Ubiquitin, proteasomes & the regulation of intracellular protein degradation. Curr. Opin. Cell Biol. 7, 215-223.

Hou, D., Cenciarelli, C., Jensen, J.P., Nguygen, H.B. & Weissman, A.M. (1994). Activation-dependent ubiquitination of a T cell antigen receptor subunit on multiple intracellular lysines J. Biol. Chem. 269, 14244-14247.

Huibregtse, J.M., Scheffner, M. & Howley, P.M. (1991). A cellular protein mediates association of p53 with the E6 oncoprotein of human papillomavirus types 16 or 18. Embo J. 10, 4129-4135.

Hunter, T. & Pines, J. (1994). Cyclins and cancer. II: Cyclin D and CDK inhibitors come of age. Cell 79, 573-582.

Ikehata, H., Kaneda, S., Yamao, F., Seno, T., Ono, T. & Hanaoka, F. (1997). Incubation at the nonpermissive temperature induces deficiencies in UV resistance and mutagenesis in mouse mutant cells expressing a temperature-sensitive ubiquitin-activating enzyme (E1). Mol. Cell. Biol. 17, 1484-1489.

Imajoh-Ohmi, S., Kawaguchi, T., Sugiyama, S., Tanaka, K., Omura, S. & Kikuchi, H. (1995). Lactacystin, a specific inhibitor of the proteasome, induces apoptosis in human monoblast U937 cells. Biochem. Biophys. Res. Commun. 217, 1070-1077.

Jariel-Encontre, I., Pariat, M., Martin, F., Carillo, S., Salvat, C. & Piechaczyk, M. (1995). Ubiquitinylation is not an absolute requirement for degradation of c- Jun protein by the 26 S proteasome. J. Biol. Chem. 270, 11623-11627.

Jensen, D.E., Proctor, M., Marquis, S.T., Gardner, H.P., Ha, S.I., Chodosh, L.A., Ishov, A.M., Tommerup, N., Vissing, H., Sekido, Y., Minna, J., Borodovsky, A., Schultz, D.C., Wilkinson, K.D., Maul, G.G., Barlev, N., Berger, S.L., Prendergast, G.C. & Rauscher, F. J. r. (1998). BAP1: a novel ubiquitin hydrolase which binds to the BRCA1 RING finger and enhances BRCA1-mediated cell growth suppression. Oncogene 16, 1097-1112.

Johnsen, A., France, J., Sy, M.S. & Harding, C.V. (1998). Down-regulation of the transporter for antigen presentation, proteasome subunits & class I major histocompatibility complex in tumor cell lines. Cancer Res. 58, 3660-3667.

Jones, S.N., Roe, A.E., Donehower, L.A. & Bradley, A. (1995). Rescue of embryonic lethality in Mdm2-deficient mice by absence of p53. Nature 378, 206-208.

Karin, M., Liu, Z. & Zandi, E. (1997). AP-1 function and regulation. Curr. Opin. Cell Biol. 9, 240-246.

Keyomarsi, K., Conte, D., Jr., Toyofuku, W. & Fox, M.P. (1995). Deregulation of cyclin E in breast cancer. Oncogene 11, 941-950.

Keyomarsi, K., O'Leary, N., Molnar, G., Lees, E., Fingert, H.J. & Pardee, A.B. (1994). Cyclin E, a potential prognostic marker for breast cancer. Cancer Res. 54, 380-385.

Keyomarsi, K. & Pardee, A.B. (1993). Redundant cyclin overexpression and gene amplification in breast cancer cells. Proc. Natl. Acad. Sci. U. S. A.90, 1112-1116.

King, R.W. (1996). How proteolysis drives the cell cycle. Science 274, 1652-1659.

King, R.W., Peters, J.M., Tugendreich, S., Rolfe, M., Hieter, P. & Kirschner, M.W. (1995). A 20S complex containing CDC27 and CDC16 catalyzes the mitosis-specific conjugation of ubiquitin to cyclin B. Cell 81, 279-288.

Koike, K., Uchiumi, T., Ohga, T., Toh, S., Wada, M., Kohno, K. & Kuwano, M. (1997). Nuclear translocation of the Y-box binding protein by ultraviolet irradiation. FEBS Lett. 417, 390-394.

Krappmann, D., Wulczyn, F.G. & Scheidereit, C. (1996). Different mechanisms control signal-induced degradation and basal turnover of the NF-kappaB inhibitor IkappaB alpha in vivo. EMBO J. 15, 6716-26.

Kubbutat, M.H., Jones, S.N. & Vousden, K.H. (1997). Regulation of p53 stability by Mdm2. Nature 387, 299-303.

Kuckelkorn, U., Frentzel, S., Kraft, R., Kostka, S., Groettrup, M. & Kloetzel, P.M. (1995). Incorporation of major histocompatibility complex--encoded subunits LMP2 and LMP7 changes the quality of the 20S proteasome polypeptide processing products independent of interferon-gamma. Eur. J. Immunol. 25, 2605-2611.

Kudo, N., Khochbin, S., Nishi, K., Kitano, K., Yanagida, M., Yoshida, M. & Horinouchi, S. (1997). Molecular cloning and cell cycle-dependent expression of mammalian CRM1, a protein involved in nuclear export of proteins. J. Biol. Chem. 272, 29742-29751.

Kumada, K., Yanagida, M. & Toda, T. (1996). Caffeine-resistance in fission yeast is caused by mutations in a single essential gene, crm1+. Mol. Gen. Genet. 250, 59-68.

Lam, Y.A., Xu, W., DeMartino, G.N. & Cohen, R.E. (1997). Editing of ubiquitin conjugates by an isopeptidase in the 26S proteasome. Nature 385, 737-740.

Lee, H.W., Smith, L., Pettit, G.R., Vinitsky, A. & Smith, J.B. (1996). Ubiquitination of protein kinase C-alpha and degradation by the proteasome. J. Biol. Chem. 271, 20973-20976.

Levine, A. (1997). p53, the cellular gatekeeper for growth and division. Cell 88, 323-331.

Lillington, D.M., MacCallum, P.K., Lister, T.A. & Gibbons, B. (1993). Translocation t(6;9) (p23;q34) in acute myeloid leukemia without myelodysplasia or basophilia: two cases and a review of the literature. Leukemia 7, 527-531.

List, A.F. (1996). The role of multidrug resistance and its pharmacological modulation in acute myeloid leukemia. Leukemia 10, Suppl 1, S36-38.

List, A.F., Spier, C.S., Grogan, T.M., Johnson, C., Roe, D.J., Greer, J.P., Wolff, S.N., Broxterman, H.J., Scheffer, G.L., Scheper, R.J. & Dalton, W.S. (1996). Overexpression of the major vault transporter protein lung-resistance protein predicts treatment outcome in acute myeloid leukemia. Blood 87, 2464-2469.

Loda, M., Cukor, B., Tam, S.W., Lavin, P., Fiorentino, M., Draetta, G.F., Jessup, J.M. & Pagano, M. (1997). Increased proteasome-dependent degradation of the cyclin-dependent kinase inhibitor p27 in aggressive colorectal carcinomas. Nat. Med. 3, 231-234.

Löwe, J., Stock, D., Jap, B., Zwickl, P., Baumeister, W. & Huber, R. (1995). Crystal structure of the 20S proteasome from the archaeon T. acidophilum at 3.4 A resolution. Science 268, 533-539.

Maki, C.G., Huibregtse, J.M. & Howley, P.M. (1996). In vivo ubiquitination and proteasome-mediated degradation of p53 (1). Cancer Res. 56, 2649-2654.

Masutani, C., Sugasawa, K., Yanagisawa, J., Sonoyama, T., Ui, M., Enomoto, T., Takio, K., Tanaka, K., Van der Spek, P.J., Bootsma, D., et al. (1994). Purification and cloning of a nucleotide excision repair complex involving the xeroderma pigmentosum group C protein and a human homologue of yeast RAD23. EMBO J. 13, 1831-1843.

Mattson, M.P., Goodman, Y., Luo, H., Fu, W. & Furukawa, K. (1997). Activation of NF-κB protects hhippocampal neurons against oxidative stress-induced apoptosis: evidence for induction of Mn-SOD and suppression of peroxynitrite production and protein tyrosine nitration. J. Neurosci. Res. 49, 681-697.

Michalek, M.T., Grant, E.P., Gramm, C., Goldberg, A.L. & Rock, K.L. (1993). A role for the ubiquitin-dependent proteolytic pathway in MHC class I- restricted antigen presentation. Nature 363, 552-554.

Moffat, G.J., McLaren, A.W. & Wolf, C.R. (1994). Involvement of Jun and Fos proteins in regulating transcriptional activation of the human pi class glutathione S-transferase gene in multidrug-resistant MCF7 breast cancer cells. J. Biol. Chem. 269, 16397-16402.

Monney, L., Otter, I., Olivier, R., Ozer, H.L., Haas, A.L., Omura, S. & Borner, C. (1998). Defects in the ubiquitin pathway induce caspase-independent apoptosis blocked by Bcl-2. J. Biol. Chem. 11, 6121-6131.

Montes de Oca Luna, R., Wagner, D.S. & Lozano, G. (1995). Rescue of early embryonic lethality in mdm2-deficient mice by deletion of p53. Nature 378, 203-206.

Mori, S., Kanaki, H., Tanaka, K., Morisaki, N. & Saito, Y. (1995). Biochem Ligand-activated platelet-derived growth factor beta-receptor is degraded through proteasome-dependent proteolytic pathway. Biophys. Res. Commun. 217, 224-229.

Morosetti, R., Kawamata, N., Gombart, A.F., Miller, C.W., Hatta, Y., Hirama, T., Said, J.W., Tomonaga, M. & Koeffler, H.P. (1995). Alterations of the p27KIP1 gene in non-Hodgkin's lymphomas and adult T-cell leukemia/lymphoma. Blood 86, 1924-1930.

Musti, A.M., Treier, M. & Bohmann, D. (1997). Reduced ubiquitin-dependent degradation of c-Jun after phosphorylation by MAP kinases. Science 275, 400-402.

Nandi, D., Jiang, H. & Monaco, J.J. (1996). Identification of MECL-1 (LMP-10) as the third IFN-gamma-inducible proteasome subunit. J. Immunol. 156, 2361-2364.

Naviglio, S., Mattecucci, C., Matoskova, B., Nagase, T., Nomura, N., Di Fiore, P.P. & Draetta, G.F. (1998). UBPY: a growth-regulated human ubiquitin isopeptidase. EMBO J. 17, 3241-3250.

Neville, M., Stutz, F., Lee, L., Davis, L. & Rosbash, M. (1997). The importin-beta family member Crm1p bridges the interaction between Rev and the nuclear pore complex during nuclear export. Curr. Biol. 7, 767-775.

Ohga, T., Koike, K., Ono, M., Makino, Y., Itagaki, Y., Tanimoto, M., Kuwano, M. & Kohno, K. (1996). Role of the human Y box-binding protein YB-1 in cellular sensitivity to the DNA-damaging agents cisplatin, mitomycin C, and ultraviolet light. Cancer Res. 56, 4224-4228.

Oliner, J.D., Kinzler, K.W., Meltzer, P.S., George, D.L. & Vogelstein, B. (1992). Amplification of a gene encoding a p53-associated protein in human sarcomas. Nature 358, 80-83.

Oliner, J.D., Pietenpol, J.A., Thiagalingam, S., Gyuris, J., Kinzler, K.W. & Vogelstein, B. (1993). Oncoprotein MDM2 conceals the activation domain of tumor suppressor p53. Nature 362, 857-860.

Orlowski, R.Z., Eswara, J.R., Lafond-Walker, A., Grever, M.R., Orlowski, M. & Dang, C.V. (1998). Tumor growth inhibition induced in a murine model of human Burkitt's lymphoma by a proteasome inhibitor. Cancer Res. 58, 4342-4348.

Ossareh-Nazari, B., Bachelerie, F. & Dargemont, C. (1997). Evidence for a role of CRM1 in signal-mediated nuclear protein export. Science 278, 141-144.

Pagano, M., Tam, S.W., Theodoras, A.M., Beer Romero, P., Del Sal, G., Chau, V., Yew, P.R., Draetta, G.F.

& Rolfe, M. (1995). Role of the ubiquitin-proteasome pathway in regulating abundance of the cyclin-dependent kinase inhibitor p27. Science 269, 682-685.

Pahl, H.L. & Baeuerle, P.A. (1996). Control of gene expression by proteolysis. Curr. Opin. Cell Biol. 8, 340-347.

Palmer, A., Mason, G.G., Paramio, J.M., Knecht, E. & Rivett, A.J. (1994). Changes in proteasome localization during the cell cycle. Eur. J. Cell Biol. 64, 163-175.

Palmer, A., Rivett, A.J., Thomson, S., K.B.H., Butcher, G.W., Fuertes, G. & Knecht, E. (1996). Subpopulations of proteasomes in rat liver nuclei, microsomes and cytosol. Biochem. J. 316, 401-407.

Palombella, V.J., Rando, O.J., Goldberg, A.L. & Maniatis, T. (1994). The ubiquitin-proteasome pathway is required for processing the NF- kappa B1 precursor protein and the activation of NF-kappa B. Cell 78, 773-785.

Papa, F.R. & Hochstrasser, M. (1993). The yeast DOA4 gene encodes a deubiquitinating enzyme related to a product of the human tre-2 oncogene. Nature 366, 313-319.

Peters, J.M. (1994). Proteasomes: protein degradation machines of the cell. Trends Biochem. Sci. 19, 377-382.

Peters, J.M., Cejka, Z., Harris, J.R., Kleinschmidt, J.A. & Baumeister, W. (1993). Structural features of the 26 S proteasome complex. J. Mol. Biol. 234, 932-937.

Philip, P.A., Rea, D., Thavasu, P., Carmichael, J., Stuart, N.S., Rockett, H., Talbot, D.C., Ganesan, T., Pettit, G.R., Balkwill, F. & et al. (1993). Phase I study of bryostatin 1: assessment of interleukin 6 and tumor necrosis factor alpha induction in vivo. The Cancer Research Campaign Phase I Committee. J. Natl. Cancer Inst. 85, 1812-1818.

Pinkus, R., Weiner, L.M. & Daniel, V. (1996). Role of oxidants and antioxidants in the induction of AP-1, NF-kappaB, and glutathione S-transferase gene expression. J. Biol. Chem. 271, 13422-13429.

Polyak, K. (1994). p27Kip1, a cyclin-Cdk inhibitor, links transforming growth factor-beta and contact inhibition to cell cycle arrest. Genes Dev. 8, 9-22.

Pomerantz, J., Schreiber-Agus, N., Liegeois, N., Silverman, A., Alland, L., Chin, L., Potes, J., Chen, K., Orlow, I., Lee, H., Cordon-Cardo, C. & DePinho, R. (1998). The Ink4a tumor suppressor gene product, p19Arf, interacts with MDM2 and neutralizes MDM2's inhibition of p53. Cell 92, 713-723.

Porter, P.L., Malone, K.E., Heagerty, P.J., Alexander, G.M., Gatti, L.A., Firpo, E.J., Daling, J.R. & Roberts, J.M. (1997). Expression of cell-cycle regulators p27Kip1 and cyclin E, alone and in combination, correlate with survival in young breast cancer patients. Nat. Med. 3, 222-225.

Reits, E.A.J., Benham, A.M., Plougastel, B., Neefjes, J. & Trowsdale, J. (1997). Dynamics of proteasome distribution in living cells.EMBO J 16, 6087-6094.

Restifo, N.P., Esquivel, F., Kawakami, Y., Yewdell, J.W., Mule, J.J., Rosenberg, S.A. & Bennink, J.R. (1993). Identification of human cancers deficient in antigen processing. J. Exp. Med. 177, 265-272.

Rinaldi, T., Bolotin-Fukuhara, M. & Frontali, L. (1995). A Saccharomyces cerevisiae gene essential for viability has been conserved in evolution. Gene 160, 135-136.

Ritke, M.K., Bergoltz, V.V., Allan, W.P. & Yalowich, J.C. (1994). A Increased c-jun/AP-1 levels in etoposide-resistant human leukemia K562 cells. Biochem. Pharmacol. 48, 525-533.

Rock, K.L., Gramm, C., Rothstein, L., Clark, K., Stein, R., Dick, L., Hwang, D. & Goldberg, A.L. (1994). Inhibitors of the proteasome block the degradation of most cell proteins and the generation of peptides presented on MHC class I molecules. Cell 78, 761-771.

Rolfe, M., Beer Romero, P., Glass, S., Eckstein, J., Berdo, I., Theodoras, A., Pagano, M. & Draetta, G. (1995). Reconstitution of p53-ubiquitinylation reactions from purified components: the role of human ubiquitin-conjugating enzyme UBC4 and E6- associated protein (E6AP). Proc. Natl. Acad. Sci. USA

92, 3264-3268.

Rotem-Yehudar, R., Groettrup, M., Soza, A., Kloetzel, P. M. & Ehrlich, R. (1996). LMP-associated proteolytic activities and TAP-dependent peptide transport for class 1 MHC molecules are suppressed in cell lines transformed by the highly oncogenic adenovirus 12. J. Exp. Med. 183, 499-514.

Roth, J., Dobbelstein, M., Freedman, D., Shenk, T. & Levine, A.J. (1998). Nucleo-cytoplasmic shuttling of the hdm2 oncoprotein regulates the levels of the p53 protein via a pathway used by the human immunodeficiency virus rev protein. EMBO J. 17, 554-564.

Rubin, D. & Finley, D. (1995). Proteolysis. The proteasome: a protein-degrading organelle? Curr. Biol. 5, 854-858.

Sadoul, R., Fernandez, P.A., Quiquerez, A.L., Martinou, I., Maki, M., Schroter, M., Becherer, J.D., Irmler, M., Tschopp, J. & Martinou, J.C. (1996). Involvement of the proteasome in the programmed cell death of NGF- deprived sympathetic neurons. EMBO J. 15, 3845-3852.

Schauber, C., Chen, L., Tongaonkar, P., Vega, I., Lambertson, D., Potts, W. & Madura, K. (1998). Rad23 links DNA repair to the ubiquitin/proteasome pathway. Nature 391, 715-718.

Scheffner, M., Huibregtse, J.M., Vierstra, R. D. & Howley, P. M. (1993). The HPV-16 E6 and E6-AP complex functions as a ubiquitin-protein ligase in the ubiquitination of p53. Cell 75, 495-505.

Scheffner, M., Werness, B.A., Huibregtse, J.M., Levine, A.J. & Howley, P.M. (1990). The E6 oncoprotein encoded by human papillomavirus types 16 and 18 promotes the degradation of p53. Cell 63, 1129-1136.

Schreiber, M., Baumann, B., Cotten, M., Angel, P. & Wagner, E.F. (1995). Fos is an essential component of the mammalian UV response. EMBO J. 14, 5338-5349.

Seeger, M., Ferrell, K. & Dubiel, W. (1997). The 26S proteasome: a dynamic structure. Mol. Biol. Rep. 24, 83-88.

Seeger, M., Kraft, R., Ferrell, K., Bech-Otschir, D., Dumdey, R., Schade, R., Gordon, C., Naumann, M. & Dubiel, W. (1998). A novel protein complex involved in signal transduction possessing similarities to 26S proteasome subunits. FASEB J. 12, 469-478.

Seliger, B., Harders, C., Lohmann, S., Momburg, F., Urlinger, S., Tampe, R. & Huber, C. (1998). Down-regulation of the MHC class I antigen-processing machinery after oncogenic transformation of murine fibroblasts. Eur. J. Immunol. 28, 122-133.

Seliger, B., Hohne, A., Knuth, A., Bernhard, H., Meyer, T., Tampe, R., Momburg, F. & Huber, C. (1996). Analysis of the major histocompatibility complex class I antigen presentation machinery in normal and malignant renal cells: evidence for deficiencies associated with transformation and progression. Cancer Res. 56, 1756-1760.

Sepp-Lorenzino, L., Ma, Z., Lebwohl, D.E., Vinitsky, A. & Rosen, N. (1995). Herbimycin A induces the 20 S proteasome- and ubiquitin-dependent degradation of receptor tyrosine kinases. J. Biol. Chem. 270, 16580-16587.

Shaeffer, J.R. & Cohen, R.E. (1996). Differential effects of ubiquitin aldehyde on ubiquitin and ATP-dependent protein degradation. Biochem. 35, 10886-10893.

Shimanuki, M., Saka, Y., Yanagida, M. & Toda, T. (1995). A novel essential fission yeast gene pad1+ positively regulates pap1(+)- dependent transcription and is implicated in the maintenance of chromosome structure. J. Cell Sci. 108, 569-579.

Shinohara, K., Tomioka, M., Nakano, H., Tone, S., Ito, H. & Kawashima, S. (1996). Apoptosis induction resulting from proteasome inhibition. Biochem. J. 317, 385-388.

Sibille, C., Gould, K.G., Willard Gallo, K., Thomson, S., Rivett, A.J., Powis, S., Butcher, G.W. & De Baetselier, P. (1995). LMP2+ proteasomes are required for the presentation of specific antigens to cytotoxic T lymphocytes. Curr. Biol. 5, 923-930.

Singh, S.P., Lipman, J., Goldman, H., Ellis, F.H.J., Aizenman, L., Cangi, M.G., Signoretti, S., Chiaur, D.S., Pagano, M. & Loda, M. (1998). Loss or altered subcellular localization of p27 in Barrett's associated adenocarcinoma. Cancer Res. 58, 1730-1735.

Slingerland, J.M. (1994). A novel inhibitor of cyclin-Cdk activity detected in transforming growth factor beta-arrested epithelial cells. Mol. Cell. Biol. 14, 3683-3694.

Spataro, V., Bosshard, G., Scudiero, D.A., Harris, A.L., Myers, T.G. & Norbury, C.J. (1998). In "Proceedings of the American Association of Cancer Research," p. 170.

Spataro, V., Toda, T., Craig, R., Seeger, M., Dubiel, W., Harris, A.L. & Norbury, C. (1997). Resistance to diverse drugs and ultraviolet light conferred by overexpression of a novel human 26 S proteasome subunit. J. Biol. Chem. 272, 30470-30475.

Stade, K., Ford, C., Guthrie, C. & Weis, K. (1997). Exportin 1 (Crm1p) is an essential nuclear export factor. Cell 90, 1041-1050.

Sundseth, R., MacDonald, G., Ting, J. & King, A.C. (1997). DNA elements recognizing NF-Y and Sp1 regulate the human multidrug- resistance gene promoter. Mol. Pharmacol.6, 963-971.

Toda, T., Shimanuki, M., Saka, Y., Yamano, H., Adachi, Y., Shirakawa, M., Kyogoku, Y. & Yanagida, M. (1992). Fission yeast pap1-dependent transcription is negatively regulated by an essential nuclear protein, crm1. Mol. Cell. Biol. 12, 5474-5484.

Toone, W.M., Kuge, S., Samuels, M., Morgan, B.A., Toda, T. & Jones, N. (1998). Regulation of the fission yeast transcription factor Pap1 by oxidative stress: requirement for the nuclear export factor Crm1 (Exportin) and the stress-activated MAP kinase Sty1/Spc1. Genes Dev. 12, 1453-1463.

Treier, M., Staszewski, L.M. & Bohmann, D. (1994). Ubiquitin-dependent c-Jun degradation in vivo is mediated by the delta domain. Cell 78, 787-798.

Tsihlias, J., Kapusta, L.R., DeBoer, G., Morava-Protzner, I., Zbieranowski, I., Bhattacharya, N., Catzavelos, G.C., Klotz, L.H. & Slingerland, J.M. (1998). Loss of cyclin-dependent kinase inhibitor p27Kip1 is a novel prognostic factor in localized human prostate adenocarcinoma. Cancer Res. 58, 542-548.

Tsurumi, C., DeMartino, G.N., Slaughter, C.A., Shimbara, N. & Tanaka, K. (1995a). cDNA cloning of p40, a regulatory subunit of the human 26S proteasome, and a homolog of the Mov-34 gene product. Biochem. Biophys. Res. Commun. 210, 600-608.

Tsurumi, C., Ishida, N., Tamura, T., Kakizuka, A., Nishida, E., Okumura, E., Kishimoto, T., Inagaki, M., Okazaki, K., Sagata, N. & et al. (1995b). Mol cDNA cloning of p40, a regulatory subunit of the human 26S proteasome, and a homolog of the Mov-34 gene product. Cell Biol. 15, 5682-5687.

Van Antwerp, D.J., Martin, S.J., Kafri, T., Green, D. & Verma, I.M. (1996). Suppression of TNF-alpha-induced apoptosis by NF-kappaB. Science 274, 787-789.

Verovski, V.N., Van den Berge, D.L., Delvaeye, M.M., Scheper, R.J., De Neve, W.J. & Storme, G.A. (1996). Low-level doxorubicin resistance in P-glycoprotein-negative human pancreatic tumor PSN1/ADR cells implicates a brefeldin A-sensitive mechanism of drug extrusion. Br. J. Cancer 73, 596-602.

Wang, C.Y., Mayo, M.W. & Baldwin, A.S.J. (1996). TNF- and cancer therapy-induced apoptosis: potentiation by inhibition of NF-kappaB. Science 274, 784-787.

Watkins, J.F., Sung, P., Prakash, L. & Prakash, S. (1993). The Saccharomyces cerevisiae DNA repair gene RAD23 encodes a nuclear protein containing a ubiquitin-like domain required for biological function. Mol. Cell. Biol. 13, 7757-7765.

Wei, N., Tsuge, T., Serino, G., Dohmae, N., Takio, K., Matsui, M. & Deng, X. (1998). The COP9 complex is conserved between plants and mammals and is related to the 26S proteasome regulatory complex. Curr. Biol. 8, 919-922.

Weinstein, J.N., Myers, T.G., O'Connor, P.M., Friend, S.H., Fornace, A.J., Kohn, K.W., Fojo, T., Bates, S.,

Rubinstein, L.V., Anderson, N.L., Buolamwini, J.K., van Osdol, W.W., Nomks, A.P., Scudiero, D.A., Sausville, E.A., Zaharevitz, D.W., Bunow, B., Viswanadan, V.N., Johnson, G.S., Wittes, R.E. & Paull, K.D. (1997). An information-intensive approach to the molecular pharmacology of cancer. Science 275, 343-349.

Wilkinson, C.R., Wallace, M., Seeger, M., Dubiel, W. & Gordon, C. (1997). Mts4, a non-ATPase subunit of the 26 S protease in fission yeast is essential for mitosis and interacts directly with the ATPase subunit Mts2. J. Biol. Chem. 272, 25768-25777.

Yang, R.M., Naitoh, J., Murphy, M., Wang, H.J., Phillipson, J., deKernion, J.B., Loda, M. & Reiter, R. E. (1998). Low p27 expression predicts poor disease-free survival in patients with prostate cancer. J. Urol. 159, 941-945.

Yang, Y., Fruh, K., Ahn, K. & Peterson, P.A. (1995). In vivo assembly of the proteasomal complexes, implications for antigen processing. J. Biol. Chem. 270, 27687-94.

Yao, K.S., Godwin, A.K., Johnson, S.W., Ozols, R.F., O'Dwyer, P.J. & Hamilton, T.C. (1995). Evidence for altered regulation of gamma-glutamylcysteine synthetase gene expression among cisplatin-sensitive and cisplatin-resistant human ovarian cancer cell lines. Cancer Res. 55, 4367-4374.

Zhang, Y., Xiong, Y. & Yarbrough, W. (1998). ARF promotes MDM2 degradation and stabilizes p53: ARF-INK4a locus deletion impairs both the Rb and p53 tumor suppression pathways. Cell 92, 725-734.

Zhou, M., Wu, X. & Ginsberg, H.N. (1996). ARF promotes MDM2 degradation and stabilizes p53: ARF-INK4a locus deletion impairs both the Rb and p53 tumor suppression pathways. J. Biol. Chem. 271, 24769-24775.

CONTRIBUTOR ADDRESSES

MOSTAFA BADR, Dr. University of Missouri-Kansas City, 2411 Holmes Street, M3-115, Kansas City, MO 64108-2792
Tel: (816) 235-1842, Fax: (816) 235-1776
E-mail: badrmz@cctr.umkc.edu

MATTHIAS F. BAUER, Institute of Clinical Chemistry, Molecular Diagnostics and Mitochondrial Genetics & Diabetes Research Group, Academic Hopital Munich-Schwabing, Ludwig-Maximilians University of Munich, Kölner Platz 1, 80804 Munich, Germany
Tel.: (49) 89 3068 2670, Fax: (49) 89 3068 3911
E-mail: bauer@bio.med.uni-muenchen.de

EVERLY CONWAY DE MACARIO, PhD., Professor; Research Scientist, Wadsworth Center, Room B-749 B, Division of Molecular Medicine, New York State Department of Health; and, Department of Biomedical Sciences, School of Public Health, The University at Albany (SUNY), Empire State Plaza, P.O. Box 509, Albany, New York 12201-0509, USA
Phone/Fax: (518) 474-1213, Fax: (518) 474-8590 -7992
E-mail: everlym@wadsworth.org
Website: http://www.wadsworth.org/BMS/SCBlinks/stress.html

MICHAEL FOSSEL, MD, PhD, Michigan State University, Box 630, Ada, Michigan 49301
Voice: 616-676-4099, Fax: 616-676-4099

KLAUS D. GERBITZ, Institute of Clinical Chemistry, Molecular Diagnostics and Mitochondrial Genetics & Diabetes Research Group, Academic Hopital Munich-Schwabing, Ludwig-Maximilians University of Munich, Kölner Platz 1, 80804 Munich, GERMANY
Tel.: (49) 89 3068 2670, Fax: (49) 89 3068 3911
E-mail: bauer@bio.med.uni-muenchen.de

GORDON W. GLAZNER, Department of Pharmacology, The University of Manitoba, 753 McDermot Avenue, Winnipeg, Manitoba, Canada R3E OT6
Phone: (204) 789-3542, Fax: (204) 789-3932)

SABINE HOFMANN, Institute of Clinical Chemistry, Molecular Diagnostics and Mitochondrial Genetics & Diabetes Research Group, Academic Hopital Munich-Schwabing, Ludwig-Maximilians University of Munich, Kölner Platz 1, 80804 Munich, GERMANY
Tel.: (49) 89 3068 2670, Fax: (49) 89 3068 3911

JEFFREY N. KELLER, Sanders-Brown Building Rm 131, University of Kentucky, Lexington, KY 40536-0230
Phone (606) 325-1412 ext 249, Fax (606) 323-2866

ALBERTO J.L. MACARIO, MD, Professor; Research Physician, Wadsworth Center, Room B-749, Division of Molecular Medicine, New York State Department of Health; and, Department of Biomedical Sciences, School of Public Health, The University at Albany (SUNY), Empire State Plaza, P.O. Box 509, Albany, New York 12201-0509, USA
Phone: (518) 474-2781, Fax: (518) 474-1213; -8590; -7992
E-mail: macario@wadsworth.org
Website: http://www.wadsworth.org/BMS/SCBlinks/stress.html

MARK P. MATTSON, Chief, Laboratory of Neurosciences, National Institute on Aging, GRC 4F02, 5600 Nathan Shock Drive, Baltimore, MD 21224
Phone: (410) 558-8463, Fax: (410) 558-8465
E-mail: mattsonm@grc.nia.nih.gov
Website: http://www.grc.nia.nih.gov/Branches/Ins/httl/httl/htdocs/index.html

CHRIS J. NORBURY, PhD, Imperial Cancer Research Fund Molecular Oncology Laboratory, University of Oxford Institute of Molecular Medicine, Headley Way, The Radcliffe, Oxford OX3 9DS, UK
Tel.: (44) 1865-222415, Fax: (44) 1865-222431

THOMAS J. PRESTON, Hamilton Regional Cancer Centre, 699 Concession Street, Hamilton, Ontario, CANADA, L8V 5C2
Tel: (905) 387 9711, Ext. 7007, Fax: (905) 575 6330
E-mail: prestotj@fhs.csu.mcmaster.ca

GURMIT SINGH, Hamilton Regional Cancer Centre, 699 Concession Street, Hamilton, Ontario, CANADA, L8V 5C2
Tel: (905) 387 9711, Ext. 7007, Fax: (905) 575 6330
E-mail: singhg@fhs.csu.mcmaster.ca

VITO J. SPATARO, MD, Servizio Oncologico Ticinese, Ospedale San Giovanni, CH-6500 Bellinzona, Switzerland
Tel.: (41) 91-8209111, Fax: (41) 91-8209044

JIHAN YOUSSEF, University of Missouri-Kansas City, 2411 Holmes Street, M3-115, Kansas City, MO 64108-2792
Tel: (816) 235-1842, Fax: (816) 235-1776

Advances in Cell Aging and Gerontology

Series Editor: Mark P. Mattson

URL: http://www.elsevier.nl/locate/series/acag

Aims and Scope:

Advances in Cell Aging and Gerontology (ACAG) is dedicated to providing timely review articles on prominent and emerging research in the area of molecular, cellular and organismal aspects of aging and age-related disease. The average human life expectancy continues to increase and, accordingly, the impact of the dysfunction and diseases associated with aging are becoming a major problem in our society. The field of aging research is rapidly becoming the niche of thousands of laboratories worldwide that encompass expertise ranging from genetics and evolution to molecular and cellular biology, biochemistry and behavior. ACAG consists of edited volumes that each critically review a major subject area within the realms of fundamental mechanisms of the aging process and age-related diseases such as cancer, cardiovascular disease, diabetes and neurodegenerative disorders. Particular emphasis is placed upon: the identification of new genes linked to the aging process and specific age-related diseases; the elucidation of cellular signal transduction pathways that promote or retard cellular aging; understanding the impact of diet and behavior on aging at the molecular and cellular levels; and the application of basic research to the development of lifespan extension and disease prevention strategies. ACAG will provide a valuable resource for scientists at all levels from graduate students to senior scientists and physicians.

Books Published:

1. P.S. Timiras, E.E. Bittar, *Some Aspects of the Aging Process,* 1996, 1-55938-631-2
2. M.P. Mattson, J.W. Geddes, *The Aging Brain,* 1997, 0-7623-0265-8
3. M.P. Mattson, *Genetic Aberrancies and Neurodegenerative Disorders,* 1999, 0-7623-0405-7
4. B.A. Gilchrest, V.A. Bohr, *The Role of DNA Damage and Repair in Cell Aging,* 2001, 0-444-50494-X
5. M.P. Mattson, S. Estus, V. Rangnekar, *Programmed Cell Death, Volume I,* 2001, 0-444-50493-1
6. M.P. Mattson, S. Estus, V. Rangnekar, *Programmed Cell Death, Volume II,* 2001, 0-444-50730-2
7. M.P. Mattson, *Interorganellar Signaling in Age-Related Disease,* 2001, 0-444-50495-8